D0782376

COUNTEREXAMPLES IN PROBABILITY AND REAL ANALYSIS

COUNTEREXAMPLES IN PROBABILITY AND REAL ANALYSIS

GARY L. WISE
Department of Electrical and Computer Engineeering
and
Department of Mathematics
The University of Texas at Austin
Austin, Texas

ERIC B. HALL
Department of Electrical Engineering
Southern Methodist University
Dallas, Texas

New York Oxford
OXFORD UNIVERSITY PRESS
1993

Oxford University Press

Oxford New York Toronto
Delhi Bombay Calcutta Madras Karachi
Kuala Lumpur Singapore Hong Kong Tokyo
Nairobi Dar es Salaam Cape Town
Melbourne Auckland Madrid

and associated companies in
Berlin Ibadan

Published by Oxford University Press, Inc.,
200 Madison Avenue, New York, New York 10016

Library of Congress Cataloging-in-Publication Data
Wise, Gary L., 1945–
Counterexamples in probability and real analysis /
Gary L. Wise, Eric B. Hall.
p. cm.
Includes bibliographical references and index.
ISBN 0-19-507068-2
1. Probabilities. 2. Mathematical analysis.
I. Hall, Eric B., 1963–
II. Title. QA273.W67 1993 515—dc20 93-22411

2 4 6 8 9 7 5 3

Printed in the United States of America
on acid-free paper

In loving memory of
my late father,
Calder Lamar Wise,
and to Stella, of course,
and to Tanna, too.

To my parents,
Jess and Millie Hall.

Preface

When we began 5 years ago to collect the almost 350 examples in this book, we were immediately faced with an important question. What is a counterexample? In this book a counterexample is any example or result that is counter to commonly held beliefs or intuition. In preparing and collecting these examples, we have tried to avoid a common technique for obtaining counterexamples. In particular, we have seldom felt the need to point out that necessary conditions are not sufficient, that sufficient conditions are not necessary, or that a theorem's result need not hold when its hypotheses are ignored. On occasion, we have presented results that, while not counterexamples in themselves, are useful in the development of other counterexamples.

Throughout this work we adopt the Zermelo–Fraenkel (ZF) axioms of set theory with the Axiom of Choice, commonly abbreviated as ZFC. It follows from the work of Gödel and Cohen that if the ZF axioms are consistent, the Axiom of Choice can be neither proved nor disproved from the ZF axioms. The Axiom of Choice is deeply embedded in standard mathematics, and we use it without apology. Indeed, much of standard mathematics falls when the Axiom of Choice is not assumed or when models of ZF are used in which the Axiom of Choice fails: non-Lebesgue measurable subsets of the reals may cease to exist, there may exist infinite sets of real numbers without countably infinite subsets, there may exist a decomposition of the real line into a countable union of countable sets, there may exist a vector space without a basis, etc. We say that a set is transitive if each element of it is a subset of it. An ordinal is a transitive set well ordered by set membership. An initial ordinal is any ordinal that is not equipotent to any smaller ordinal. If α is an ordinal, we will let $\alpha + 1$ denote the ordinal successor of α. For any set A, the cardinality of A is the smallest ordinal equipotent to A. Any initial ordinal is its own cardinal, and any cardinal is an initial ordinal. Hence, initial ordinals and cardinals are the same thing. We let $\aleph_0$ denote the cardinality of the positive integers. We let $\aleph_1$ denote the cardinality of the first uncountable ordinal. Hence, $\aleph_1$ is the smallest cardinal after $\aleph_0$. We let c denote the cardinality of the reals.

We will occasionally use the Continuum Hypothesis, commonly abbreviated as CH, which is the assertion that $\aleph_1 = c$. We know that c is some ordinal not less than $\aleph_1$, and CH pins down for us just which ordinal it is. It was Cantor's conjecture that CH followed from ZFC, and Hilbert later published an incorrect proof of Cantor's conjecture. Later, Gödel showed that if ZFC were consistent, then CH could not be disproved from ZFC. Later still, Cohen showed that CH could not be proved from ZFC. Any time we use CH, we will explicitly mention it. In the absence of our mentioning CH, we are making no assumption regarding CH.

Acknowledgments

While we, of course, must take the blame for any errors this book may contain, we benefited greatly from the advice and encouragement of many friends and colleagues. In particular, David Blackwell was a constant source of encouragement during GLW's two visits to the Department of Statistics at the University of California at Berkeley. We would also like to thank GLW's faculty sponsors at Berkeley, Lucien Le Cam and Terry Speed, as well as Steve Evans, Peter Bickel, Michael Klass, Leo Breiman, David Aldous, Warry Millar, Jim Pitman, Lester Dubins, and GLW's office mate at Berkeley, Robert Dalang. In addition, we would like to express our appreciation to Alan Wessel who provided many useful suggestions and to David Drumm who read the manuscript several times and also provided many useful suggestions.

Contents

Preface vii

Notation xi

I. The Counterexamples 3

II. The Counterexamples with Proofs 31

1. The Real Line 33

2. Real–Valued Functions 53

3. Differentiation 71

4. Measures 81

5. Integration 103

6. Product Spaces 121

7. Basic Probability 137

8. Conditioning 159

9. Convergence in Probability Theory 171

10. Applications of Probability 183

Index 209

Notation

$\alpha+1$	$\alpha \cup \{\alpha\}$ for an ordinal α
$\aleph_0$	the first infinite ordinal
$\aleph_1$	the first uncountable ordinal
c	the cardinality of the real line
CH	the continuum hypothesis (i.e., that $\aleph_1 = c$)
$\mathbb{R}$	the set of all real numbers
$\mathbb{Z}$	the set of all integers
$\mathbb{Z}^+$	the set of nonnegative integers
$\mathbb{N}$	the set of all positive integers
$\mathbb{Q}$	the set of all rational numbers
$\imath$	the imaginary unit
z^*	the complex conjugate of the complex number z
$\mathbb{P}(S)$	the set of all subsets of the set S
$\mathrm{card}(A)$	the cardinality of the set A
2^α	$\mathrm{card}(\mathbb{P}(\alpha))$ for an ordinal α
I_A	the indicator function of the set A
A^c	the complement of the set A
$A \oplus B$	the set $\{a+b\colon a \in A, b \in B\}$ for $A, B \subset \mathbb{R}$
$A \oplus b$	the set $A \oplus \{b\}$ for $A \subset \mathbb{R}$ and $b \in \mathbb{R}$
$A \ominus B$	the set $\{a-b\colon a \in A, b \in B\}$ for $A, B \subset \mathbb{R}$
$A \ominus b$	the set $A \ominus \{b\}$ for $A \subset \mathbb{R}$ and $b \in \mathbb{R}$
$A \triangle B$	the symmetric difference of the sets A and B
$A \setminus B$	the set of points in A that are not in B
$-A$	$\{-x\colon x \in A\}$ for $A \subset \mathbb{R}$
B^A	the set of all functions mapping A to B
m_*	the inner measure on $\mathbb{P}(\Omega)$ induced by the measure m on $(\Omega, \mathcal{S})$
m^*	the outer measure on $\mathbb{P}(\Omega)$ induced by the measure m on $(\Omega, \mathcal{S})$
$C(K)$	the set of all continuous functions mapping K into $\mathbb{R}$
$C_b(K)$	the set of all bounded continuous functions mapping K into $\mathbb{R}$
$\mathbf{L}_p(\Omega, \mathcal{F}, \mu)$	the set of all μ-equivalence classes of functions $f\colon(\Omega, \mathcal{F}) \to (\mathbb{R}, \mathcal{B}(\mathbb{R}))$ such that $\int_\Omega \lvert f\rvert^p \, d\mu < \infty$
$\mathcal{B}(T)$	the Borel subsets of the topological space T
$\mathcal{M}(A)$	the Lebesgue measurable subsets of $A \in \mathcal{B}(\mathbb{R})$
λ	Lebesgue measure on $\mathcal{M}(\mathbb{R})$
$\sigma(\{A_i\colon i \in I\})$	the smallest σ-algebra including $\{A_i\colon i \in I\}$
$\sigma(\{X_i\colon i \in I\})$	the smallest σ-algebra for which X_i is measurable for all $i \in I$
$f\vert_A$	the function f restricted to A
f^+	$\max\{f, 0\}$ for a real-valued function f
f^-	$-\min\{f, 0\}$ for a real-valued function f

$\nu \ll \mu$	the measure ν is absolutely continuous with respect to the measure μ
S_x	the x section of the subset S of $\mathbb{R}^2$
S^y	the y section of the subset S of $\mathbb{R}^2$
a.e. $[\mu]$	almost everywhere with respect to the measure μ
a.s.	almost surely

COUNTEREXAMPLES IN PROBABILITY AND REAL ANALYSIS

Part I
The Counterexamples

The Counterexamples

1. The Real Line

1.1. *There exists a subset of* $\mathbb{R}$ *such that the infimum of the set is greater than the supremum of the set.*

1.2. *Given any nonzero real number* a*, there exists a metric on* $\mathbb{R}$ *such that* a *is closer to 0 than to any other real number.*

1.3. *There exists an uncountable family of distinct infinite subsets of* $\mathbb{N}$ *such that the intersection of any two distinct sets from this family is finite.*

1.4. *Given any positive number* ε*, there exists an open subset* U *of* $\mathbb{R}$ *such that* U *contains each of the rationals, yet the Lebesgue measure of* U *is less than* ε*.*

1.5. *There exists a subset* S *of* $\mathbb{R}$ *such that* $\inf\{\varepsilon > 0\colon S \oplus \varepsilon = S^c\} = 0$.

1.6. *There do not exist Lebesgue measurable subsets of* $\mathbb{R}$ *with positive Lebesgue measure and with cardinality less than* c*.*

1.7. *There exists an uncountable collection of distinct subsets of* $\mathbb{N}$ *that are totally ordered by the inclusion relation.*

1.8. *For any closed interval* I *of* $\mathbb{R}$ *such that* $\lambda(I) > 0$ *and any* $\alpha \in [0, \lambda(I))$*, there exists a Cantor-like subset of* I *having Lebesgue measure* α*.*

1.9. *Each element in the Cantor ternary set may be written in the form* $\sum_{n=1}^{\infty} x_n/3^n$*, where* $x_n \in \{0, 2\}$ *for each* n*.*

1.10. *If* C *denotes the Cantor ternary set, then* $C \oplus C = [0, 2]$.

1.11. *If* C *denotes the Cantor ternary set, then* $C \ominus C = [-1, 1]$.

1.12. *If* F *is a first category subset of* $\mathbb{R}$*, then* $F^c \oplus F^c = \mathbb{R}$.

1.13. *There exists a subset of* $\mathbb{R}$ *that is dense, has cardinality* c*, and is of the first category.*

1.14. *The image of a Lebesgue measurable subset of the real line under a continuous transformation need not be Lebesgue measurable.*

1.15. *There exists a continuous function* f *mapping* $[0, 1]$ *into* $[0, 1]$ *and a Lebesgue measurable set* L *such that* $f^{-1}(L)$ *is not Lebesgue measurable.*

1.16. *The image of a Lebesgue null subset of the real line under a continuous transformation may be Lebesgue measurable yet not have zero Lebesgue measure.*

1.17. *If* A *is a Lebesgue measurable subset of* $\mathbb{R}$ *having positive Lebesgue measure, then* $A \ominus A$ *contains an interval of the form* $(-\alpha, \alpha)$ *for some* $\alpha > 0$.

1.18. *There exists a subset* V *of* $\mathbb{R}$ *such that* $(V \ominus V) \cap \mathbb{Q} = \{0\}$ *yet such that* $\mathbb{R} = \bigcup_{q \in \mathbb{Q}} q \oplus V$.

1.19. *There exists a set $G \in \mathcal{B}(\mathbb{R})$ such that $I \cap G$ and $I \cap G^c$ each have positive Lebesgue measure for any nonempty open interval I.*

1.20. *Any uncountable G_δ subset of $\mathbb{R}$ contains a nowhere dense closed subset of Lebesgue measure zero that can be mapped continuously onto $[0, 1]$.*

1.21. *There exist two Lebesgue null subsets A and B of $\mathbb{R}$ such that $A \oplus B = \mathbb{R}$.*

1.22. *There exists a set B of real numbers such that B and B^c each have at least one point in common with every uncountable closed subset of $\mathbb{R}$.*

1.23. *The intersection of a Bernstein set with any Lebesgue measurable set of positive Lebesgue measure is not Lebesgue measurable.*

1.24. *A Bernstein set lacks the property of Baire.*

1.25. *There exists a nowhere dense, non-Lebesgue measurable subset of the real line.*

1.26. *There exists an uncountable subset B of $\mathbb{R}$ such that every closed (in $\mathbb{R}$) subset of B is countable.*

1.27. *For any integer $N > 1$ there exist disjoint subsets $B_1, \ldots, B_N$ of $\mathbb{R}$ such that B_i and B_i^c have a nonempty intersection with every uncountable closed subset of $\mathbb{R}$ for each positive integer $i \leq N$.*

1.28. *Any subset of $\mathbb{R}$ of the second category has a subset that does not possess the property of Baire.*

1.29. *There exists a Hamel basis for $\mathbb{R}$ over $\mathbb{Q}$.*

1.30. *The inner Lebesgue measure of any Hamel basis for $\mathbb{R}$ over $\mathbb{Q}$ is zero.*

1.31. *A Lebesgue measurable Hamel basis for $\mathbb{R}$ over $\mathbb{Q}$ exists.*

1.32. *Assuming the continuum hypothesis, there exists a countable collection of Hamel bases of $\mathbb{R}$ over $\mathbb{Q}$ that covers $\mathbb{R} \setminus \{0\}$.*

1.33. *Assuming the continuum hypothesis, there exists a non-Lebesgue measurable Hamel basis for $\mathbb{R}$ over $\mathbb{Q}$.*

1.34. *If H is a Hamel basis for $\mathbb{R}$ over $\mathbb{Q}$, if $h \in H$, and if A is the set of all real numbers that are expressible as a finite linear combination of elements from $H \setminus \{h\}$ with rational coefficients, then A is not Lebesgue measurable.*

1.35. *There exists a subset Λ of $\mathbb{R}$ such that $\mathbb{R}$ may be partitioned into a countably infinite number of translations of Λ, each of which is dense in $\mathbb{R}$.*

1.36. *Any Hamel basis for $\mathbb{R}$ over $\mathbb{Q}$ has cardinality c.*

1.37. *There exists a Hamel basis for $\mathbb{R}$ over $\mathbb{Q}$ that contains a nonempty perfect subset.*

1.38. *There exists a proper subset N of $\mathbb{R}$ such that the family $\{N \oplus ny: n \in \mathbb{N}\}$ contains only a finite number of different subsets of $\mathbb{R}$ for any fixed $y \in \mathbb{R}$.*

1.39. *No Hamel basis for $\mathbb{R}$ over $\mathbb{Q}$ is an analytic set.*

1.40. *There exists a proper subset T of $\mathbb{R}$ such that the set $\{T \oplus t\colon t \in \mathbb{R}\}$ contains only countably many distinct elements.*

1.41. *There exists a partition of the positive reals into two nonempty sets, each of which is closed under addition.*

1.42. *The Minkowski sum of two bounded Lebesgue null subsets of the reals can be non-Lebesgue measurable.*

1.43. *Assuming the continuum hypothesis, there exists an uncountable subset of the real line that has a countable intersection with every subset of $\mathbb{R}$ of the first category.*

1.44. *Assuming the continuum hypothesis, there exists a subset of the real line of the second category that does not include any uncountable sets of the first category.*

1.45. *There exists a partition of the real line into a Lebesgue null set and a set of the first category.*

1.46. *For any closed subset K of $\mathbb{R}$ having positive Lebesgue measure and any cardinal β such that $2 \le \beta \le c$, there exists a partition $\{B_j\colon j < \beta\}$ of K such that $\lambda^*(B_j) = \lambda(K)$ and $\lambda_*(B_j) = 0$ for each $j < \beta$.*

1.47. *For any closed subset K of $\mathbb{R}$ having positive Lebesgue measure and any nonnegative real number $r \le \lambda(K)$, there exists a partition $\{B_j\colon j < c\}$ of K such that $\lambda^*(B_j) = r$ and $\lambda_*(B_j) = 0$ for each $j < c$.*

1.48. *There exists a saturated non-Lebesgue measurable subset S of $\mathbb{R}$ such that the set $\{S \oplus y\colon y \in I\}$ is a partition of $\mathbb{R}$ into saturated non-Lebesgue measurable sets congruent under translation, where card(I) is any infinite cardinal less than or equal to c.*

2. Real–Valued Functions

2.1. *There exists a function $f\colon \mathbb{R} \to \mathbb{R}$ that is nowhere continuous yet is equal a.e. $[\lambda]$ to a continuous function.*

2.2. *There exists a function $f\colon \mathbb{R} \to \mathbb{R}$ that is continuous at all but one point yet is not equal a.e. $[\lambda]$ to any continuous function.*

2.3. *A sequence of Borel measurable functions on $\mathbb{R}$ can converge almost everywhere (with respect to Lebesgue measure) to a function that is not Borel measurable.*

2.4. *The composition of a Lebesgue measurable function mapping $\mathbb{R}$ to $\mathbb{R}$ with a continuous function mapping $\mathbb{R}$ to $\mathbb{R}$ need not be Lebesgue measurable.*

2.5. *A non-Lebesgue measurable function mapping the reals into the reals can be such that the image under this function of any Lebesgue measurable set is a Lebesgue measurable set.*

2.6. *There exists a continuous function f: $[0, 1] \to \mathbb{R}$ such that $f(0) = 0$ and $f(1) = 1$ and such that f is constant on each of the open intervals removed in the construction of the Cantor ternary set.*

2.7. *There exists a strictly increasing, continuous, singular function g: $\mathbb{R} \to \mathbb{R}$.*

2.8. *Lower semicontinuity of each element of the set $\{f_n: n \in \mathbb{N}\}$ of functions mapping $\mathbb{R}$ into $\mathbb{R}$ does not imply that $\lim_{n\to\infty} f_n$ is lower semicontinuous.*

2.9. *Any real-valued function of three real variables is expressible as a combination of two real-valued functions of two real variables.*

2.10. *There exists a real-valued function of a real variable that is continuous at every irrational point, discontinuous at every rational point, right continuous, strictly increasing, and that converges to zero at $-\infty$ and one at ∞.*

2.11. *If f: $\mathbb{R} \to \mathbb{R}$ is a pointwise limit of a sequence of continuous functions mapping $\mathbb{R}$ into $\mathbb{R}$, then f is continuous except at a set of points of the first category.*

2.12. *There exists a continuous function f: $[0, 1] \to [0, 1]$ such that it is zero on the Cantor ternary set and nonzero off the Cantor ternary set.*

2.13. *If $(\Omega, \mathcal{F}, \mu)$ is a measure space such that all singletons are measurable, then it need not follow that the function f: $\Omega \to \mathbb{R}$ via $f(x) = \mu(\{x\})$ is measurable.*

2.14. *Given any three real numbers α, β, and γ that are linearly independent over $\mathbb{Q}$, there exist real-valued (not identically constant) functions f and g defined on $\mathbb{R}$ such that f is periodic with period α, g is periodic with period β, and $f + g$ is periodic with period γ.*

2.15. *There does not exist a metric d on $C([0, 1])$ such that $\lim_{n\to\infty} d(f_n, f_0) = 0$ if and only if the sequence $\{f_n\}_{n\in\mathbb{N}}$ converges pointwise to f_0.*

2.16. *There exists a periodic function f: $\mathbb{R} \to \mathbb{R}$ that has no chord of irrational length even though, for any $h \in \mathbb{R}$, $f(x+h) - f(x)$ is a constant function of x.*

2.17. *A real-valued midpoint convex function defined on the reals need not be continuous.*

2.18. *Any real-valued function on $[0, 1]$ may be expressed as the composition of a Lebesgue measurable function with a Borel measurable function.*

2.19. *There exists a nonnegative real-valued Lebesgue integrable function f defined on $\mathbb{R}$ such that for any $a < b$ and any $R > 0$ the set $(a, b) \cap \{x \in \mathbb{R}: f(x) \geq R\}$ has positive Lebesgue measure.*

2.20. *There exists a real-valued function defined on $[0, 1]$ whose set of discontinuities has measure zero yet has an uncountable intersection with every nonempty open subinterval.*

2.21. *Given any nonempty subset M of $\mathbb{R}$, there exists a function $g\colon \mathbb{R} \to \mathbb{R}$ such that g is constant a.e. with respect to Lebesgue measure and $g(\mathbb{R}) = M$.*

2.22. *There exists a function $f\colon \mathbb{R} \to \mathbb{R}$ such that $f(x) = 0$ holds for some value of x, such that f is everywhere discontinuous, such that f is unbounded and non-Lebesgue measurable on any nonempty open interval, yet such that, for any real number x_1 and the corresponding sequence $\{x_n\}_{n\in\mathbb{N}}$ defined via $x_{n+1} = f(x_n)$, $f(x_n) = 0$ for all $n \geq 3$.*

2.23. *There exist two real-valued functions f and g of a real variable such that each is continuous almost everywhere with respect to Lebesgue measure and yet $f \circ g$ is nowhere continuous.*

2.24. *There exists a Lebesgue measurable function $f\colon [0, 1] \to \mathbb{R}$ such that, given any auteomorphism h on $[0, 1]$, there is no function of the first class of Baire that agrees with $f \circ h$ almost everywhere with respect to Lebesgue measure.*

2.25. *There exists a nondecreasing continuous function f mapping $[0, 1]$ to $[0, 1]$ such that $|f'(x)| \leq 1$ off a Lebesgue null set and yet such that f is not a Lipschitz function.*

2.26. *There exists an absolutely continuous function $f\colon [0, 1] \to \mathbb{R}$ that is monotone in no nonempty open interval.*

2.27. *There exists a net $\{f_t\}_{t\in\mathbb{R}}$ of Borel measurable real-valued functions defined on $(0, 1]$ such that the set of numbers $x \in (0, 1]$ for which $\lim_{t\to\infty} f_t(x) = 0$ is not a Borel set.*

2.28. *If f is a Lebesgue measurable real-valued function of a real variable, then the function of the real variable x given by $\sup_{t\in\mathbb{R}} |f(x+t) - f(x-t)|$ can be non-Lebesgue measurable.*

2.29. *There exists a real-valued function defined on the reals that takes on every real value infinitely many times in every nonempty open interval yet is equal to zero a.e. with respect to Lebesgue measure.*

2.30. *There exists a function $f\colon \mathbb{R} \to \mathbb{R}$ such that f takes on every real value $\mathfrak{c}$ times on every nonempty perfect subset of $\mathbb{R}$.*

2.31. *For any function $f\colon \mathbb{R} \to \mathbb{R}$ there exists a Darboux function $g\colon \mathbb{R} \to \mathbb{R}$ such that the set $\{x \in \mathbb{R}\colon f(x) \neq g(x)\}$ is a set of Lebesgue measure zero.*

2.32. *There exists a Darboux function mapping $\mathbb{R}$ into $\mathbb{R}$ that is non-Lebesgue measurable on every subset of $\mathbb{R}$ having positive Lebesgue measure.*

2.33. *If f is any real-valued function of a real variable, then there exist two Darboux functions g and h mapping the reals into the reals such that $f = g + h$.*

2.34. *There exists a sequence $\{f_n\}_{n\in\mathbb{N}}$ of Lebesgue measurable functions mapping the reals into the reals such that f_n converges pointwise to zero* *and such that for any positive integer n, f_n takes on each real value in any nonempty open set.*

2.35. *For any real-valued function f defined on the reals, there exists a sequence $\{g_n\}_{n\in\mathbb{N}}$ of measurable Darboux functions mapping the reals into the reals such that g_n converges pointwise to f.*

2.36. *There exists a Darboux function $f: \mathbb{R} \to \mathbb{R}$ such that f, as a subset of $\mathbb{R}^2$, is not connected.*

2.37. *If f is a Darboux function mapping the reals into the reals and if a is a real number, the function g mapping the reals into the reals via $g(x) = f(x) + ax$ need not be a Darboux function.*

2.38. *For each $h \in [0, 1)$, there exists an almost everywhere continuous function f_h: $[0, 1) \to [0, 1]$ such that for each $x \in [0, 1)$, $f_h(x) \to 0$ as $h \to 0$, yet this convergence is not uniform on any Lebesgue measurable set of positive Lebesgue measure.*

2.39. *There exists a symmetric non-Lebesgue measurable, nonnegative definite function mapping $\mathbb{R}$ to $\mathbb{R}$.*

2.40. *Assuming the continuum hypothesis, there exists a sequence of real-valued functions defined on the reals that converges on an uncountable subset but converges uniformly on none of its uncountable subsets.*

3. Differentiation

3.1. *If f: $\mathbb{R} \to \mathbb{R}$ is a function having a Taylor's series at a real number a, then there exist c distinct functions mapping $\mathbb{R}$ into $\mathbb{R}$ having the same Taylor's series at the real number a.*

3.2. *There exists a function f: $\mathbb{R} \to \mathbb{R}$ that is everywhere differentiable yet whose derivative is not Lebesgue integrable.*

3.3. *A differentiable function mapping the reals into the reals need not be absolutely continuous.*

3.4. *An absolutely continuous function mapping the reals into the reals need not be differentiable.*

3.5. *There exists a function f: $[0, 1] \to \mathbb{R}$ that is differentiable with a bounded derivative but whose derivative is not Riemann integrable.*

3.6. *There exists a continuously differentiable function defined on $[0, 1]$ whose set of critical values is uncountable.*

3.7. *A sequential limit of derivatives of real-valued functions of a real variable need not be the derivative of a real-valued function of a real variable.*

3.8. *The product of the derivatives of two differentiable real-valued functions of a real variable need not be a derivative.*

3.9. *For any Lebesgue null set A of $\mathbb{R}$, there exists a continuous nondecreasing function $f: \mathbb{R} \to \mathbb{R}$ such that each Dini derivative is equal to ∞ at each point of A.*

3.10. *The Dini derivatives of a continuous function $f: \mathbb{R} \to \mathbb{R}$ need not have the Darboux property.*

3.11. *There exists a real-valued function defined on $(0, 1)$ whose derivative vanishes on a dense set and yet does not vanish everywhere.*

3.12. *Let F_1 and F_2 be continuous, real-valued functions defined on an interval $[a, b]$ and assume that F_2 is a.e. differentiable. Given any $\varepsilon > 0$, there exists a continuous a.e. differentiable function G: $[a, b] \to \mathbb{R}$ such that $G' = F_2'$ a.e. and such that $|F_1(x) - G(x)| \leq \varepsilon$ for all $x \in [a, b]$.*

3.13. *Let $g: \mathbb{R} \to \mathbb{R}$ be Lebesgue integrable on $[a, b]$ and let $\varepsilon > 0$. There exists a continuous a.e. differentiable function $G: \mathbb{R} \to \mathbb{R}$ such that $G' = g$ a.e., $G(a) = G(b) = 0$, and $|G(x)| \leq \varepsilon$ for all $x \in [a, b]$.*

3.14. *Let f be any real-valued function defined on $\mathbb{R}$ and let $\{h_n\}_{n \in \mathrm{N}}$ be any sequence of nonzero numbers converging to zero. There exists a function $F: \mathbb{R} \to \mathbb{R}$ such that $\lim_{n\to\infty}[F(x+h_n) - F(x)]/h_n = f(x)$ for all $x \in \mathbb{R}$.*

3.15. *Given any closed subinterval $[a, b]$ of $\mathbb{R}$ with $a < b$ and any sequence $\{h_n\}_{n \in \mathrm{N}}$ of nonzero real numbers converging to zero, there exists a continuous function F: $[a, b] \to \mathbb{R}$ such that for any Lebesgue measurable function f: $[a, b] \to \mathbb{R}$ there exists a subsequence $\{h_{n_k}\}_{k \in \mathrm{N}}$ of $\{h_n\}_{n \in \mathrm{N}}$ such that*

$$\lim_{k\to\infty} \frac{F(x + h_{n_k}) - F(x)}{h_{n_k}} = f(x) \text{ a.e. on } [a, b].$$

4. Measures

4.1. *A σ-subalgebra of a countably generated σ-algebra need not be countably generated.*

4.2. *If $\mathcal{F}$ is a family of subsets of a set Ω and if A is an element of $\sigma(\mathcal{F})$, then there exists a countable subset $\mathcal{C}$ of $\mathcal{F}$ such that $A \in \sigma(\mathcal{C})$.*

4.3. *If Ω is a set having cardinality $\aleph_1$, then there does not exist a diffuse probability measure μ defined on the power set of Ω.*

4.4. *Assuming the continuum hypothesis, there does not exist a diffuse probability measure on the power set of any interval of real numbers having a positive length.*

4.5. *A σ-finite measure μ on $\mathcal{B}(\mathbb{R}^k)$ can be such that any Borel set with a nonempty interior has infinite measure.*

4.6. *A σ-finite measure on $\mathcal{B}(\mathbb{R}^k)$ need not be a Borel measure.*

4.7. *A measure μ on $\mathcal{B}(\mathbb{R}^k)$ can be σ-finite but not locally finite.*

4.8. *A Borel measure on a Hausdorff space need not be locally finite.*

4.9. *A locally finite Borel measure on a locally compact Hausdorff space need not be σ-finite.*

4.10. *For a family $\mathcal{F}$ of open intervals of real numbers, the fact that the endpoints of the intervals in $\mathcal{F}$ are dense in $\mathbb{R}$ does not imply that $\sigma(\mathcal{F}) = \mathcal{B}(\mathbb{R})$.*

4.11. *The σ-algebra generated by the open balls of a metric space need not be the family of Borel subsets.*

4.12. *If H is a Hausdorff space and if B is a closed subset of H with the subspace topology, then the Borel sets of B need not be a subset of the Borel sets of H.*

4.13. *There does not exist any σ-algebra whose cardinality is $\aleph_0$.*

4.14. *If $\{\mathcal{F}_n\}_{n\in\mathbb{N}}$ is a sequence of σ-algebras on a given set such that for each $n \in \mathbb{N}$, $\mathcal{F}_n$ is a proper subset of $\mathcal{F}_{n+1}$, then $\bigcup_{n\in\mathbb{N}} \mathcal{F}_n$ is not a σ-algebra on the given set.*

4.15. *A sequence of decreasing measures on $\mathcal{B}(\mathbb{R}^k)$ may be such that the limiting set function is not a measure.*

4.16. *The existence of a measure space $(\Omega, \mathcal{F}, \mu)$ and a sequence $\{A_n\}_{n\in\mathbb{N}}$ of measurable sets such that $A_n \supset A_{n+1}$ for each positive integer n and such that $\bigcap_{n\in\mathbb{N}} A_n = \emptyset$ does not imply that $\lim_{n\to\infty} \mu(A_n) = 0$.*

4.17. *A countably additive, real-valued function on an algebra $\mathcal{A}$ need not be extendable to a signed measure on $\sigma(\mathcal{A})$.*

4.18. *There exists an algebra $\mathcal{A}$ on a set Ω and a σ-finite measure μ on $\sigma(\mathcal{A})$ such that μ is not σ-finite on $\mathcal{A}$.*

4.19. *There exists an algebra $\mathcal{A}$ on a set Ω and a σ-finite measure μ on $\sigma(\mathcal{A})$ such that there are sets in $\sigma(\mathcal{A})$ that cannot be well approximated by sets in the algebra $\mathcal{A}$.*

4.20. *There exists an algebra $\mathcal{A}$ on a set Ω and an uncountable collection of distinct σ-finite measures on $\sigma(\mathcal{A})$ that each agree on the algebra $\mathcal{A}$, yet disagree on $\sigma(\mathcal{A})$.*

4.21. *There exists a σ-finite measure on $\mathcal{B}(\mathbb{R})$ that is not a Lebesgue–Stieltjes measure.*

4.22. *Given a σ-subalgebra $\mathcal{A}_0$ of $\mathcal{B}(\mathbb{R})$, there need not exist a function $f\colon \mathbb{R} \to \mathbb{R}$ such that $f^{-1}(\{B \subset \mathbb{R}\colon f^{-1}(B) \in \mathcal{B}(\mathbb{R})\}) = \mathcal{A}_0$.*

4.23. *There exists a measure ν on $\mathcal{M}(\mathbb{R})$ that is absolutely continuous with respect to Lebesgue measure, and yet is such that $\nu(A) = \infty$ for any Lebesgue measurable set A with a nonempty interior.*

4.24. *A σ-finite measure ν may be absolutely continuous with respect to a finite measure μ yet not be such that for every positive real number ε there* *exists a positive real number δ such that if $\mu(A) < \delta$ for some measurable set A, then $\nu(A) < \varepsilon$.*

4.25. *There exist two measures μ and ν on $\mathcal{B}(\mathbb{R}^k)$ such that μ is nonatomic and ν is atomic, yet μ is absolutely continuous with respect to ν.*

4.26. *For any positive integer k, Lebesgue measure restricted to a σ-subalgebra of the Lebesgue measurable subsets of $\mathbb{R}^k$ need not be σ-finite.*

4.27. *If $(\Omega, \mathcal{F}, m)$ is a measure space, A_1 and A_2 are subsets of Ω, and C_1 and C_2 are measurable covers of A_1 and A_2, respectively, then $C_1 \cap C_2$ need not be a measurable cover of $A_1 \cap A_2$.*

4.28. *For a measure space $(\Omega, \mathcal{F}, \mu)$, the completion of $\mathcal{F}$ with respect to μ need not be the σ-algebra of sets measurable with respect to the outer measure on $\mathbb{P}(\Omega)$ generated by μ.*

4.29. *A measure induced by an outer measure need not be saturated.*

4.30. *If $(\Omega, \mathcal{F}, m)$ is a measure space and if W is any subset of Ω such that $m^*(W) < \infty$, then m^* is a finite measure on the σ-algebra given by $\{F \cap W\colon F \in \mathcal{F}\}$.*

4.31. *If $(\Omega, \mathcal{F}, \mu)$ is a measure space and if $\mathcal{G}$ is the σ-algebra of all locally measurable sets, there may exist different saturated extensions of μ to $\mathcal{G}$.*

4.32. *Let $(\Omega, \mathcal{F})$ be a measurable space and let μ and ν be σ-finite measures on $(\Omega, \mathcal{F})$. Let $\mathcal{P}$ be a collection of elements in $\mathcal{F}$ such that $\Omega \in \mathcal{P}$ and such that the intersection of any two sets in $\mathcal{P}$ is in $\mathcal{P}$. If $\mu(A) = \nu(A)$ for $A \in \mathcal{P}$, then $\mu(A)$ need not equal $\nu(A)$ for $A \in \sigma(\mathcal{P})$.*

4.33. *A normalized regular Borel measure on a completely normal locally compact Hausdorff space can have an empty support.*

4.34. *A measure can be diffuse and atomic.*

4.35. *For a measure space $(\Omega, \mathcal{F}, \mu)$ such that there exists an atom A, a measurable subset B of A may be such that neither B nor $A \setminus B$ is a null set.*

4.36. *The sum of an atomic measure and a nonatomic measure, each defined on a given measurable space, can be atomic.*

4.37. *A normalized Borel measure on a completely normal compact Hausdorff space can have a nonempty support that has zero measure.*

4.38. *A normalized Borel measure ν on a compact completely normal Hausdorff space can have a nonempty support S such that the support of ν restricted to S is a nonempty proper subset of S.*

4.39. *A Borel measure on a compact completely normal Hausdorff space can fail to be outer regular or strongly inner regular.*

4.40. *A normalized Borel measure on a Hausdorff space can be τ-smooth but not outer regular.*

4.41. *A normalized regular Borel measure on a locally compact completely normal Hausdorff space can fail to be τ-smooth.*

4.42. *A Borel measure on a Hausdorff space can be inner regular, but fail to be compact inner regular, or outer regular.*

4.43. *A regular Borel measure on a locally compact metric space may be neither compact inner regular nor σ-finite.*

4.44. *A Borel measure on a completely normal Hausdorff space might not be outer regular, compact inner regular, or inner regular.*

4.45. *A Borel measure on a locally compact metric space may be compact inner regular but not be outer regular and not be σ-finite.*

4.46. *If $(\Omega, \mathcal{F})$ is a measurable space and μ and ν are finite measures on $(\Omega, \mathcal{F})$, it need not be true that the smallest measure not less than μ or ν is* $\max\{\mu(A), \nu(A)\}$.

4.47. *There exists a compact subset K of the reals that is either null or else non-σ-finite for any translation-invariant measure defined on the real Borel sets.*

4.48. *There exists a compact Hausdorff space H and a probability measure P on $\mathcal{B}(H)$ and a $\mathcal{B}(H)$-measurable real-valued function f such that for any continuous real-valued function g, $P(\{x \in H\colon f(x) \neq g(x)\}) \geq 1/2$.*

4.49. *There exists a compact metric space Ω and a probability measure P on $\mathcal{B}(\Omega)$ such that for any countable set $\mathcal{C}$ of disjoint closed balls of radius not exceeding unity, $\sum_{B \in \mathcal{C}} P(B) < 1/2$.*

4.50. *If B is a Bernstein subset of $\mathbb{R}$, then B is not measurable for any nonatomic, $\mathcal{B}(\mathbb{R})$ regular outer measure μ that is not identically zero.*

4.51. *There exists a compact Hausdorff space that is not the support of any compact inner regular Borel measure.*

5. Integration

5.1. *A nondecreasing sequence of functions mapping $[0, 1]$ into $[0, 1]$ may be such that each term in the sequence is Riemann integrable and such that the limit of the resulting sequence of Riemann integrals exists, but the limit of the sequence of functions is not Riemann integrable.*

5.2. *A function $f\colon \mathbb{R} \to \mathbb{R}$ that has a finite improper Riemann integral need not be bounded.*

5.3. *If f and g map $[0, 1]$ into $[0, 1]$ and f is continuous and g is Riemann integrable, $g \circ f$ need not be Riemann integrable.*

5.4. *There exists a bounded infinitely differentiable function that is not Lebesgue integrable yet whose improper Riemann integral exists and is finite.*

5.5. *There exists a bounded Lebesgue integrable function f: $[0, 1] \to \mathbb{R}$ such that there does not exist a Riemann integrable function g: $[0, 1] \to \mathbb{R}$ that is equal to f almost everywhere with respect to Lebesgue measure.*

5.6. *Integration by parts may fail for Stieltjes integrals.*

5.7. *There exist monotone nondecreasing functions f and g mapping $[-1, 1]$ into $[-1, 1]$ such that the Riemann–Stieltjes integrals*

$$\int_{-1}^{0} f(x)\, dg(x)$$

and

$$\int_{0}^{1} f(x)\, dg(x)$$

exist but the Riemann–Stieltjes integral

$$\int_{-1}^{1} f(x)\, dg(x)$$

does not exist.

5.8. *If f maps $[0, 1]$ into $\mathbb{R}$ and is continuous and nondecreasing, then $\int_{[0,1]} df$ need not equal $\int_{[0,1]} f'\, dx$.*

5.9. *If f and g map $[0, 1]$ into $\mathbb{R}$, f is nondecreasing and continuous, and g is a polynomial, then $\int_{[0,1]} fg'\, dx$ need not equal $f(1)g(1) - f(0)g(0) - \int_{[0,1]} gf'\, dx$.*

5.10. *If f and g map $\mathbb{R}$ into $\mathbb{R}$, f is strictly increasing, and g is continuously differentiable with compact support, then $\int_{\mathbb{R}} fg'\, dx$ need not equal $-\int_{\mathbb{R}} gf'\, dx$.*

5.11. *If f: $\mathbb{R} \to \mathbb{R}$ is such that $\int_{[-1,0)} f\, d\lambda$ exists and is such that $\int_{[0,1)} f\, d\lambda$ exists, then the integral of f over $[-1, 1)$ need not exist.*

5.12. *The limit of a uniformly convergent sequence of integrable functions mapping $\mathbb{R}$ into $\mathbb{R}$ need not be integrable.*

5.13. *There exists a sequence $\{f_n\}_{n\in\mathbb{N}}$ ofcontinuous real-valued functions defined on $[0, 1]$ such that $f_1(x) \geq f_2(x) \geq \cdots \geq 0$ for all $x \in [0, 1]$, such that the only continuous function f for which $f_n(x) \geq f(x) \geq 0$ for all $x \in [0, 1]$ and all $n \in \mathbb{N}$ is the identically zero function, and such that $\int_0^1 f_n(x)\, dx \not\to 0$ as $n \to \infty$.*

5.14. *The existence of a sequence of nonnegative Borel measurable functions $\{f_n\}_{n\in\mathbb{N}}$ defined on the same measure space $(\Omega, \mathcal{F}, \mu)$ such that for $n \in \mathbb{N}$ and some positive real constant K, $K \geq f_n \geq f_{n+1} \geq 0$, does not imply that $\lim_{n\to\infty} \int_\Omega f_n(\omega)\, d\mu = \int_\Omega \lim_{n\to\infty} f_n(\omega)\, d\mu$.*

5.15. *If p is a real number not less than one, I is any interval of real numbers having positive length, μ is Lebesgue measure restricted to $\mathcal{M}(I)$,* *and g is any element of $\mathbf{L}_p(I, \mathcal{M}(I), \mu)$, then the subset S of $\mathbf{L}_p(I, \mathcal{M}(I), \mu)$ consisting of all points f that are almost everywhere not greater than g on any subinterval of positive length is a closed nowhere dense subset of $\mathbf{L}_p(I, \mathcal{M}(I), \mu)$.*

5.16. *The set of all nonnegative Lebesgue integrable functions defined up to almost everywhere equivalence is a nowhere dense closed subset of $\mathbf{L}_1(\mathbb{R}, \mathcal{M}(\mathbb{R}), \lambda)$.*

5.17. *A sequence of real-valued functions defined on [0, 1] exists that converges in $\mathbf{L}_p([0, 1], \mathcal{M}([0, 1]), \lambda|_{\mathcal{M}([0, 1])})$ to a function in $\mathbf{L}_p([0, 1], \mathcal{M}([0, 1]), \lambda|_{\mathcal{M}([0, 1])})$ for any $p > 0$, yet converges pointwise at no point.*

5.18. *There exists a sequence of bounded measurable functions defined on [0, 1] that converges pointwise, yet for any $p > 0$, does not converge in $\mathbf{L}_p([0, 1], \mathcal{M}([0, 1]), \lambda|_{\mathcal{M}([0, 1])})$.*

5.19. *There exists a sequence $\{f_n\}_{n\in\mathbb{N}}$ of bounded Lebesgue integrable functions mapping $\mathbb{R}$ into $\mathbb{R}$ such that*

$$\liminf_{n\to\infty} \int_{\mathbb{R}} f_n \, d\lambda < \int_{\mathbb{R}} \liminf_{n\to\infty} f_n \, d\lambda.$$

5.20. *If $(\Omega, \mathcal{F}, \mu)$ is a finite measure space and if $\{f_\alpha\}_{\alpha\in A}$ is an increasing net of integrable functions taking values in [0, 1], then*

$$\lim_\alpha \int_\Omega f_\alpha \, d\mu$$

need not equal

$$\int_\Omega \lim_\alpha f_\alpha \, d\mu.$$

5.21. *If $(\Omega, \mathcal{F}, \mu)$ is a finite measure space and if $\{f_\alpha\}_{\alpha\in A}$ is a net of integrable functions taking values in [0, 1] for which there exists an integrable function g such that $|f_\alpha| \le g$, then*

$$\lim_\alpha \int_\Omega f_\alpha \, d\mu$$

need not equal

$$\int_\Omega \lim_\alpha f_\alpha \, d\mu.$$

5.22. *If $(\Omega, \mathcal{F}, \mu)$ is a finite measure space and if $\{f_\alpha\}_{\alpha\in A}$ is a net of integrable functions taking values in [0, 1], then*

$$\liminf_\alpha \int_\Omega f_\alpha \, d\mu$$

need not be lower bounded by

$$\int_\Omega \liminf_\alpha f_\alpha \, d\mu.$$

5.23. *There exists an integrable function mapping $\mathbb{R}$ into $\mathbb{R}$ that is not square integrable on (a, b) for any real numbers a and b such that $a < b$.*

5.24. *There exists a real-valued Lebesgue integrable function that is not bounded over any neighborhood of any point.*

5.25. *A bounded continuous function on $(0, \infty)$ may vanish at infinity yet not be in $\mathbf{L}_p((0, \infty), \mathcal{B}((0, \infty)), \lambda|_{\mathcal{B}((0,\infty))})$ for any $p > 0$.*

5.26. *For any positive integer k, there exists a σ-finite measure μ on $\mathcal{B}(\mathbb{R}^k)$ such that there exists no continuous nonidentically zero real-valued function defined on $\mathbb{R}^k$ that is integrable.*

5.27. *For any real number $p > 1$, there exists a measure space $(\Omega, \mathcal{F}, \mu)$ and a measurable, nonnegative, real-valued function f such that fg is in $\mathbf{L}_1(\Omega, \mathcal{F}, \mu)$ for all g in $\mathbf{L}_p(\Omega, \mathcal{F}, \mu)$, and yet f is not in $\mathbf{L}_q(\Omega, \mathcal{F}, \mu)$ for any real positive q.*

5.28. *For any real number $p > 1$, there exists a measure space $(\Omega, \mathcal{F}, \mu)$ such that for any two functions f and g in $\mathbf{L}_p(\Omega, \mathcal{F}, \mu)$ for which $\|f\|_{\mathbf{L}_p} \neq 0$ and $\|g\|_{\mathbf{L}_p} \neq 0$,*

$$2^{-|2-p|/p} \leq \frac{\|f+g\|^2_{\mathbf{L}_p} + \|f-g\|^2_{\mathbf{L}_p}}{2\left(\|f\|^2_{\mathbf{L}_p} + \|g\|^2_{\mathbf{L}_p}\right)} \leq 2^{|2-p|/p},$$

and such that for $p \neq 2$, these bounds are not achievable by any two functions f and g in $\mathbf{L}_p(\Omega, \mathcal{F}, \mu)$ for which $\|f\|_{\mathbf{L}_p} \neq 0$ and $\|g\|_{\mathbf{L}_p} \neq 0$.

5.29. *There exists a measure space $(\Omega, \mathcal{F}, \mu)$ such that $\mathbf{L}_1(\Omega, \mathcal{F}, \mu) = \mathbf{L}_1^*(\Omega, \mathcal{F}, \mu) \neq \mathbf{L}_\infty(\Omega, \mathcal{F}, \mu)$.*

5.30. *There exists a measure space $(\Omega, \mathcal{F}, \mu)$ and an element $L \in \mathbf{L}_1^*(\Omega, \mathcal{F}, \mu)$ such that there exists no $g \in \mathbf{L}_\infty(\Omega, \mathcal{F}, \mu)$ for which $L(f) = \int_\Omega fg\,d\mu$.*

5.31. *If $(\Omega, \mathcal{F}, \mu)$ is a normalized measure space and S is a subset of $\mathbf{L}_1(\Omega, \mathcal{F}, \mu)$ that is uniformly integrable, there need not exist $h \in \mathbf{L}_1(\Omega, \mathcal{F}, \mu)$ such that $|f| \leq |h|$ for all $f \in S$.*

5.32. *There exists a compact, completely normal Hausdorff space Ω and a nonnegative linear function L defined on the continuous real-valued functions on Ω such that there are two distinct normalized measures on $\mathcal{B}(\Omega)$, μ and ν, such that $L(f) = \int_\Omega f\,d\mu = \int_\Omega f\,d\nu$ for all continuous real-valued functions f defined on Ω and such that μ and ν each have the same nonempty support S, and yet $\mu(S) = 0$ and $\nu(S) = 1$.*

5.33. *If H is a compact Hausdorff space and m is a normalized measure on $\mathcal{B}(H)$, $C(H)$ need not be dense in $\mathbf{L}_1(H, \mathcal{B}(H), m)$.*

5.34. *There exists a compact Hausdorff space H and a finite nonidentically zero signed measure m on $\mathcal{B}(H)$ such that for any $f \in C(H)$, the integral over H of f with respect to m is zero.*

5.35. *A Radon–Nikodym derivative may exist yet not be real-valued.*

5.36. *The existence of two measures on a measurable space, the first absolutely continuous with respect to the second, does not imply the existence of a Radon–Nikodym derivative of the first with respect to the second.*

5.37. *The result of the Radon–Nikodym Theorem may hold even when neither measure of interest is σ-finite.*

5.38. *There exists a measurable space $(\Omega, \mathcal{F})$ and two equivalent measures μ_1 and μ_2 defined on $(\Omega, \mathcal{F})$ such that the Radon–Nikodym derivative of μ_2 with respect to μ_1 exists, but a Radon–Nikodym derivative of μ_1 with respect to μ_2 does not exist.*

5.39. *An integrable, square-integrable, band-limited function may be everywhere discontinuous.*

5.40. *The Fourier transform of a Lebesgue integrable function $f\colon \mathbb{R} \to [0, \infty)$ need not be integrable.*

5.41. *If $f \in \mathbf{L}_1(\mathbb{R}, \mathcal{M}(\mathbb{R}), \lambda)$ has a Fourier transform F such that $F(\omega) = 0$ for $\omega > W$, where W is a positive real number, then f need not be equal to*

$$\sum_{n=1}^{\infty} f\left(\frac{n\pi}{w}\right) \frac{\sin(n\pi - wt)}{wt},$$

where w is a real number greater than W.

5.42. *Multidimensional convolution need not be associative.*

5.43. *For any positive integer k, the multidimensional convolution of two real-valued, bounded, integrable, nowhere zero functions defined on $\mathbb{R}^k$ can be identically equal to zero.*

5.44. *A linear system mapping $\mathbf{L}_2(\mathbb{R}, \mathcal{M}(\mathbb{R}), \lambda)$ to $\mathbf{L}_2(\mathbb{R}, \mathcal{M}(\mathbb{R}), \lambda)$ described via convolution with a fixed point in $\mathbf{L}_2(\mathbb{R}, \mathcal{M}(\mathbb{R}), \lambda)$ that takes on each real value in any nonempty open subset of $\mathbb{R}$ may map each element of $\mathbf{L}_2(\mathbb{R}, \mathcal{M}(\mathbb{R}), \lambda)$ into the identically zero function.*

6. Product Spaces

6.1. *Given any bounded subsets X and Y of $\mathbb{R}^3$ having nonempty interiors, there exists a positive integer n, a partition $\{X_i\colon 1 \le i \le n\}$ of X, and a partition $\{Y_i\colon 1 \le i \le n\}$ of Y such that X_i is congruent to Y_i for each i.*

6.2. *For any integer $k > 1$, there exist non-Borel measurable convex subsets of $\mathbb{R}^k$.*

6.3. *For any integer $k > 1$, there exists a bounded real-valued function defined on a convex subset of $\mathbb{R}^k$ that is convex yet is not Borel measurable.*

6.4. *A real-valued function defined on $\mathbb{R}^2$ that is nondecreasing in each variable need not be Borel measurable.*

6.5. *There exists a real-valued function defined on the plane that is discontinuous and yet such that both partial derivatives exist and are continuous.*

6.6. *If f and g are real-valued functions of a real variable, if $h: \mathbb{R}^2 \to \mathbb{R}$ via $h(x, y) = f(x)g(y)$, if $\lim_{x\to 0} h(x, y)$ exists and is uniform in y, if the limit as $y \to 0$ of this limit exists, if $\lim_{y\to 0} h(x, y)$ exists and is uniform in x, and if the limit as $x \to 0$ of this limit exists, then it need not be true that*

$$\lim_{(x,y)\to(0,0)} h(x, y)$$

exists.

6.7. *If A and B are nonempty sets, $\mathcal{A}$ and $\mathcal{B}$ are σ-algebras on A and B, respectively, and $f: A \to B$ is an $(\mathcal{A} \times \mathcal{B})$-measurable subset of $A \times B$, then f need not be a measurable function.*

6.8. *If A and B are nonempty sets, $\mathcal{A}$ and $\mathcal{B}$ are σ-algebras on A and B, respectively, and $f: A \to B$ is a measurable function, then f need not be an $(\mathcal{A} \times \mathcal{B})$-measurable subset of $A \times B$.*

6.9. *For any positive integer k and any integer $N > 1$, there exist N subsets $S_1, S_2, \ldots, S_N$ of $\mathbb{R}^k$ that partition $\mathbb{R}^k$ and that are such that, for each positive integer $j \leq N$, S_j is a saturated non-Lebesgue measurable set.*

6.10. *A subset of $\mathbb{R}^2$ such that each vertical line intersects it in at most one point need not be a Lebesgue null set.*

6.11. *Assuming the continuum hypothesis, there exists a countable cover of $\mathbb{R}^2$ into sets of the form $\{(x, f(x)): x \in \mathbb{R}$ and $f: \mathbb{R} \to \mathbb{R}\}$ or $\{(f(x), x): x \in \mathbb{R}$ and $f: \mathbb{R} \to \mathbb{R}\}$.*

6.12. *Assuming the continuum hypothesis, there exists a subset S of $[0, 1]^2$ such that S_x is countable for all $x \in [0, 1]$, such that S^y is the complement of a countable set for all $y \in [0, 1]$, and such that S is non-Lebesgue measurable.*

6.13. *A real-valued, bounded, Borel measurable function f defined on $\mathbb{R}^2$ such that for each real number x, $f(\cdot, x)$ and $f(x, \cdot)$ are integrable and each integrates over $\mathbb{R}$ to zero may be such that f is not integrable over $\mathbb{R}^2$ with respect to Lebesgue measure on $\mathcal{B}(\mathbb{R}^2)$.*

6.14. *The product of two complete measure spaces need not be complete.*

6.15. *If T is a locally compact completely normal Hausdorff space, $\mathcal{B}(T) \times \mathcal{B}(T)$ need not equal $\mathcal{B}(T \times T)$.*

6.16. *The product of two Borel measures on locally compact completely normal Hausdorff spaces need not be a Borel measure on the product of the Hausdorff spaces.*

6.17. *If H is a locally compact Hausdorff space, if $S \subset H \times H$, and if for each compact $K \subset H$, the union over x in K of S_x is a Borel subset of H and the union over y in K of S^y is a Borel subset of H, then S need not be an element of $\mathcal{B}(H) \times \mathcal{B}(H)$.*

6.18. *A measurable set in a product of two σ-finite measure spaces can have positive measure and yet not include any measurable rectangle of positive measure.*

6.19. *For two locally compact metric spaces there may exist a closed set in the product space that is not in any product σ-algebra on the product space.*

6.20. *A continuous real-valued function defined on the product of two locally compact metric spaces need not be measurable with respect to the product of the Borel sets.*

6.21. *There exists a sequence $\{f_n\}_{n\in\mathbb{N}}$ of continuous functions mapping $[0, 1]$ into $[0, 1]^{[0,1]}$ with the product topology that converges pointwise to a function $f\colon [0, 1] \to [0, 1]^{[0,1]}$ such that, for the Borel sets of $[0, 1]^{[0,1]}$, f is not Lebesgue measurable.*

6.22. *There exists a measurable space $(\Omega, \mathcal{F})$ with a finite measure μ and a real-valued bounded function f defined on the product space that is measurable in each coordinate and yet*

$$\int_\Omega \int_\Omega f(x, y)\, d\mu(x)\, d\mu(y) \neq \int_\Omega \int_\Omega f(x, y)\, d\mu(y)\, d\mu(x).$$

6.23. *There exist two measures μ and ν on $\mathcal{B}([0, 1])$ such that if D is the diagonal in $[0, 1]^2$, then*

$$\int_{[0,1]} \left[\int_{[0,1]} \mathrm{I}_D\, d\mu\right] d\nu = 0$$

and

$$\int_{[0,1]} \left[\int_{[0,1]} \mathrm{I}_D\, d\nu\right] d\mu = 1.$$

6.24. *If for each $x \in \mathbb{R}$, I_x denotes a copy of $[0, 1]$ and U_x denotes a copy of $(0, 1]$ included therein, then $\prod_{x\in\mathbb{R}} U_x$ is not a Borel set in the compact Hausdorff space $\prod_{s\in\mathbb{R}} \mathrm{I}_x$.*

7. Basic Probability

7.1. *A nonnegative Riemann integrable function or a nonnegative Lebesgue integrable function that integrates to one need not be a probability density function.*

7.2. *The set of all probability density functions defined up to almost everywhere equivalence is a nowhere dense subset of $\mathbf{L}_1(\mathbb{R}, \mathcal{B}(\mathbb{R}), \lambda|_{\mathcal{B}(\mathbb{R})})$.*

7.3. *Two absolutely continuous probability distribution functions F and G may have disjoint supports and yet be such that $\sup_{x\in\mathbb{R}} |F(x) - G(x)|$ is less than any preassigned positive real number.*

7.4. *There exists a probability distribution μ on $\mathcal{B}(\mathbb{R})$ such that $\mu(\mathcal{B}(\mathbb{R}))$ is the Cantor ternary set.*

7.5. *For a probability distribution μ on $\mathcal{B}(\mathbb{R})$, the family of μ-continuity sets need not be a σ-algebra.*

7.6. *A probability distribution on the Borel subsets of a separable metric space need not be tight.*

7.7. *Given any positive integer k, any partition $\{S_n: 1 \leq n \leq N\}$ of $\mathbb{R}^k$ into saturated non-Lebesgue measurable subsets, any probability measure P on $(\mathbb{R}^k, \mathcal{B}(\mathbb{R}^k))$ that is equivalent to Lebesgue measure, there exists a probability space $(\mathbb{R}^k, \mathcal{F}, \mu)$ such that $\mathcal{F}$ includes $\mathcal{B}(\mathbb{R}^k)$ and $\sigma(\{S_n: 1 \leq n \leq N\})$, such that μ agrees with P on $\mathcal{B}(\mathbb{R}^k)$, and such that $\mathcal{B}(\mathbb{R}^k)$ is independent of $\sigma(\{S_n: 1 \leq n \leq N\})$.*

7.8. *There exist two uncorrelated Gaussian random variables such that they are not independent and their sum is not Gaussian.*

7.9. *A Lebesgue measurable function of a random variable need not be a random variable.*

7.10. *A random variable can have a continuous probability density function and yet not have an integrable characteristic function.*

7.11. *There do not exist two strict sense, independent, identically distributed positive random variables X and Y defined on a common probability space such that $\mathrm{E}[X/Y] \leq 1$.*

7.12. *There exists a nonempty convex subset C of $\mathbb{R}^2$ and a convex real-valued function f defined on C and a random variable X taking values in C such that $f(X)$ is not a real-valued random variable.*

7.13. *There exist two independent strict sense random variables X and Y and a function $f: \mathbb{R} \to \mathbb{R}$ such that $f(X) = Y$.*

7.14. *There exist two uncorrelated positive variance Gaussian random variables X and Y and a function $g: \mathbb{R} \to \mathbb{R}$ such that $X = g(Y)$.*

7.15. *Given any integer $n > 2$, there exist n standard Gaussian random variables $\{X_1, \ldots, X_n\}$ that are not mutually independent and not mutually Gaussian yet such that the random variables in any proper subset of $\{X_1, \ldots, X_n\}$ containing at least two elements are mutually independent and mutually Gaussian.*

7.16. *If X and Y are two random variables defined on a probability space $(\Omega, \mathcal{F}, P)$, if $C: \mathbb{R}^2 \to \mathbb{R}$ is continuous at all but one point, and if $C(X, y)$ is independent of Y and has a fixed distribution D for each real number y, then it need not follow that $C(X, Y)$ is independent of Y or that $C(X, Y)$ has the distribution D.*

7.17. *There exist simple random variables X and Y defined on a common probability space that are not independent and yet the moment generating* *function of $X + Y$ is the product of the moment generating function of X and the moment generating function of Y.*

7.18. *There exists a standard Gaussian random variable X defined on a probability space $(\Omega, \mathcal{F}, P)$ such that $X(\Omega)$ is a saturated non-Lebesgue measurable set.*

7.19. *For any integer $N > 1$, there exists a probability space and N standard Gaussian random variables $X_1, X_2, \ldots, X_N$ such that $P(X_i \in A_i$ for each $i \in I) = \prod_{i \in I} P(X_i \in A_i)$ holds whenever the sets in $\{A_i: i \in I\}$ are Borel sets but does not hold for all sets $\{A_i: i \in I\}$ for which the indicated probabilities are well defined where I is any subset of $\{1, 2, \ldots, N\}$ having cardinality at least two.*

7.20. *For any integer $N > 1$ and any diffuse distribution on $\mathbb{R}$, there exists a probability space and N random variables $X_1, X_2, \ldots, X_N$, each with the given distribution such that $P(X_i \in A_i$ for each $i \in I) = \prod_{i \in I} P(X_i \in A_i)$ holds whenever the sets in $\{A_i: i \in I\}$ are Borel sets but does not hold for all sets $\{A_i: i \in I\}$ for which the indicated probabilities are well defined where I is any subset of $\{1, 2, \ldots, N\}$ having cardinality at least two.*

7.21. *There exists a sequence $\{X_n\}_{n \in \mathbb{N}}$ of mutually independent, identically distributed, nonnegative random variables defined on a common probability space such that $\sup_{n \in \mathbb{N}} \mathrm{E}[X_n^p] < \infty$ for all positive real numbers p and yet $P(\sup_{j \in \mathbb{N}} X_{n_j} < \infty) = 0$ whenever $\{n_j\}_{j \in \mathbb{N}}$ is a subsequence of the positive integers.*

7.22. *There exists a non-Markovian random process that satisfies the Chapman–Kolmogorov equation.*

7.23. *If $\{X(t): t \in T\}$ is a stochastic process defined on some underlying probability space $(\Omega, \mathcal{F}, P)$, where T is any uncountable index set, then for any random variable Y defined on this probability space such that Y is $\sigma(\{X(t): t \in T\})$-measurable, there exists a countable subset K of T such that Y is $\sigma(\{X(t): t \in K\})$-measurable.*

7.24. *If $\{X(t): t \in \mathbb{R}\}$ is a stochastic process defined on a complete probability space $(\Omega, \mathcal{F}, P)$ and if almost all sample paths of $\{X(t): t \in \mathbb{R}\}$ are continuous, then the stochastic process need not be $(\mathcal{B}(\mathbb{R}) \times \mathcal{F})$-measurable.*

7.25. *If $\{X(t): t \in \mathbb{R}\}$ is a stochastic process defined on a complete probability space $(\Omega, \mathcal{F}, P)$ and if each sample path of $\{X(t): t \in \mathbb{R}\}$ is a.e. $[\lambda]$ continuous, then the stochastic process need not be $(\mathcal{B}(\mathbb{R}) \times \mathcal{F})$-measurable.*

7.26. *There exists a random process $\{X(t): t \in \mathbb{R}\}$ whose sample functions are each positive and Lebesgue integrable yet whose mean is not integrable.*

7.27. *A modification $\{Y(t): t \in [0, \infty)\}$ of a random process $\{X(t): t \in [0, \infty)\}$ that is adapted to a filtration $\{\mathcal{F}_t: t \in [0, \infty)\}$ need not itself be adapted to this filtration even if $\mathcal{F}_0$ is complete with respect to the underlying probability measure.*

7.28. *The σ-algebra generated by a random process $\{X(t)\colon t \in \mathbb{R}\}$ defined on a probability space such that $X(t) = 0$ a.s. need not be countably generated.*

7.29. *The sample-path continuity of a random process does not imply the continuity of its generated filtration.*

7.30. *There exists a partition $\{S_t\colon t \in \mathbb{R}\}$ of the real line, a probability space $(\Omega, \mathcal{F}, P)$, and a stationary Gaussian random process $\{X(t)\colon t \in \mathbb{R}\}$ defined thereon composed of mutually independent standard Gaussian random variables such that for any real number t, $X(t, \Omega) = S_t$.*

7.31. *There exists a random process $\{X(t)\colon t \in \mathbb{R}\}$ such that every sample path is non-Lebesgue measurable on any Lebesgue measurable set having positive Lebesgue measure, every sample path is nonzero in each uncountable closed subset of the reals, and $P(X(t) = 0) = 1$ for all real numbers t.*

7.32. *For any random process indexed by the reals, there exists a probability space and a random process indexed by the reals and defined on that probability space such that the two random processes have the same family of finite-dimensional distributions yet such that almost all of the sample paths of the second random process are not locally integrable.*

7.33. *There exists a zero mean wide sense stationary random process that does not possess a spectral representation.*

7.34. *There exists a wide sense stationary random process having periodic sample paths yet whose autocorrelation function is not periodic.*

7.35. *For a zero mean, wide sense stationary random process, the existence of an ordinary derivative pointwise on the underlying probability space does not imply the existence of a mean-square derivative.*

7.36. *Almost sure continuity does not imply almost sure sample continuity of a random process.*

7.37. *If $(\Omega, \mathcal{F}, P)$ is a probability space such that $\mathcal{F}$ is separable with respect to the pseudometric d defined via $d(A, B) = P(A \,\triangle\, B)$, then no random process of strict sense mutually independent random variables defined on this probability space indexed by an uncountable parameter set exists.*

7.38. *There exists a Gaussian random process indexed by the reals whose autocorrelation function is almost everywhere zero yet whose variance takes on every nonnegative real number in any nonempty open interval of real numbers.*

7.39. *There exists a strictly stationary, mean-square continuous, zero mean Gaussian random process $\{X(t)\colon t \in \mathbb{R}\}$ whose spectral measure has compact support and yet is such that no sample path is Lebesgue measurable on any Lebesgue measurable set of positive Lebesgue measure.*

7.40. *There exists a Gaussian random process $\{X(t)\colon t \in \mathbb{R}\}$ defined on a probability space $(\Omega, \mathcal{F}, P)$ such that each sample path takes on every nonzero real number c times on any nonempty perfect set.*

7.41. *There exists a probability space $(\Omega, \mathcal{F}, P)$ and a stationary zero mean Gaussian random process $\{X(t)\colon t \in \mathbb{R}\}$ defined thereon such that for any real number t, the range of $X(t)$ is a saturated non-Lebesgue measurable set.*

7.42. *There exists a probability space $(\Omega, \mathcal{F}, P)$, and a stationary Gaussian random process $\{X(t)\colon t \in \mathbb{R}\}$ defined thereon and composed of mutually independent standard Gaussian random variables such that, given any real number v, there exists one and only one real number t for which $X(t, \omega) = v$ for some $\omega \in \Omega$.*

7.43. *There exists a probability space $(\Omega, \mathcal{F}, P)$, and a stationary Gaussian random process $\{X(t)\colon t \in \mathbb{R}\}$ defined thereon and composed of mutually independent standard Gaussian random variables such that, given any fixed $\omega \in \Omega$, there exist no distinct real numbers t_1 and t_2 such that $X(t_1, \omega) = X(t_2, \omega)$.*

7.44. *There exists a probability space $(\Omega, \mathcal{F}, P)$, and a stationary Gaussian random process $\{X(t)\colon t \in \mathbb{R}\}$ defined thereon and composed of mutually independent standard Gaussian random variables such that for any real number t, $\bigcup_{y\in\mathbb{R}\setminus\{t\}} X(y, \Omega)$ is a proper subset of the real line.*

7.45. *There exists a probability space $(\Omega, \mathcal{F}, P)$, and a stationary Gaussian random process $\{X(t)\colon t \in \mathbb{R}\}$ defined thereon and composed of mutually independent standard Gaussian random variables such that for any disjoint subsets A_1 and A_2 of $\mathbb{R}$ and any $\omega \in \Omega$, it follows that $X(A_1, \omega) \cap X(A_2, \omega) = \emptyset$.*

7.46. *A Gaussian process indexed by $[0, \infty)$ with continuous sample paths and with stationary, mutually independent increments need not be a Brownian motion.*

7.47. *The Skorohod imbedding theorem may fail for $\mathbb{R}^2$-valued Brownian motion.*

7.48. *There exists a Gaussian random process $\{X(t)\colon t \in [0, \infty)\}$ taking values in $\mathbb{R}^2$ such that the inner product of $X(t)$ with any point in $\mathbb{R}^2$ other than the origin is a standard real-valued Brownian motion, yet $\{X(t)\colon t \in [0, \infty)\}$ is not standard Brownian motion.*

8. Conditioning

8.1. *Conditional probability and knowledge need not be related.*

8.2. *Conditional probabilities need not be measures.*

8.3. *Regular conditional probabilities need not exist.*

8.4. *Independence need not imply conditional independence.*

8.5. *Conditional independence need not imply independence.*

8.6. *Conditional expectation need not be a smoothing operator.*

8.7. *The existence of* $\mathrm{E}[\mathrm{E}[X|Y]]$ *does not imply that the mean of* X *exists.*

8.8. *Conditional expectation of an integrable random variable need not be obtainable from a corresponding conditional probability distribution.*

8.9. $\mathrm{E}[X \mid \mathcal{F}]$ *and* $\mathrm{E}[Y \mid \mathcal{F}]$ *need not be independent if* X *and* Y *are independent.*

8.10. *There exist random variables* X, Y_1, *and* Y_2 *such that* $\mathrm{E}[X \mid Y_1] = \mathrm{E}[X \mid Y_2] = 0$, *and yet* $\mathrm{E}[X \mid Y_1, Y_2] = X$.

8.11. *Conditional expectation need not minimize mean-square error.*

8.12. *The existence of a joint probability density function need not imply that the regression function satisfies any regularity property beyond Borel measurability.*

8.13. *A* σ*-subalgebra that includes a sufficient* σ*-subalgebra need not be sufficient.*

8.14. *There exists a sufficient statistic that is a function of a random variable that is not a sufficient statistic.*

8.15. *The smallest sufficient* σ*-algebra including two sufficient* σ*-algebras need not exist.*

8.16. *Uniform integrability and almost sure convergence of a sequence of random variables need not imply convergence of the corresponding conditional expectations, and Fatou's lemma need not hold for conditional expectations.*

8.17. *There exists a martingale* $\{X_n, \mathcal{F}_n: n \in \mathbb{N}\}$ *defined on a probability space* $(\Omega, \mathcal{F}, P)$ *and an almost surely finite stopping time* T *relative to the filtration* $\{\mathcal{F}_n: n \in \mathbb{N}\}$ *such that* $T \geq 1$ *a.s. and yet such that* $\mathrm{E}[X_T \mid \mathcal{F}_1]$ *is not almost surely equal to* X_1.

9. Convergence in Probability Theory

9.1. *A sequence* $\{Y_1, Y_2, \ldots\}$ *of integrable random variables may be mutually independent yet not be such that* $\mathrm{E}[Y_1 Y_2 \cdots]$ *is equal to* $\mathrm{E}[Y_1]\mathrm{E}[Y_2]\cdots$.

9.2. *For any positive real number* B, *there exists a sequence* $\{X_n\}_{n \in \mathbb{N}}$ *of integrable random variables and an integrable random variable* X *such that* X_n *converges to* X *pointwise and yet* $\mathrm{E}[X_n] = -B$ *and* $\mathrm{E}[X] = B$.

9.3. *If a random variable* X *and a sequence of random variables* $\{X_n\}_{n \in \mathbb{N}}$ *are defined on a common probability space, if* X_n *has a unique median for each positive integer* n, *and if* $X_n \to X$ *almost surely as* $n \to \infty$, *then it need not follow that the sequence of medians of the* X_n*'s converges.*

9.4. *There exists a sequence of probability distribution functions* $\{F_n: n \in \mathbb{N}\}$ *such that the corresponding sequence of characteristic functions* $\{C_n$:

$n \in \mathbb{N}\}$ converges to a function $C(t)$ for all t, yet $C(t)$ is not a characteristic function.

9.5. *A density of a centered normalized sum of a sequence of second-order, mutually independent, identically distributed random variables may exist yet not converge to a standard Gaussian probability density function.*

9.6. *There exists a sequence $\{\mu\}_{n\in\mathrm{N}}$ of probability distributions and a probability distribution μ such that each of these probability distributions has a probability density function and such that μ_n converges in the weak* topology to μ and yet the probability density function of μ_n does not converge to that of μ as $n \to \infty$.*

9.7. *There exists a sequence $\{\mu\}_{n\in\mathrm{N}}$ of probability distributions and a probability distribution μ such that each of these probability distributions has an infinitely differentiable probability density function, such that μ_n converges in the weak* topology to μ and yet for any $\varepsilon \in (0, 1)$ there exists a closed set C such that $\mu(C) = 1-\varepsilon$ and such that $\mu_n(C)$ does not converge to $\mu(C)$ as $n \to \infty$.*

9.8. *There exists a sequence of atomic probability distributions that converge in the weak* topology to an absolutely continuous probability distribution.*

9.9. *If $\{X_n\}_{n\in\mathrm{N}}$ is a sequence of mutually independent identically distributed random variables defined on the same probability space such that the common probability distribution function has at least two points of increase, then there does not exist any random variable X defined on that probability space such that X_n converges to X in probability.*

9.10. *If $\{X_n\}_{n\in\mathrm{N}}$ is a sequence of mutually independent random variables defined on a common probability space, then the almost sure unconditional convergence of*

$$\sum_{n=1}^{\infty} X_n$$

to a real-valued random variable need not imply the almost sure convergence of

$$\sum_{n=1}^{\infty} |X_n|$$

to a real-valued random variable.

9.11. *There exists a sequence $\{X_n\}_{n\in\mathrm{N}}$ of mutually independent, zero mean random variables defined on a common probability space such that*

$$\frac{1}{n}\sum_{i=1}^{n} X_i$$

converges to $-\infty$ with probability one.

9.12. *The necessary conditions of a weak law of large numbers due to Kolmogorov are not necessary.*

9.13. *There exists a martingale with a constant positive mean that converges almost surely to zero in finite time and yet with positive probability exceeds any real number.*

9.14. *A last element of a martingale need not exist.*

9.15. *A martingale may converge without being* $\mathbf{L}_1$ *bounded.*

9.16. *There exists a nonnegative martingale* $\{(X_n, \mathcal{F}_n): n \in \mathbb{N}\}$ *such that* $\mathrm{E}[X_n] = 1$ *for all* $n \in \mathbb{N}$ *and such that* X_n *converges to zero pointwise as* $n \to \infty$ *yet such that* $\sup_{n \in \mathbb{N}} X_n$ *is not integrable.*

9.17. *A martingale that is indexed by a directed set can be* $\mathbf{L}_2$ *bounded and converge in* $\mathbf{L}_2$ *and yet not converge for any point off a null set.*

9.18. *There exists a martingale* $\{X(t), \mathcal{F}_t: t \in \mathbb{R}\}$ *such that for any positive real number* p, $\mathrm{E}[|X(t)|^p] < \infty$ *for all real numbers* t *and yet for almost no point on the underlying probability space does* $X(t)$ *converge as* $t \to \infty$.

9.19. *If* $\{X(t): t \in \mathbb{R}\}$ *is a random process defined on a probability space such that for each* $t \in \mathbb{R}$, $X(t) = 0$ *a.s., the set on which* $\lim_{t\to\infty} X(t) = 0$ *need not be an event.*

9.20. *If* $\{Y_n\}_{n \in \mathbb{N}}$ *is a sequence of random variables defined on a common probability space such that* $Y_n \to 0$ *in probability as* $n \to \infty$, *and if* $\{X(t) : t \in [0, \infty)\}$ *is an* $\mathbb{N}$*-valued random process defined on the same probability space such that* $X(t) \to \infty$ *a.s. as* $t \to \infty$, *then* $Y_{X(t)}$ *need not converge to zero in probability as* $t \to \infty$.

10. Applications of Probability

10.1. *There exist two complex-valued, uncorrelated, mutually Gaussian random variables that are not independent.*

10.2. *There exists a linear filter described via convolution with a nowhere zero function that is a no-pass filter for a certain family of input random processes.*

10.3. *There exists a bounded random variable* X *and a symmetric, bounded function* C *mapping* $\mathbb{R}$ *into* $\mathbb{R}$ *that is nondecreasing on* $[0, \infty)$ *such that there exists no one-level quantizer* Q *with the property that* $\mathrm{E}[C(X-Q(X))]$ *is minimized over all one-level quantizers.*

10.4. *A random variable* X *having a symmetric probability density function may be such that no two-level quantizer* $\tilde{Q}$ *such that* $\tilde{Q}(x) = -\tilde{Q}(-x)$ *for all real x minimizes the quantity* $\mathrm{E}[[X - Q(X)]^2]$ *over all two-level quantizers* Q.

10.5. *For any positive integer* k *and any integer* $N > 1$, *a probability space, a Gaussian random vector* X *defined on the space taking values in* $\mathbb{R}^k$ *and* *having a positive definite covariance matrix, and an N-level quantizer Q exist such that the random vector $Q(X)$ takes on each of the N values in its range with equal probability and such that X and $Q(X)$ are independent.*

10.6. *The martingale convergence theorem need not be a useful estimation technique even when every random variable of concern is Gaussian.*

10.7. *For any positive integer N, a perfectly precise Kalman filtering scheme may fail completely when based upon observations subjected to an N-level maximum entropy quantizer.*

10.8. *A Kalman filter may fail to provide a useful estimate based upon observations subjected to an arbitrarily high-level quantizer even when the quantity being estimated may be reconstructed precisely from a single observation.*

10.9. *For an integer $n > 2$, there exists a set $\{Y_1, \ldots, Y_n\}$ of random variables such that Y_n is a strict sense random variable that is equal to a polynomial function of Y_i for some positive integer $i < n$, yet the best $\mathbf{L}_2$ linear unbiased estimator of Y_n based on $\{Y_1, \ldots, Y_{n-1}\}$ is zero.*

10.10. *There exists a probability density function for which conventional importance sampling is a useless variance reduction technique.*

10.11. *There exists a probability density function for which improved importance sampling is a useless variance reduction technique.*

10.12. *There exists a situation involving a Gaussian signal of interest and mutually Gaussian observations in which methods of fusion are useless.*

10.13. *A popular Markovian model for bit error rate estimation may lead to misleading estimates in practical applications.*

10.14. *Sampling a strictly stationary random process at a random time need not preserve measurability.*

10.15. *There exists a random process $\{X(t)\colon t \in \mathbb{R}\}$ that is not stationary yet is such that for any positive real number Δ the random process $\{X(n\Delta)\colon n \in \mathbb{Z}\}$ is strictly stationary.*

10.16. *Estimating an autocorrelation function via a single sample path can be futile.*

10.17. *A method of moments estimator need not be unbiased.*

10.18. *A method of moments estimator need not be a reasonable estimator.*

10.19. *An unbiased estimator need not exist.*

10.20. *A situation exists in which a biased constant estimator of the success probability in a binomial distribution has a smaller mean-square error than an unbiased relative frequency estimator.*

10.21. *A minimum variance unbiased estimator need not exist.*

10.22. *A unique minimum variance unbiased estimator need not be a reasonable estimator.*

10.23. *There exists a situation in which a single observation from a distribution with a finite variance is a better estimator of the mean than is a sample mean of two such observations.*

10.24. *Ancillary statistics need not be location invariant.*

10.25. *The Fisher information may stress small values of a parameter yet stress large values when estimating a strictly increasing function of the parameter.*

10.26. *There exists a sequence of estimators with an asymptotic variance that is never greater than the Cramer–Rao lower bound and that is less than that bound for certain values of the parameter of interest.*

10.27. *A maximum likelihood estimator need not exist.*

10.28. *A maximum likelihood estimator need not be unique.*

10.29. *A maximum likelihood estimator need not be unbiased or admissible.*

10.30. *A maximum likelihood estimator need not be consistent.*

10.31. *A maximum likelihood estimator of a real parameter may be strictly less than the parameter with probability one.*

10.32. *The computational demands of a maximum likelihood estimator may increase dramatically with the number of observations.*

10.33. *A maximum likelihood estimator may be unique for an odd number of observations yet fail to be unique for an even number of observations.*

10.34. *A maximum likelihood estimator of a real parameter may increase without bound as the number of observations upon which the estimate is based approaches infinity.*

10.35. *There exists a probability space* $(\Omega, \mathcal{F}, P)$, *a stationary Gaussian random process* $\{X(t): t \in \mathbb{R}\}$ *defined thereon and composed of mutually independent standard Gaussian random variables, and a function* $f: \mathbb{R}^2 \to \mathbb{R}$ *such that for any fixed but unknown real number* θ, *any real number* t, *and any point* ω *from* Ω, $f(t, X(t - \theta, \omega)) = \theta$.

10.36. *There exists a probability space* $(\Omega, \mathcal{F}, P)$, *and a stationary Gaussian random process* $\{X(t): t \in \mathbb{R}\}$ *defined thereon and composed of mutually independent standard Gaussian random variables, and a function* $f: \mathbb{R} \to \mathbb{R}$ *such that for any* $t \in \mathbb{R}$ *and any* $\omega \in \Omega$, $f(X(t, \omega)) = t$.

10.37. *A most powerful statistical hypothesis test need not correspond to an optimal detection procedure.*

10.38. *A Neyman–Pearson detector for a known positive signal corrupted by additive noise possessing a unimodal and symmetric probability density* *function may announce the absence of the signal for large values of the observation.*

10.39. *A uniformly most powerful test of a simple hypothesis against a composite alternative may exist for one size yet not exist for another size.*

Part II
The Counterexamples with Proofs

1. The Real Line

The real line is the playground of real analysis. Although many introductory treatments of the real line might lead one to believe that the real line is rather lackluster and dull, such is not the case. In this chapter we will explore several pathological properties and subsets of the real line. Before proceeding, however, we will first present some useful definitions.

A subset P of $\mathbb{R}$ is said to be perfect if it is closed and if every point of P is a limit point. A subset A of $\mathbb{R}$ is said to be nowhere dense if its closure has an empty interior. A subset of $\mathbb{R}$ is said to be of the first category if it is a countable union of nowhere dense sets and otherwise is said to be of the second category. A subset of $\mathbb{R}$ is said have the property of Baire if it is given by the symmetric difference of an open set and a set of the first category.

A Hamel basis H for $\mathbb{R}$ over $\mathbb{Q}$ is a subset of $\mathbb{R}$ such that each nonzero real number may be uniquely represented as a finite linear combination of distinct elements from H with nonzero rational coefficients. That such a set exists is a consequence of the axiom of choice and is shown in Example 1.29.

A subset C of $\mathbb{R}$ is said to be a Cantor-like set if C is an uncountable, perfect, nowhere dense subset of $\mathbb{R}$. Such sets are constructed in Example 1.8. The Cantor ternary set, also given in Example 1.8, is an uncountable, perfect, nowhere dense, Lebesgue null subset of $\mathbb{R}$. In Example 1.9 we show that each element of the Cantor ternary set may be expressed as a sum of the form $\sum_{n=1}^{\infty} c_n/3^n$, where $c_n \in \{0, 2\}$ for each n. That the cardinality of the Cantor ternary set is c follows as an immediate consequence of this last property.

Consider a measure m on $(\mathbb{R}, \mathcal{B}(\mathbb{R}))$ and let m_* denote the inner measure on $(\mathbb{R}, \mathbb{P}(\mathbb{R}))$ induced by m. That is, for any subset A of $\mathbb{R}$, $m_*(A) = \sup\{m(B)\colon A \supset B \in \mathcal{B}(\mathbb{R})\}$. A subset S of $\mathbb{R}$ is said to be a saturated non-m-measurable subset if $m_*(S) = m_*(S^c) = 0$. In Example 1.46 we show for any cardinal β with $2 \leq \beta \leq c$ that the real line may be partitioned into β saturated non-Lebesgue measurable subsets.

A subset B of $\mathbb{R}$ is said to be a Bernstein set if $B \cap C$ and $B^c \cap C$ are nonempty for any uncountable closed subset C of $\mathbb{R}$. In Example 1.22 we construct a Bernstein set, and in Example 1.46 we show that a Bernstein set is a saturated non-Lebesgue measurable set.

A topological space that is metrizable by a metric for which the metric space is complete and separable is called a Polish space. A subset A of $\mathbb{R}$ is said to be an analytic set if it is of the form $f(B)$ for some continuous function f mapping a Borel subset B of some Polish space X into $\mathbb{R}$. Clearly any Borel subset of $\mathbb{R}$ is analytic. In addition, any analytic subset of $\mathbb{R}$ is Lebesgue measurable. In Example 1.39 we show that no Hamel basis for $\mathbb{R}$ over $\mathbb{Q}$ is analytic.

Example 1.1. *There exists a subset of $\mathbb{R}$ such that the infimum of the set is greater than the supremum of the set.*

Proof: Consider the empty set. The infimum of the empty set is ∞, and the supremum of the empty set is $-\infty$. □

Example 1.2. *Given any nonzero real number a, there exists a metric on $\mathbb{R}$ such that a is closer to 0 than to any other real number.*

Proof: Define a metric d on $\mathbb{R}$ via

$$d(x, y) = \begin{cases} 0, & \text{if } x = y \\ 1, & \text{if } x = 0 \text{ and } y = a \text{ or if } x = a \text{ and } y = 0 \\ 2, & \text{if } x = 0 \text{ and } y \neq a \text{ or if } x = a \text{ and } y \neq 0 \\ 3, & \text{if neither } x \text{ nor } y \text{ is either 0 or } a. \end{cases}$$

The desired result follows immediately. □

Example 1.3. *There exists an uncountable family of distinct infinite subsets of $\mathbb{N}$ such that the intersection of any two distinct sets from this family is finite.*

Proof: Consider a fixed enumeration of the rationals. For each number t in $[0, 1]$, fix an infinite sequence of distinct rationals that converges to t. Now, define a subset N_t of $\mathbb{N}$ as the indices of this sequence of rational numbers. Obviously, N_t is an infinite set. Further, note that for distinct elements t_1 and t_2 from $[0, 1]$, $N_{t_1} \cap N_{t_2}$ is a finite set since the corresponding sequences are converging to distinct real numbers. □

Example 1.4. *Given any positive number ε, there exists an open subset U of $\mathbb{R}$ such that U contains each of the rationals, yet the Lebesgue measure of U is less than ε.*

Proof: Let $\{r_n: n \in \mathbb{N}\}$ be an enumeration of the rationals and for each positive integer n define an open interval U_n via

$$U_n = \left(r_n - \frac{\varepsilon}{2^{n+2}}, r_n + \frac{\varepsilon}{2^{n+2}}\right).$$

Note that U_n contains r_n and that the Lebesgue measure of U_n is equal to $\varepsilon/2^{n+1}$. Thus, the set $U = \bigcup_{n\in\mathbb{N}} U_n$ is an open subset of $\mathbb{R}$ that contains every rational number. Further, via countable subadditivity, the Lebesgue measure of U is not greater than $\sum_{n=1}^{\infty}(\varepsilon/2^{n+1}) = \varepsilon/2$, which is less than ε. □

Example 1.5. *There exists a subset S of $\mathbb{R}$ such that* $\inf\{\varepsilon > 0: S \oplus \varepsilon = S^c\} = 0$.

Proof: Let $D = \{(m/3^n): m \in \mathbb{Z}, n \in \mathbb{N} \cup \{0\}\}$ and note that the relation $\sim$ defined by $x \sim y$ if $x - y \in D$ is an equivalence relation on $\mathbb{R}$. Let E be any subset of $\mathbb{R}$ containing exactly one point from each set in the partition of $\mathbb{R}$ provided in the usual manner by this equivalence relation. Let $\{P_x: x \in E\}$ denote this partition of $\mathbb{R}$ into equivalence classes. For each $x \in E$, we may partition P_x into two sets A_x and B_x, where $A_x = \{x + (m/3^n): m \in \mathbb{Z}, n \in \mathbb{N} \cup \{0\}, m \text{ even}\}$ and $B_x = \{x + (m/3^n): m \in \mathbb{Z}, n \in \mathbb{N} \cup \{0\}, m \text{ odd}\}$. Now, let n be a nonnegative integer and let $\alpha = 1/3^n$. Note that $A_x \oplus \alpha = B_x$ for each $x \in E$, since, if $y \in A_x \oplus \alpha$, then for some even integer m and nonnegative integer k

$$y = x + \frac{m}{3^k} + \frac{1}{3^n} = x + \frac{m3^n + 3^k}{3^{k+n}} \in B_x,$$

and if $y \in B_x$, then for some odd integer m and nonnegative integer k

$$y = x + \frac{m}{3^k} = x + \left(\frac{m3^n - 3^k}{3^{k+n}}\right) + \frac{1}{3^n} \in A_x \oplus \alpha.$$

Thus, if $S = \bigcup_{x \in E} A_x$, then

$$S \oplus \frac{1}{3^n} = \bigcup_{x \in E} A_x \oplus \frac{1}{3^n} = \bigcup_{x \in E} B_x = S^c$$

for each nonnegative integer n, and hence $\inf\{\varepsilon > 0\colon\ S \oplus \varepsilon = S^c\} = 0$. □

Example 1.6. *There do not exist Lebesgue measurable subsets of* $\mathbb{R}$ *with positive Lebesgue measure and with cardinality less than c.*

Proof: This example is from Briggs and Schaffter [1979]. We will show that if A is a Lebesgue measurable set with positive Lebesgue measure, then the cardinality of A is c. Let F be a closed subset of A having positive Lebesgue measure, and let I_0 and I_1 be disjoint closed intervals such that the Lebesgue measure of each is less than that of F and such that the intersection of either with F has positive Lebesgue measure. Note that we can do this since for any Lebesgue measurable set L of positive Lebesgue measure, there exists a real number x such that $L \cap (x, \infty)$ and $L \cap (-\infty, x)$ each have positive Lebesgue measure. Now, let I_{00} and I_{01} be disjoint closed subintervals of I_0 such that each has Lebesgue measure less than the Lebesgue measure of $F \cap I_0$ and such that each intersects F with positive Lebesgue measure. Similarly, let I_{10} and I_{11} be subintervals of I_1 with analogous properties. Continuing, we obtain, at the nth step, 2^n disjoint closed intervals $I_{j_1,\dots,j_n}$ where j_i equals zero or one for each i and where each interval intersects F with positive Lebesgue measure. Note that any infinite sequence $\{j_i\}_{i \in \mathbb{N}}$ taking values in $\{0, 1\}$ corresponds to an infinite sequence of nested closed subintervals whose nth term is $I_{j_1,\dots,j_n}$. Note that the intersection of the elements of such a sequence of intervals is nonempty. We say that the points in this nonempty set correspond to the sequence that produced it in this matter. Now, simply note that distinct infinite sequences of zeros and ones correspond to disjoint subsets of F. Since there are c such sequences, the result now follows. □

Example 1.7. *There exists an uncountable collection of distinct subsets of* $\mathbb{N}$ *that are totally ordered by the inclusion relation.*

Proof: This example is from Golomb and Gilmer [1974]. Let $f\colon \mathbb{N} \to \mathbb{Q}$ be one-to-one and, for each $t \in \mathbb{R}$, let $N_t = \{n \in \mathbb{N}\colon\ f(n) < t\}$. Note that the collection $\{N_t\colon\ t \in \mathbb{R}\}$ contains uncountably many distinct sets since if $t \neq s$, then $N_t \neq N_s$. Further, if $s \leq t$, then $N_s \subset N_t$. That is, the elements in $\{N_t\colon\ t \in \mathbb{R}\}$ are totally ordered by the inclusion relation. □

Example 1.8. *For any closed interval I of* $\mathbb{R}$ *such that $\lambda(I) > 0$ and any $\alpha \in [0, \lambda(I))$, there exists a Cantor-like subset of I having Lebesgue measure α.*

Proof: Consider a closed subinterval I of $\mathbb{R}$ such that $\lambda(I) > 0$ and let α be a real number such that $0 \leq \alpha < \lambda(I)$. Further, let $\gamma = \lambda(I) - \alpha$, and let

$\{\beta_n\}_{n\in\mathbb{N}}$ be a sequence of positive numbers such that $\sum_{n=1}^{\infty} 2^{n-1}\beta_n = \gamma$. Next, from the middle of the interval I remove the open subinterval of length β_1, from the middle of each of the remaining two closed intervals remove the open subinterval of length β_2, and from the middle of each of the remaining four closed intervals remove the open subinterval of length β_3. Continuing this procedure, we obtain a countable collection of disjoint open subintervals of I whose union forms an open set A such that $\lambda(A) = \gamma$. The set $I \setminus A$ has Lebesgue measure α and is a called a Cantor-like subset of I. As we will now show, it also follows that $I \setminus A$ is a perfect, nowhere dense subset of I.

Now, note that at the nth stage of the above construction, the lengths of each of the remaining 2^n closed intervals are no more than $2^{-n}\lambda(I)$. Hence, the resulting Cantor-like set includes no nonempty open interval, and thus, since it is closed, we see that it is nowhere dense. Finally, for any point $x \in I \setminus A$, let A_n be the remaining closed subinterval at the nth stage of the above construction that contains x. Since the right endpoints of the A_n's are elements of $I \setminus A$ that converge to x, we see that $I \setminus A$ is perfect.

If $I = [0, 1]$ and $\beta_n = 3^{-n}$, then $I \setminus A$ is called the Cantor ternary set and is an uncountable, perfect, nowhere dense, Lebesgue null set. □

Example 1.9. *Each element in the Cantor ternary set may be written in the form $\sum_{n=1}^{\infty} x_n/3^n$, where $x_n \in \{0, 2\}$ for each n.*

Proof: Recall that each point $x \in [0, 1]$ has a ternary expansion of the form $x = \sum_{n=1}^{\infty} x_n/3^n$, where $x_n \in \{0, 1, 2\}$ for each n, and that this expansion is unique unless x has a finite expansion of the form $x = \sum_{n=1}^{m} x_n/3^n$, where $x_m \in \{1, 2\}$. If, in such a case, $x_m = 2$, then we will express x via this finite expansion, but if $x_m = 1$, then we will express x as

$$x = \frac{x_1}{3} + \cdots + \frac{x_{m-1}}{3^{m-1}} + \sum_{n=m+1}^{\infty} \frac{2}{3^n}.$$

In this way each $x \in [0, 1]$ has a unique ternary expansion.

Next, let C_n denote the union of the 2^n closed intervals remaining at the nth stage of the construction of the Cantor ternary set C and recall that $C = \bigcap_{n=1}^{\infty} C_n$. Since the intervals that make up C_{n+1} are obtained by removing the middle thirds from the intervals that make up C_n, we see that

$$C_n = \left\{ x \in [0, 1]: x = \sum_{m=1}^{\infty} \frac{x_m}{3^m}, \text{ with } x_i \in \{0, 2\} \text{ for } 1 \le i \le n \right\}.$$

Thus, $x = \sum_{n=1}^{\infty} x_n/3^n \in C = \bigcap_{n=1}^{\infty} C_n$ if and only if $x_n \in \{0, 2\}$ for each positive integer n. □

Example 1.10. *If C denotes the Cantor ternary set, then $C \oplus C = [0, 2]$.*

Proof: Recall from Example 1.9 that C is given by the set of all real numbers expressible as

$$\sum_{n=1}^{\infty} \frac{c_n}{3^n},$$

where for each positive integer n, $c_n \in \{0, 2\}$. Hence, $C \oplus C$ is given by all real numbers of the form

$$\sum_{n=1}^{\infty} \frac{b_n}{3^n} + \sum_{n=1}^{\infty} \frac{c_n}{3^n},$$

where for each positive integer n, b_n and c_n belong to $\{0, 2\}$. Now since the preceding sum can be written as

$$2\sum_{n=1}^{\infty} \frac{(b_n + c_n)/2}{3^n},$$

we see that $C \oplus C = [0, 2]$. □

Example 1.11. *If C denotes the Cantor ternary set, then $C \ominus C = [-1, 1]$.*

Proof: Recall from Example 1.9 that C is given by the set of all reals numbers expressible as

$$\sum_{n=1}^{\infty} \frac{c_n}{3^n},$$

where for each positive integer n, $c_n \in \{0, 2\}$. Hence, $C \ominus C$ is given by all real numbers of the form

$$\sum_{n=1}^{\infty} \frac{b_n}{3^n} - \sum_{n=1}^{\infty} \frac{c_n}{3^n},$$

where for each positive integer n, b_n and c_n belong to $\{0, 2\}$. Now since the preceding difference can be written as

$$2\sum_{n=1}^{\infty} \frac{(b_n - c_n)/2}{3^n},$$

we see that $C \ominus C = [-1, 1]$. □

Example 1.12. *If F is a first category subset of $\mathbb{R}$, then $F^c \oplus F^c = \mathbb{R}$.*

Proof: Let F be a first category subset of $\mathbb{R}$, and let a be any real number. Note that $a \ominus F^c$ is the complement of a first category subset of $\mathbb{R}$, and note that $F^c \cap (a \ominus F^c)$ is nonempty. Let b be an element of this intersection and note that $b = a - x$ for some x in F^c. □

Example 1.13. *There exists a subset of $\mathbb{R}$ that is dense, has cardinality c, and is of the first category.*

Proof: For such a set, consider the union of the Cantor ternary set and the rationals. □

Example 1.14. *The image of a Lebesgue measurable subset of the real line under a continuous transformation need not be Lebesgue measurable.*

Proof: Let f denote the Cantor–Lebesgue function on $[0, 1]$ that is given in Example 2.6. Note that $f(0) = 0$, $f(1) = 1$, f is nondecreasing and continuous on $[0, 1]$, and f is constant on every interval removed in the construction of the Cantor ternary subset C of $[0, 1]$. Since there are only countably many intervals removed during the construction of C, it is clear that $f([0, 1] \setminus C)$ is a countable subset of $[0, 1]$. Hence, $f(C)$ is Lebesgue measurable with Lebesgue measure one. Recall that any set of positive Lebesgue measure has a subset that is not Lebesgue measurable. Let A be such a subset of the set $f(C)$. Then $f^{-1}(A)$ is a subset of the null set C and hence is Lebesgue measurable with Lebesgue measure zero. Thus, f is continuous yet maps the measurable set $f^{-1}(A)$ onto the nonmeasurable set A. □

Example 1.15. *There exists a continuous function f mapping $[0, 1]$ into $[0, 1]$ and a Lebesgue measurable set L such that $f^{-1}(L)$ is not Lebesgue measurable.*

Proof: Consider the function g and the set $h(N)$ constructed in Example 2.4. The set $h(N)$ is Lebesgue measurable and g is continuous, yet $g(h(N)) = N$ is not Lebesgue measurable. □

Example 1.16. *The image of a Lebesgue null subset of the real line under a continuous transformation may be Lebesgue measurable yet not have zero Lebesgue measure.*

Proof: Adopting the notation of Example 1.14, it follows that the function f is continuous yet maps C, which has Lebesgue measure zero, to $f(C)$, which has Lebesgue measure 1. Thus, the image of a null set under a continuous transformation may be measurable yet not have zero measure. In fact, as demonstrated in Example 1.14, the image of a measurable set under a continuous transformation need not even be measurable. □

Example 1.17. *If A is a Lebesgue measurable subset of $\mathbb{R}$ having positive Lebesgue measure, then $A \ominus A$ contains an interval of the form $(-\alpha, \alpha)$ for some $\alpha > 0$.*

Proof: Assume without loss of generality that $\lambda(A)$ is finite. Let U be an open subset of $\mathbb{R}$ such that $A \subset U$ and such that $\lambda(U \setminus A) \le \lambda(A)/3$. Let $\{I_n\}_{n\in\Lambda}$ for $\Lambda \subset \mathbb{N}$ be a collection of disjoint, nonempty open intervals such that $\bigcup_{n\in\Lambda} I_n = U$. Note that

$$\sum_{n\in\Lambda} \lambda(I_n \cap A) = \lambda(A) \ge \frac{3}{4}\lambda(U) = \frac{3}{4}\sum_{n\in\Lambda} \lambda(I_n).$$

Thus, there exists some $m \in \Lambda$ such that $\lambda(I_m \cap A) \ge 3\lambda(I_m)/4$. Let $B = I_m \cap A$. If $\alpha \in \mathbb{R}$ and $(\alpha \oplus B) \cap B$ is empty, then $2\lambda(B) = \lambda((\alpha \oplus B) \cup B) \le \lambda((\alpha \oplus I_m) \cup I_m)$ via the translation invariance of Lebesgue measure, countable additivity, and since $B \subset I_m$. Let β and γ denote the left and right endpoints of I_m, respectively. If $\alpha \geq 0$, then

$$(\alpha \oplus I_m) \cup I_m \subset (\beta,\ \alpha + \gamma) = I_m \cup [\gamma,\ \gamma + \alpha),$$

and if $\alpha < 0$, then

$$(\alpha \oplus I_m) \cup I_m \subset (\alpha + \beta,\ \gamma) = (\alpha + \beta,\ \beta] \cup I_m.$$

Thus, $\lambda((\alpha \oplus I_m) \cup I_m) \leq |\alpha| + \lambda(I_m)$ for any $\alpha \in \mathbb{R}$, and hence if $(\alpha \oplus B) \cap B = \emptyset$, then $2\lambda(B) \leq |\alpha| + \lambda(I_m)$, which implies that $|\alpha| \geq \lambda(I_m)/2$. That is, if $|\alpha| < \lambda(I_m)/2$, then $(\alpha \oplus B) \cap B$ is not empty. Thus, if $\alpha \in (-\lambda(I_m)/2,\ \lambda(I_m)/2)$, then $\alpha = x - y$ for some x and y from A. □

Example 1.18. *There exists a subset V of $\mathbb{R}$ such that $(V \ominus V) \cap \mathbb{Q} = \{0\}$ yet such that $\mathbb{R} = \bigcup_{q \in \mathbb{Q}} q \oplus V$.*

Proof: Consider an equivalence relation $\sim$ on $\mathbb{R}$ defined by setting $x \sim y$ if $y - x$ is rational. For each $x \in \mathbb{R}$, let $[x]$ denote the equivalence class of x with respect to $\sim$; that is, $[x]$ is the set of $y \in \mathbb{R}$ such that $x \sim y$. Note that $[x]$ is simply $x \oplus \mathbb{Q}$. Let V be a set that contains exactly one element from each distinct such equivalence class. The desired results about the Vitale set V follow immediately. In addition, note that V is not Lebesgue measurable, since if it were, it would then follow that

$$\sum_{q \in \mathbb{Q}} \lambda([0,\ 1] \cap (q \oplus V)) = 1$$

and hence that $\lambda(([0,\ 1] \cap (r \oplus V)) > 0$ for some rational r, which implies that $\lambda(r \oplus V) > 0$. This along with Example 1.17 would imply that $(r \oplus V) \ominus (r \oplus V)$ contains a nonempty open interval about the origin, which is not possible since this difference set contains no nonzero rationals. □

Example 1.19. *There exists a set $G \in \mathcal{B}(\mathbb{R})$ such that $I \cap G$ and $I \cap G^c$ each have positive Lebesgue measure for any nonempty open interval I.*

Proof: For $1/2 < \alpha < 1$, let C be a Cantor-like subset of $[0,\ 1]$ having Lebesgue measure α and let $\{S_n\colon n \in \mathbb{N}\}$ denote the open intervals removed during the construction of C. For each $n \in \mathbb{N}$, construct another Cantor-like subset in the interval S_n having Lebesgue measure $\alpha\lambda(S_n)$, and let $\{S_{n,m}\colon m \in \mathbb{N}\}$ denote the open intervals removed during the process. For each $n \in \mathbb{N}$ and $m \in \mathbb{N}$, construct yet another Cantor-like subset in the interval $S_{n,m}$ having Lebesgue measure $\alpha\lambda(S_{n,m})$, and let $\{S_{n,m,j}\colon j \in \mathbb{N}\}$ denote the open intervals removed during the process. Continuing in this way, it follows that corresponding to any finite sequence $n_1, \ldots, n_j$ of positive integers, there exists an open subinterval $S_{n_1,\ldots,n_j}$ of $[0,\ 1]$ on which is constructed a Cantor-like subset $C_{n_1,\ldots,n_j}$ having Lebesgue measure $\alpha\lambda(S_{n_1,\ldots,n_j})$. Note that

$$\lambda\left(\bigcup_{n_1=1}^{\infty} \bigcup_{n_2=1}^{\infty} \cdots \bigcup_{n_j=1}^{\infty} S_{n_1,\ldots,n_j}\right) = (1 - \alpha)^j.$$

Let A_1, A_2, ... be an enumeration of the sets in $\{C, C_{n_1,\ldots,n_j}: n_i, j \in \mathbb{N}\}$, and let $A = \bigcup_{n=1}^{\infty} A_n$. Notice the A_i's are disjoint and that $A \in \mathcal{B}(\mathbb{R})$. Further,

$$\lambda(A) = 1 - \sum_{j=1}^{\infty}(1-\alpha)^j = \frac{2\alpha - 1}{\alpha}.$$

Now, let I be an open nonempty subinterval of [0, 1] and assume that $I \cap A_i$ is empty for each $i \in \mathbb{N}$. It thus follows that for any $j \in \mathbb{N}$ there exist positive integers $n_1, n_2, \ldots, n_j$ such that $I \subset S_{n_1,n_2,\ldots,n_j}$ and hence that I has zero Lebesgue measure. This contradiction implies that $I \cap A_i$ is nonempty for some $i \in \mathbb{N}$. Let $B_i = I \cap A_i$, let $D_i = I \cap ([0, 1] - A_i)$, and note that D_i is nonempty. Let $x_1 \in B_i$, and note that since A_i is perfect there exists some point $x_2 \neq x_1$ that is also an element of B_i. Assume without loss of generality that $x_1 < x_2$. By construction there exists a nonempty open interval (a, b) disjoint from A_i such that $x_1 < a$ and $b > x_2$. It thus follows that D_i has positive Lebesgue measure. Further, $\lambda(D_i) < 1$, since $\lambda(D_i) \leq 1 - \lambda(A_i) \leq \alpha < 1$. Thus, since $\lambda(B_i) + \lambda(D_i) = 1$, it follows that B_i also has positive Lebesgue measure.

Finally, let $G = \bigcup_{n \in \mathbb{Z}} A \oplus n$ and note that $\lambda(G \cap I)$ and $\lambda(G^c \cap I)$ are each positive for any nonempty open subinterval I of $\mathbb{R}$. □

Example 1.20. *Any uncountable G_δ subset of $\mathbb{R}$ contains a nowhere dense closed subset of Lebesgue measure zero that can be mapped continuously onto* [0, 1].

Proof: This example is from Oxtoby [1980]. Let E be an uncountable G_δ subset of $\mathbb{R}$ such that $E = \bigcap_{n=1}^{\infty} G_n$, with G_n open for each n. Let F denote the set of all of the condensation points of E that are elements of E. Note that F is nonempty and that every point in F is a limit point of F. Let $I(0)$ and $I(1)$ be disjoint closed subintervals of the open set G_1 such that $\lambda(I(0)) \leq 1/3$, $\lambda(I(1)) \leq 1/3$, and such that the interiors of $I(0)$ and $I(1)$ each have at least one point in common with F, where λ denotes Lebesgue measure. Further, let $I(0, 0)$, $I(0, 1)$, $I(1, 0)$, and $I(1, 1)$ be disjoint closed intervals such that, for each i and j from $\{0, 1\}$, $\lambda(I(i, j)) \leq 1/3^2$, $I(i, j) \subset G_2 \cap I(i)$, and the interior of $I(i, j)$ has at least one point in common with F. Continuing in this manner, we obtain a family of closed intervals $\{I(i_1, i_2, \ldots, i_n): n \in \mathbb{N}, i_j \in \{0, 1\}\}$ such that each has at least one point in common with F and such that, for each $n \in \mathbb{N}$, $\lambda(I(i_1, i_2, \ldots, i_n)) \leq 1/3^n$ and $I(i_1, i_2, \ldots, i_{n+1})$ is a subset of $G_{n+1} \cap I(i_1, i_2, \ldots, i_n)$. Next, for each $n \in \mathbb{N}$, let

$$C_n = \bigcup_{i_1=0}^{1} \bigcup_{i_2=0}^{1} \cdots \bigcup_{i_n=0}^{1} I(i_1, i_2, \ldots, i_n)$$

and, also, let $C = \bigcap_{n=1}^{\infty} C_n$. Note that C is a closed, nowhere dense, Lebesgue null subset of $\mathbb{R}$. (It follows that C is nowhere dense since it is a closed subset that contains no intervals and that C is a Lebesgue null set since $\lambda(C_n) \to 0$ as $n \to \infty$.) Notice, also, that for each $x \in C$ there exists a unique sequence $\{i_j\}_{j \in \mathbb{N}}$ of 0's and 1's such that $x \in I(i_1, \ldots, i_n)$ for each $n \in \mathbb{N}$. Further, every such sequence corresponds to some point from C. That is, for a sequence $\{i_j\}_{j \in \mathbb{N}}$ let $D_n = I(i_1, \ldots, i_n)$ and note that $D_{n+1} \subset D_n$ for each n, which implies that the intersection of the nonempty bounded closed sets D_n over all $n \in \mathbb{N}$ is nonempty. Finally, for $x \in C$ and the corresponding sequence $\{i_j\}_{j \in \mathbb{N}}$, let $f(x)$ denote the element of [0, 1] whose binary expansion is given by $0.i_1 i_2 \ldots$. Note that the function f so defined maps C onto [0, 1], which thus implies that C has cardinality c. Note that f is continuous since $|f(x) - f(y)| \leq 1/2^n$ whenever x and y are in $C \cap I(i_1, \ldots, i_n)$. □

Example 1.21. *There exist two Lebesgue null subsets A and B of $\mathbb{R}$ such that $A \oplus B = \mathbb{R}$.*

Proof: Let C denote the Cantor ternary subset of [0, 1]. Define a subset A of $\mathbb{R}$ via $A = \bigcup_{n \in \mathbb{Z}} C \oplus n$ and let $B = C$. Recalling Example 1.10, we see that the result follows. □

Example 1.22. *There exists a set B of real numbers such that B and B^c each have at least one point in common with every uncountable closed subset of $\mathbb{R}$.*

Proof: This example is from Oxtoby [1980]. Note that there are at most c open subsets of $\mathbb{R}$, since any open set may be expressed as a countable union of open intervals with rational endpoints and hence that there are at most c closed subsets of $\mathbb{R}$. Also, there are at least c closed uncountable sets, since there are c closed uncountable intervals. Thus, there are exactly c uncountable closed subsets of $\mathbb{R}$. Let $\mathcal{F} = \{F_\alpha\colon \alpha < c\}$ denote a well ordering of the family of closed uncountable intervals. Note from Example 1.20 that each element of $\mathcal{F}$ has cardinality c since each closed subset of $\mathbb{R}$ is a G_δ subset of $\mathbb{R}$. Consider a well ordering of $\mathbb{R}$ and let p_1 and q_1 denote the first two elements of F_1 with respect to that ordering. Let p_2 and q_2 denote the first two elements of F_2 that are different from p_1 and q_1. In general, if $1 < \alpha < c$ and if p_β and q_α have been defined in this way for all $\beta < \alpha$, then let p_α and q_α be the first two elements from H_α, where $H_\alpha = F_\alpha \setminus \bigcup_{\beta < \alpha} \{p_\beta, q_\alpha\}$. Note that H_α has cardinality c for each $\alpha < c$, since F_α has cardinality c. In this way, p_α and q_α are defined for all $\alpha < c$. Let $B = \{p_\alpha\colon \alpha < c\}$ and the desired result follows. Such a set B is called a Bernstein set. □

Example 1.23. *The intersection of a Bernstein set with any Lebesgue measurable set of positive Lebesgue measure is not Lebesgue measurable.*

Proof: This example is from Oxtoby [1980]. Consider a Bernstein set B and let A be a Lebesgue measurable subset of B. If F is a closed subset contained in A, then F must be countable since every uncountable closed subset of $\mathbb{R}$ has at least one point in common with B^c. Thus, any closed subset of A must have Lebesgue measure zero. Since the inner Lebesgue measure of A is given by the supremum of $\lambda(F)$ over all closed and bounded subsets F of A, it thus follows that $\lambda(A) = 0$; that is, any Lebesgue measurable subset of B has Lebesgue measure zero. Similarly, it follows that any Lebesgue measurable subset of B^c has Lebesgue measure zero. Thus, if B is Lebesgue measurable, then both B and B^c must have Lebesgue measure zero, which thus implies that B is not Lebesgue measurable. Further, since any Lebesgue measurable subset of B or B^c must have zero Lebesgue measure, the desired result follows immediately.

Finally, as a consequence of this example, note that any Lebesgue measurable subset of the real line with positive Lebesgue measure contains a non-Lebesgue measurable subset. That is, if A be a subset of $\mathbb{R}$ with positive Lebesgue measure, then this example implies that $A \cap B$ and $A \cap B^c$ cannot both be Lebesgue measurable. □

Example 1.24. *A Bernstein set lacks the property of Baire.*

Proof: This example is from Oxtoby [1980]. Consider a Bernstein set B and let A be a subset of B that has the property of Baire. That is, assume that A is a subset of B such that $A = G \cup F$, where G is a G_δ set and F is of the first category. Note that the set G must be countable since by Example 1.20 every uncountable G_δ subset of $\mathbb{R}$ contains an uncountable closed set and hence has at least one point in common with B^c. Thus it follows that A must be of the first category. Similarly, it follows that any subset of B^c that has the property of Baire must be of the first category. Thus, if B has the property of Baire, then B and B^c must each be of the first category. However, if B has the property of Baire, then B^c is dense. This contradiction implies that B does not possess the property of Baire. □

Example 1.25. *There exists a nowhere dense, non-Lebesgue measurable subset of the real line.*

Proof: Consider a Bernstein set B as constructed in Example 1.22 and consider a Cantor-like set C of positive Lebesgue measure as constructed in Example 1.8. The set $B \cap C$ is clearly nowhere dense and, in light of Example 1.23, is not Lebesgue measurable. □

Example 1.26. *There exists an uncountable subset B of $\mathbb{R}$ such that every closed (in $\mathbb{R}$) subset of B is countable.*

Proof: Recall from Example 1.22 that the Bernstein set B is such that the intersection of B and B^c with any uncountable closed subset of $\mathbb{R}$ is each nonempty. It follows immediately that neither B nor B^c can entirely contain an uncountable closed subset. □

Example 1.27. *For any integer $N > 1$ there exist disjoint subsets $B_1, \ldots, B_N$ of $\mathbb{R}$ such that B_i and B_i^c have a nonempty intersection with every uncountable closed subset of $\mathbb{R}$ for each positive integer $i \le N$.*

Proof: Let $\mathcal{F} = \{F_\alpha\colon \alpha < c\}$ denote the family of closed uncountable intervals. By Example 1.6 it follows that each element in $\mathcal{F}$ has cardinality c. Consider a well ordering of $\mathbb{R}$ and let $x_{1,1}, x_{1,2}, \ldots, x_{1,N}$ denote the first N elements of F_1 with respect to that ordering. Next, let $x_{2,1}, x_{2,2}, \ldots, x_{2,N}$

denote the first N elements of F_2 that are different from $x_{1,j}$ for each nonnegative integer $j \leq N$. In general, if $1 < \alpha < c$ and if $x_{\beta,1}, \ldots, x_{\beta,N}$ have been defined as above for all $\beta < \alpha$, then let $x_{\alpha,1}, \ldots, x_{\alpha,N}$ be the first N elements from the set $F_\alpha \setminus \bigcup_{\beta<\alpha}\{x_{\beta,1}, \ldots, x_{\beta,N}\}$. In this way, $x_{\alpha,1}, \ldots, x_{\alpha,N}$ are defined for each $\alpha < c$. Let $B_i = \{x_{\alpha,i}\colon \alpha < c\}$ for each nonnegative integer $i \leq N$, and the desired result follows. □

Example 1.28. *Any subset of $\mathbb{R}$ of the second category has a subset that does not possess the property of Baire.*

Proof: Consider a Bernstein set B and let A be a subset of $\mathbb{R}$ that is of second category. If $A \cap B$ and $A \cap B^c$ each possess the property of Baire, then, by the development of Example 1.24, they must be of the first category. Thus, their union A must also be of the first category. This contradiction implies that either $A \cap B$ or $A \cap B^c$ does not possess the property of Baire. □

Example 1.29. *There exists a Hamel basis for $\mathbb{R}$ over $\mathbb{Q}$.*

Proof: Let $\{x_\alpha\colon \alpha < c\}$ denote a well ordering of $\mathbb{R}$. If $x_1 \neq 0$, then put x_1 in H, and otherwise put x_2 in H. For a given α assume that we have decided whether or not to put x_β in H for all $\beta < \alpha$. Then, put x_α in H if and only if x_α is not of the form $q_1x_{\beta_1} + \cdots + q_nx_{\beta_n}$, where $n \in \mathbb{N}$, $q_i \in \mathbb{Q}$, $x_{\beta_i} \in H$, and $\beta_i < \alpha$ for each nonnegative integer $i \leq n$. The definition of H now follows from transfinite induction.

To show that H is indeed a Hamel basis for $\mathbb{R}$ over $\mathbb{Q}$, consider any real number x_α. By the definition of H, we know that either x_α is in H or x_α is expressible as a finite linear combination of elements from H with rational coefficients. Now, let $\hat{H}$ be a proper subset of H, let $x_\alpha \in H \cap \hat{H}^c$, and assume that $x_\alpha = c_1x_{\alpha_1} + \cdots + c_nx_{\alpha_n}$, where $n \in \mathbb{N}$, $c_i \in \mathbb{Q}$ and $x_{\alpha_i} \in \hat{H}$ for each nonnegative integer $i \leq n$. We know that α_k cannot be less than α for each choice of k, for if such were true, then x_α would not be in H. Thus, there exists some k for which $\alpha_k > \alpha$ and $\alpha_k > \alpha_i$ for each $i \neq k$. Yet, x_{α_k} cannot be in H, since it is expressible as a finite linear combination with rational coefficients of elements that precede it. This contradiction implies that no proper subset of H possesses the property that any real number may be expressed as a linear combination of its elements over $\mathbb{Q}$.

Finally, let $x \in \mathbb{R}$ and assume that $x = r_1x_{\alpha_1} + \cdots + r_nx_{\alpha_n}$ and that $x = q_1x_{\beta_1} + q_mx_{\beta_m}$, where $n, m \in \mathbb{N}$, $r_i \in \mathbb{Q}$, $q_j \in \mathbb{Q}$, $x_{\alpha_i} \in H$, and $x_{\beta_j} \in H$ for each nonnegative integer $i \leq n$ and each nonnegative integer $j \leq m$. Subtracting, we obtain an expansion for zero of the form $y_1x_{\gamma_1} + \cdots + y_kx_{\gamma_k}$, where $y_i \in \mathbb{Q}$ and $x_{\gamma_i} \in H$ for each nonnegative integer $i \leq k$. But y_i must equal zero for each i, since otherwise an element from H would be expressible as a finite linear combination of other elements from H with rational coefficients. Thus, it follows that every real number has a unique representation as a linear combination of elements from H with rational coefficients. □

Example 1.30. *The inner Lebesgue measure of any Hamel basis for $\mathbb{R}$ over $\mathbb{Q}$ is zero.*

Proof: As in Hahn and Rosenthal [1948], let H be a Hamel basis for $\mathbb{R}$ over $\mathbb{Q}$ and let h be an element from H. Let H' denote the set $\{x \in \mathbb{R}: x = a/h$ for $a \in H\}$. Let λ_* denote inner Lebesgue measure on $(\mathbb{R}, \mathbb{P}(\mathbb{R}))$. Note that if $\lambda_*(H) > 0$, then $\lambda_*(H') > 0$. Further, by Example 1.17, if $\lambda_*(H) > 0$, then there exist two points h_1' and h_2' from H' such that $h_1' - h_2'$ is equal to some positive rational q. Thus, we see that there exist corresponding elements h_1 and h_2 from H such that $h_1 = h_2 + qh$. But since such a relation violates the uniqueness of the Hamel basis representation, we conclude that $\lambda_*(H) = 0$. □

Example 1.31. *A Lebesgue measurable Hamel basis for $\mathbb{R}$ over $\mathbb{Q}$ exists.*

Proof: As in Example 1.21, let A and B be two Lebesgue null sets such that every real number x is of the form $a + b$ for some $a \in A$ and some $b \in B$ and define $C = A \cup B$. Also, let $\{x_\alpha: \alpha < c\}$ denote a well ordering of C. Let c_1 be an element of H and, for a given α, assume that it has been decided whether or not to put c_β in H for all $\beta < \alpha$. Then, put c_α in H if and only if c_α is not of the form $q_1 c_{\beta_1} + \cdots + q_n c_{\beta_n}$, where $n \in \mathbb{N}$, $q_i \in \mathbb{Q}$, $c_{\beta_i} \in H$, and $\beta_i < \alpha$ for each nonnegative integer $i \leq n$. The definition of H now follows from transfinite induction.

To show that H is indeed a Hamel basis for $\mathbb{R}$ over $\mathbb{Q}$, consider any real number x. By definition, $x = a + b$ for some $a \in A$ and some $b \in B$. By the definition of H, we know that either a is in H or a is expressible as a finite linear combination of elements from H with rational coefficients. Similarly, either b is in H or b is expressible as a finite linear combination of elements from H with rational coefficients. Thus, x is expressible as a finite linear combination of elements from H with rational coefficients. As in Example 1.29, it follows that H is a Hamel basis for $\mathbb{R}$ over $\mathbb{Q}$. Finally, since H is a subset of the Lebesgue null set C, it follows that H is itself a Lebesgue null set. □

Example 1.32. *Assuming the continuum hypothesis, there exists a countable collection of Hamel bases of $\mathbb{R}$ over $\mathbb{Q}$ that covers $\mathbb{R} \setminus \{0\}$.*

Proof: This example is based on an argument in Harazisvili [1980]. Let H be a Hamel basis for $\mathbb{R}$ over $\mathbb{Q}$ and recall from Example 1.36 that the cardinality of H is c. Let $\{h_\alpha: \alpha < c\}$ denote a well ordering of H. Further, for each $\alpha < c$ and for each nonzero rational q, let $B_{\alpha,q}$ denote the set of all real numbers expressible in the form $q_1 h_{\alpha_1} + \cdots + q_n h_{\alpha_n} + q h_\alpha$, where $n \in \mathbb{N}$, where q_i is rational for each i, and where $\alpha_i < \alpha$ for each i. Since $B_{\alpha,q}$ is countable for each α and q, we may enumerate the elements of the set as $\{b^n_{\alpha,q}\}_{n \in \mathbb{N}}$. For each positive integer k and each nonzero rational q, let $C_{k,q}$ denote the set $\bigcup_{\alpha<c}\{b^k_{\alpha,q}\}$. Fix a positive integer k and a nonzero rational q and let $c_1, \ldots, c_n$ be a collection of distinct elements from $C_{k,q}$. Let $q_1, \ldots, q_n$ be rationals such that q_i is nonzero for at least one value of i. Let j be the largest value of i for which q_i is nonzero and note that the Hamel basis representation of c_j (with respect to H) includes a term of the form qh_α, where q is nonzero and h_α does not appear in the Hamel basis representation for c_i if $i < j$. Thus $q_1 c_1 + \cdots + q_n c_n$ is nonzero, and we conclude that the set $C_{k,q}$ is linearly independent over the rationals. Note that, via the manner suggested by Example 1.29, there exists a Hamel basis $H_{k,q}$ for $\mathbb{R}$ over $\mathbb{Q}$ that such that $C_{k,q} \subset H_{k,q}$. Since any nonzero real number is in $B_{\alpha,q}$ for some $\alpha < c$ and nonzero rational q, it follows under CH that the countable collection of Hamel bases $\{H_{k,q}$: $k \in \mathbb{N}$ and $q \in \mathbb{Q} \setminus \{0\}\}$ covers $\mathbb{R} \setminus \{0\}$. □

Example 1.33. *Assuming the continuum hypothesis, there exists a non-Lebesgue measurable Hamel basis for* $\mathbb{R}$ *over* $\mathbb{Q}$.

Proof: Recall from Example 1.30 that the inner Lebesgue measure of any Hamel basis for $\mathbb{R}$ over $\mathbb{Q}$ is zero and from Example 1.32 that there exists under CH a countable collection of Hamel bases that covers $\mathbb{R}$ over $\mathbb{Q}$. The desired result now follows immediately. □

Example 1.34. *If* H *is a Hamel basis for* $\mathbb{R}$ *over* $\mathbb{Q}$, *if* $h \in H$, *and if* A *is the set of all real numbers that are expressible as a finite linear combination of elements from* $H \setminus \{h\}$ *with rational coefficients, then* A *is not Lebesgue measurable.*

Proof: Assume that A is a Lebesgue null set. Then the translation $B_q = A \oplus qh$ for some fixed rational q is also a Lebesgue null set. But this results in a contradiction since $\mathbb{R}$ is equal to the (countable) union of B_q over all $q \in \mathbb{Q}$. Thus, A is not a Lebesgue null set.

Next, assume that A is Lebesgue measurable with positive measure and let C denote the set $\{(x/h)$: $x \in A\}$. It follows that C is also Lebesgue measurable with positive measure. Further, there exist two points x and y in C such that $x - y = q \in \mathbb{Q} \setminus \{0\}$. (See Example 1.17.) Hence, $hx \in A$ and $hy \in A$ with $hx = hy + qh$, which contradicts the definition of A. Thus, A is not Lebesgue measurable. □

Example 1.35. *There exists a subset* Λ *of* $\mathbb{R}$ *such that* $\mathbb{R}$ *may be partitioned into a countably infinite number of translations of* Λ, *each of which is dense in* $\mathbb{R}$.

Proof: This example is from Rubin [1967]. Let H be a Hamel basis for $\mathbb{R}$ over $\mathbb{Q}$ and let h be some fixed element of H. Let Λ denote the set of all $x \in \mathbb{R}$ such that x is expressible as a finite linear combination of elements from $H \setminus \{h\}$ with rational coefficients. Due to the uniqueness of Hamel basis representations, it follows that the sets in $\{\Lambda \oplus qh$: $q \in \mathbb{Q}\}$ provide a partition of $\mathbb{R}$ into a countable collection of sets. Finally, let $q \in \mathbb{Q}$ and let x and y be real numbers such that $x < y$. Since, given any $h \in H$ and any $\varepsilon > 0$, there exists some $r \in \mathbb{Q}$ such that $0 < rh < \varepsilon$, it follows that there exists some $\lambda \in \Lambda$ such that $0 < \lambda < y - x$. Thus, there exists some $n \in \mathbb{N}$ such that $x < qh + n\lambda < y$. Since, $qh + n\lambda$ is an element of $\Lambda \oplus qh$, it follows that $\Lambda \oplus qh$ is dense in $\mathbb{R}$. □

Example 1.36. *Any Hamel basis for* $\mathbb{R}$ *over* $\mathbb{Q}$ *has cardinality* c.

Proof: Consider Example 1.35, in which a set representable as a countable union of scaled Hamel bases (each with one point removed) is used to partition the real line into a countable partition. Since such is possible for any Hamel basis for $\mathbb{R}$ over $\mathbb{Q}$, it follows that any such basis has cardinality c. □

Example 1.37. *There exists a Hamel basis for $\mathbb{R}$ over $\mathbb{Q}$ that contains a nonempty perfect subset.*

Proof: This example is from Jones [1942]. Let $\{r_1, r_2, \ldots\}$ be an enumeration of the rational numbers with $r_1 = 0$. In addition, let I_{11} be a closed interval of real numbers that does not contain the origin. Note that I_{11} contains two closed intervals $I_{21} = [\alpha_{21}, \beta_{21}]$ and $I_{22} = [\alpha_{22}, \beta_{22}]$ such that $\beta_{21} < \alpha_{22}$ and such that no number of the form $n_1x_1 + n_2x_2$ different from x_1 and x_2 belongs to $I_2 = I_{21} \cup I_{22}$, where $x_i \in I_2$ and $n_i \in \{r_1, r_2\}$ for $i = 1, 2$. That is, if $r_1 \neq 1$, then such intervals may be obtained by making $\beta_{21} - \alpha_{21}$ and $\beta_{22} - \alpha_{22}$ sufficiently small and positive. Similarly, I_{21} contains two closed intervals $I_{31} = [\alpha_{31}, \beta_{31}]$ and $I_{32} = [\alpha_{32}, \beta_{32}]$ with $\beta_{31} < \alpha_{32}$, and I_{22} contains two closed intervals $I_{33} = [\alpha_{33}, \beta_{33}]$ and $I_{34} = [\alpha_{34}, \beta_{34}]$ with $\beta_{33} < \alpha_{34}$ such that no number of the form $n_1x_1 + n_2x_2 + n_3x_3$ different from x_1, x_2, and x_3 belongs to $I_3 = I_{31} \cup I_{32} \cup I_{33} \cup I_{34}$, where $x_i \in I_3$ and $n_i \in \{r_1, r_2, r_3\}$ for $i = 1, 2, 3$. Continuing this procedure, let $G_k = \{I_{k1}, \ldots, I_{kq}\}$ for each $k \in \mathbb{N}$, where $q = 2^{k-1}$. Note, for each $k \in \mathbb{N}$, that the intervals in G_k are disjoint and that each interval in G_k includes two intervals from G_{k+1}. Further, if for each k we let $H_k = I_{k1} \cup \cdots \cup I_{kq}$, where $q = 2^{k-1}$, and if $x_i \in H_k$ for $1 \leq i \leq k$, then no number of the form $n_1x_1 + \cdots + n_kx_k$ different from all of the x_i's belongs to H_k, where $n_i \in \{r_1, \ldots, r_k\}$ for each positive integer $i \leq k$.

Next, let $M = H_1 \cap H_2 \cap \cdots$ and note that M is nonempty since it is expressible as a countable intersection of decreasing nonempty compact sets. Let $x \in M$ and note that, for each $k \in \mathbb{N}$, $x \in J_k$, where $J_k = I_{kj}$ for some positive integer $j \leq 2^{k-1}$. Let $\delta > 0$ be given and note that there exists a decreasing sequence $\{N_k\}_{k\in\mathbb{N}}$ of closed intervals such that $N_k = I_{kj}$ for each k and for some j, such that $N_k \neq J_k$ for all k greater than some positive integer N, and such that $|z - y| < \delta$ whenever $z \in N_k$, $y \in J_k$, and $k > N$. Since the intersection of the N_k's is nonempty, there thus must exist some point $x' \in M$ such that $x \neq x'$ and $|x - x'| < \delta$. That is, M is perfect. Further, if $x \in M$, then x is not expressible as a finite linear combination of elements from $M \setminus \{x\}$ with rational coefficients. Next, note that if $x \in \mathbb{R}$ possesses two distinct representations as finite linear combinations of elements from M with rational coefficients, then $0 = q_1x_1 + \cdots + q_nx_n$, where $x_i \in M$ and $q_i \in \mathbb{Q} \setminus \{0\}$ for each positive integer $i \leq n$. That is, x_1 may be written as a finite linear combination of elements from $M \setminus \{x_1\}$ with rational coefficients. This contradiction implies that if $x \in \mathbb{R}$ is expressible as a finite linear combination of elements from M with nonzero rational coefficients, then that representation is unique.

Finally, let Γ be a well ordering of $\mathbb{R} \setminus M$ and let α denote the first element in Γ that is not expressible as a finite linear combination of elements from M with rational coefficients. Next, let β denote the first element in Γ that is not expressible as a finite linear combination of elements from

$M \cup \{\alpha\}$ with rational coefficients. Continuing in this manner, it follows, as in Example 1.29, via transfinite induction that $M \cup \{\alpha\} \cup \{\beta\} \cup \cdots$ is a Hamel basis for $\mathbb{R}$ over $\mathbb{Q}$ that contains the perfect set M. □

Example 1.38. *There exists a proper subset N of $\mathbb{R}$ such that the family $\{N \oplus ny\colon n \in \mathbb{N}\}$ contains only a finite number of different subsets of $\mathbb{R}$ for any fixed $y \in \mathbb{R}$.*

Proof: This example is from Rubin [1967]. Consider a Hamel basis H for $\mathbb{R}$ over $\mathbb{Q}$ and let N denote the set of all real numbers that are expressible as a finite linear combination of elements from H with integer coefficients. Let $y \in \mathbb{R} \setminus \{0\}$ and write y as $\alpha_1 h_1 + \cdots + \alpha_n h_n$, where $\alpha_i \in \mathbb{Q}$ and $h_i \in H$ for each positive integer $i \leq n$. Let β be the least common denominator of the α_i's and note that $\beta y \in N$ and that $x + \beta y \in N$ for each $x \in N$. Thus, since any element $x \in N$ may be written as $(x - \beta y) + \beta y$, where $(x - \beta y) \in N$, it follows that $N \oplus \beta y = N$ and hence that $\{N \oplus ny\colon n \in \mathbb{N}\} = \{N, N \oplus y, \ldots, N \oplus (\beta - 1)y\}$. □

Example 1.39. *No Hamel basis for $\mathbb{R}$ over $\mathbb{Q}$ is an analytic set.*

Proof: This example is from Hahn and Rosenthal [1948]. Recall that a subset A of $\mathbb{R}$ is said to be an analytic set if there exists a Polish space Y and a continuous function $f\colon Y \to \mathbb{R}$ such that $f(Y) = A$. Also, recall that every Borel subset of a Polish space is analytic, that countable unions and countable intersections of analytic sets are analytic, and that the complement of an analytic set A is itself analytic if and only if A is a Borel set. Finally, recall that if A is an analytic subset of $\mathbb{R}$ and if $f\colon \mathbb{R} \to \mathbb{R}$ is Borel measurable, then $f(A)$ is an analytic subset of $\mathbb{R}$.

Let B be a Hamel basis for $\mathbb{R}$ over $\mathbb{Q}$, fix $b \in B$, and assume that B is an analytic subset of $\mathbb{R}$. Since B is analytic, it follows that the set $B' = B - \{b\}$ is also analytic. For each $k \in \mathbb{N}$, let $B^{(k)}$ denote the set of all real numbers that are expressible in the form $q_1 b_1 + \cdots + q_k b_k$, where, for each positive integer $i \leq k$, q_i is a nonzero rational and b_i is an element of B' with $b_i \neq b_j$ when $i \neq j$. Further, for each nonzero rational r, let $B'_r = \{rx\colon x \in B'\}$. Note that B'_r is an analytic set since it is given by a Borel measurable transformation of the analytic set B'. Thus, $B^{(1)}$ is analytic since it is given by the countable union of the sets in $\{B'_r\colon r \in \mathbb{Q} - \{0\}\}$. Further, $B^{(k)}$ must be analytic for each $k \in \mathbb{N}$, since $B^{(k)} = B^{(1)} \oplus B^{(1)} \oplus \cdots \oplus B^{(1)}$ (k times), which is expressible as a continuous transformation of an analytic subset of $\mathbb{R}^k$. Finally, it follows that the countable union of the sets in $\{B^{(k)}\colon k \in \mathbb{N}\}$ is analytic. This, however, is not possible, since, by Example 1.34, this set is not Lebesgue measurable and each analytic subset of $\mathbb{R}$ must be Lebesgue measurable. □

Example 1.40. *There exists a proper subset T of $\mathbb{R}$ such that the set $\{T \oplus t\colon t \in \mathbb{R}\}$ contains only countably many distinct elements.*

Proof: Let H be a Hamel basis for $\mathbb{R}$ over $\mathbb{Q}$ and let $h \in H$. Further, let T denote the set of all real numbers whose Hamel basis expansion with respect to elements from H does not require the element h. Note first that if $t \in T$, then $T \oplus t = T$. Further, if t is of the form $s + qh$, where $s \in T$ and $q \in \mathbb{Q}$, then $T \oplus t = T \oplus qh$. Thus, the only translates of T different from T are those of the form $T \oplus qh$, where $q \in \mathbb{Q}$. That is, the set $\{T \oplus t: t \in \mathbb{R}\}$ contains only countably many elements. □

Example 1.41. *There exists a partition of the positive reals into two nonempty sets, each of which is closed under addition.*

Proof: This example is from Sánchez et al. [1976]. Let H be a Hamel basis for $\mathbb{R}$ over $\mathbb{Q}$ and let h be some fixed element of H. Let T denote the set of all positive real numbers whose Hamel basis expansion with respect to elements from H does not require the element h. Let A denote the set of all positive real numbers of the form $t+qh$, where $t \in T$ and $q \in \mathbb{Q}\setminus\{0\}$. Note that A and $[0,\infty) \setminus A$ are each nonempty and closed under addition. Notice, also, that as in the proof of Example 1.34 it follows that A is not Lebesgue measurable. □

Example 1.42. *The Minkowski sum of two bounded Lebesgue null subsets of the reals can be non-Lebesgue measurable.*

Proof: This example is taken from Rubel [1963]. Let H be a Lebesgue measurable Hamel basis included in [0, 1] that has Lebesgue measure zero. Recall that each nonzero real number x has a unique representation as a finite sum $x = \sum_{i \in F_x} r_i h_i$, where F_x is a finite set, with the r_i's nonzero rational numbers and the h_i's belonging to H. For a subset A of $\mathbb{R}$ and for a nonzero real number y, let $yA = \{x \in \mathbb{R}: x/y \in A\}$. Now, let $E_0 = H \cup (-H) \cup \{0\}$, and for positive integers n, let the sets E_n be defined inductively by $E_{n+1} = E_n \oplus E_n$. Thus, E_n consists of all sums of at most 2^n elements of E_0. Now, we claim that there exists a positive integer m such that E_m has Lebesgue measure zero and E_{m+1} has positive outer Lebesgue measure. Assume the contrary. Let

$$J = \bigcup_{n \in \mathbb{Z}^+} \bigcup_{j \in \mathbb{N}} \frac{1}{j} E_n.$$

So, if all the E_n's have Lebesgue measure zero, then J, as a countable union of sets of Lebesgue measure zero, has Lebesgue measure zero. However, note that $J = \mathbb{R}$ since for any real number x we may write all the r_i's in the Hamel basis representation for x with a common denominator to get $x = \sum_{i \in F_x} (R_i h_i)/D$, where D and the R_i's are integers, so that x is an element of $(1/D)E_n \subset J$ as soon as $2^n \geq \sum_{i \in F_x} |R_i|$.

Now, let $n_0 = m + 1$ be the first positive integer such that E_n has positive outer Lebesgue measure. Then E_{n_0} is not Lebesgue measurable. To prove this, assume the contrary. Recall from Example 1.17 that the Minkowski difference set of a set of positive Lebesgue measure includes a nonempty open interval I that contains the origin. Further, for each nonnegative integer k, $E_k = -E_k$, and thus $E_{n_0+2} = E_{n_0+1} \oplus E_{n_0+1}$ includes such an interval. However, for any such interval I, $I \oplus I \oplus I \oplus \cdots = \mathbb{R}$. This implies that in the representation of any real number via the Hamel basis, we may choose the coefficients R_i as integers. But, if $h \in H$, then $h/2$ would have a nonunique representation in terms of the Hamel basis. Hence, the set E_m is a subset of $\mathbb{R}$ having zero Lebesgue measure such that $E_m \oplus E_m$ is non-Lebesgue measurable. □

Example 1.43. *Assuming the continuum hypothesis, there exists an uncountable subset of the real line that has a countable intersection with every subset of* $\mathbb{R}$ *of the first category.*

Proof: This example is from Goffman [1953]. To begin, recall that there are at most c closed subsets of $\mathbb{R}$ and thus that there are at most c closed nowhere dense subsets of $\mathbb{R}$. Further, since each singleton set is a closed nowhere dense subset of $\mathbb{R}$, it follows that there are precisely c closed nowhere dense subsets of the real line. Let $\{C_\alpha\colon \alpha < c\}$ denote a well ordering of all closed nowhere dense subsets of $\mathbb{R}$. Since C_1 is nowhere dense, there exists some real number x_1 that is not in C_1. Further, since the set $\bigcup_{\beta\leq\alpha} C_\beta$ for $\alpha < c$ is of the first category, there must exist some real number that is not in $\bigcup_{\beta\leq\alpha} C_\beta$. Indeed, since $C_1 \cup C_2$ is nowhere dense, there exists some real number x_2 different from x_1 such that x_2 is not in $C_1 \cup C_2$. Fix $\alpha < c$ and assume that for each $\beta < \alpha$ there exists some x_β from $\mathbb{R}$ such that $x_\beta \neq x_\gamma$ for any $\gamma < \beta$ and such that x_β is not an element of the set $\bigcup_{\gamma\leq\beta} C_\gamma$. Then, since the set

$$\left(\bigcup_{\gamma\leq\alpha} C_\gamma\right) \cup \left(\bigcup_{\gamma<\alpha} \{x_\gamma\}\right)$$

is of the first category, there must exist some real number x_α not in $\bigcup_{\gamma\leq\alpha} C_\gamma$ and different from x_γ for all $\gamma < \alpha$. By transfinite induction we thus obtain an uncountable subset $\Lambda \equiv \{x_\alpha\colon \alpha < c\}$ of $\mathbb{R}$ such that $x_\alpha \neq x_\beta$ if $\alpha \neq \beta$ and such that, given any closed nowhere dense subset F_α, $x_\beta \notin F_\alpha$ for any $\beta \geq \alpha$. That is, under CH the intersection of Λ with any closed nowhere dense subset of $\mathbb{R}$ is either finite or countable. Since any nowhere dense subset of $\mathbb{R}$ is a subset of the closed nowhere dense subset given by its closure and since any set of the first category may be written as a countable union of nowhere dense sets, the desired result now follows immediately. □

Example 1.44. *Assuming the continuum hypothesis, there exists a subset of the real line of the second category that does not include any uncountable sets of the first category.*

Proof: Consider the subset Λ of $\mathbb{R}$ that was constructed in Example 1.43 and recall that, under CH, Λ has a countable intersection with any subset of the first category. If Λ is of the first category, then $\Lambda = \bigcup_{i\in M} A_i$ for some subset M of $\mathbb{N}$ and where A_i is nowhere dense for each i. Since Λ is uncountable, it follows that A_j is uncountable for some index j. Thus, we see that Λ contains the uncountable, nowhere dense subset A_j. This contradiction implies that Λ is of the second category. Finally, if B is any uncountable subset of the first category then, since $\Lambda \cap B$ is countable, it follows that Λ does not contain the set B. □

Example 1.45. *There exists a partition of the real line into a Lebesgue null set and a set of the first category.*

Proof: This example is from Oxtoby [1980]. To begin, let $\{r_n: n \in \mathbb{N}\}$ be an enumeration of the rationals. Also, for positive integers i and j let I_{ij} denote the open interval of length $2^{-(i+j)}$ centered at r_i. For each positive integer j, let $G_j = \bigcup_{i=1}^{\infty} I_{ij}$ and let $B = \bigcap_{j=1}^{\infty} G_j$. Fix $\varepsilon > 0$ and let j be a positive integer such that $2^{-j} < \varepsilon$. Then B is a subset of G_j and $\lambda(G_j) \leq \sum_{i=1}^{\infty} \lambda(I_{ij}) = \sum_{i=1}^{\infty} 2^{-(i+j)} = 2^{-j} < \varepsilon$. Hence, B is a Lebesgue null set. Finally, note that for any positive integer i, G_i is a dense open subset of $\mathbb{R}$. Thus, G_i^c is nowhere dense and B^c is of first category. □

Example 1.46. *For any closed subset K of $\mathbb{R}$ having positive Lebesgue measure and any cardinal β such that $2 \leq \beta \leq c$, there exists a partition $\{B_j: j < \beta\}$ of K such that $\lambda^*(B_j) = \lambda(K)$ and $\lambda_*(B_j) = 0$ for each $j < \beta$.*

Proof: This example is from Abian [1976]. To begin, let C be a closed subset of $\mathbb{R}$ and let B be a subset of C that meets every closed subset of K having positive Lebesgue measure. We will show that $\lambda^*(B) = \lambda(C)$. Assume that $\lambda^*(B) < \lambda(K)$. But then there exists a covering of B by pairwise disjoint open intervals such that the complement of this cover relative to C is a closed subset of C having positive Lebesgue measure. Since B does not meet this closed subset, we have a contradiction and thus we see that $\lambda^*(B) = \lambda(C)$.

Let β be a cardinal not greater than c and let $\{A_i: i < \beta\}$ be a collection of subsets of $\mathbb{R}$ each having cardinality c. We will show that there exists a collection $\{a_i: i < \beta\}$ of distinct real numbers such that $a_i \in A_i$ for $i < \beta$. Consider a well ordering of the reals. Choose an element a_1 from A_1 and for $1 < i < \beta$ let a_i be the first element of $A_i \setminus \bigcup_{j<i} A_j$. The desired result now follows by transfinite induction.

Recall from Example 1.6 that every subset of the reals having positive Lebesgue measure has cardinality c. Further note that there are precisely c closed subsets of K having positive Lebesgue measure. Let $\{P_i: i < c\}$ denote the family of all closed subsets of K having positive Lebesgue measure. Since $\beta c = c$ and since the cardinality of P_i equals c for each i, it follows that for each $i < c$, P_i may be written as a disjoint union of β subsets A_{ij} for $j < \beta$ each having cardinality c. Using the above result, it follows that for each $j < \beta$ there exists a collection $\{a_{ij}: i < c\}$ of distinct real numbers such that a_{ij} is an element of A_{ij} for each $i < c$. Let $B_0 = \{a_{i0}: i < c\} \cup (K \setminus \{a_{ij}: i < c, j < \beta\})$ and for $0 < j < \beta$ let $B_j = \{a_{ij}: i < c\}$. Note that $\{B_j: j < \beta\}$ partitions K. Further note that for each $j < \beta$, B_j meets every closed subset of K having positive Lebesgue measure. From the argument in the first paragraph, it now follows that $\lambda^*(B_j) = \lambda(K)$ for each $j < \beta$. Finally, fix $j < \beta$, let $j \neq i$, and note that $B_i \subset K \setminus B_j$. Hence it follows that $\lambda^*(B_j) = \lambda^*(K \setminus B_j) = \lambda(K)$ and thus that $\lambda_*(B_j) = 0$. Note that by choosing $K = \mathbb{R}$ one obtains a partition of $\mathbb{R}$ into β saturated non-Lebesgue measurable subsets. □

Example 1.47. *For any closed subset K of $\mathbb{R}$ having positive Lebesgue* *measure and any nonnegative real number $r \leq \lambda(K)$, there exists a partition $\{B_j\colon j < c\}$ of K such that $\lambda^*(B_j) = r$ and $\lambda_*(B_j) = 0$ for each $j < c$.*

Proof: This example is from Abian [1976]. If $r = 0$, then the result is immediate since K may be written as a union of c singleton sets. Assume that r is positive and let C be a closed subset of K having measure r. Using Example 1.46, let $\{D_i\colon i < c\}$ be a partition of C such that $\lambda^*(D_i) = r$ and $\lambda_*(D_i) = 0$ for each $i < c$. Let α denote the cardinality of $K \setminus C$ and note that $\alpha \leq c$. Let $\{x_j\colon j < \alpha\}$ be a well ordering of $K \setminus C$ and define a collection $\{B_j\colon j < c\}$ of subsets of K via

$$B_j = \begin{cases} D_j \cup \{x_j\}, & \text{if } j < \alpha \\ D_j, & \text{if } \alpha \leq j < c. \end{cases}$$

The desired result follows immediately. □

Example 1.48. *There exists a saturated non-Lebesgue measurable subset S of $\mathbb{R}$ such that the set $\{S \oplus y\colon y \in I\}$ is a partition of $\mathbb{R}$ into saturated non-Lebesgue measurable sets congruent under translation, where card(I) is any infinite cardinal less than or equal to c.*

Proof: This example is from Halperin [1951]. Note that since the reals are second countable, there are c open sets, and hence no more than c perfect sets. Further, there are clearly at least c perfect sets. Hence there are precisely c perfect subsets of $\mathbb{R}$. Let $\{P_\alpha\}_{\alpha<c}$ be a well ordering of the nonempty perfect subsets of $\mathbb{R}$. Choose an element a_1 from P_1 and let b_1 be any other element in P_1 that is not a rational multiple of a_1. For each ordinal $\alpha < c$, let a_α and b_α be elements from P_α such that the elements in the set $\{a_\beta, b_\beta\colon \beta \leq \alpha\}$ are linearly independent over the rationals. Note that a_α and b_α exist since P_α contains c elements and there are fewer than c linear combinations of the a_β's and b_β's over the rationals when $\beta < \alpha$. Hence by transfinite induction we obtain a set $\{a_\alpha, b_\alpha\colon \alpha < c\}$ consisting of real numbers that are linearly independent over the rationals. Let H be a Hamel basis that includes the set $\{a_\alpha, b_\alpha\colon \alpha < c\}$. (To construct such a basis proceed in the manner suggested by Example 1.29.) Let γ be any infinite cardinal not greater than c and let H_0 be a subset of $\{b_\alpha\colon \alpha < c\}$ such that card$(H_0) = \gamma$. Let S be the set of all real numbers expressible as finite linear combinations of elements from $H \setminus H_0$ over the rationals. Let I be the set of all real numbers expressible as finite linear combinations of elements from H_0 over the rationals, and note that card$(I) = \gamma$. Note that the sets in $\{S \oplus i\colon i \in I\}$ partition the real line into γ subsets that are congruent under translation. Finally, since S contains each of the a_α's, it follows that $S \oplus i$ meets every nonempty perfect set for each $i \in I$. Thus, as in Example 1.46, it follows that the sets in $\{S \oplus i\colon i \in I\}$ are saturated non-Lebesgue measurable. □

References

Abian, A., "Partition of nondenumerable closed sets of reals," *Czechoslovak Mathematical Journal,* Vol. 26, 1976, pp. 207–210.

Briggs, J. M., and T. Schaffter, " Measure and cardinality," *American Mathematical Monthly,* Vol. 86, December 1979, pp. 852–855.

Goffman, C., *Real Functions,* Prindle, Weber, & Schmidt, Boston, 1953.

Golomb, S. W., and R. Gilmer, "Problem 5918," *American Mathematical Monthly,* Vol. 81, September 1974, p. 789.

Hahn, H., and A. Rosenthal, *Set Functions,* The University of New Mexico Press, Albuquerque, New Mexico, 1948.

Halperin, I., "Non-measurable sets and the equation $f(x+y) = f(x)+f(y)$," *Proceedings of the American Mathematical Society,* Vol. 2, April 1951, pp. 221–224.

Harazisvili, A. B., "Nonmeasurable Hamel bases," *Soviet Mathematics Doklady,* Vol. 22, 1980, pp. 232–234.

Jones, F. B., "Measure and other properties of a Hamel basis," *Bulletin of the American Mathematical Society,* Vol. 48, 1942, pp. 472–481.

Oxtoby, J. C., *Measure and Category,* Second edition, Springer-Verlag, New York, 1980.

Rubel, L. A., "A pathological Lebesgue-measurable function," *Journal of the London Mathematical Society,* Vol. 38, pp. 1–4, 1963.

Rubin, J. E., *Set Theory for the Mathematician,* Holden–Day, San Francisco, 1967.

Sánchez, W., D. Spear, D. Borwein, and P. B. Borwein, "Problem 5971," *American Mathematical Monthly,* Vol. 83, January 1976, p. 66.

2. Real–Valued Functions

Consider a function f mapping the real line into the real line. The function f is said to be periodic with period p if there exists a positive number p such that $f(x) = f(x+p)$ for all $x \in \mathbb{R}$. If a periodic function f possesses a smallest positive period, then that period is said to be the fundamental period of f. The function f is said to be continuous if $f^{-1}(U)$ is an open subset of $\mathbb{R}$ whenever U is an open subset of $\mathbb{R}$. The function f is said to be absolutely continuous if for every $\varepsilon > 0$ there exists a $\delta > 0$ such that $\sum_{k=1}^{n} |f(b_k) - f(a_k)| < \varepsilon$ for every finite, disjoint collection $\{(a_k, b_k)\}_{k=1}^{n}$ of open intervals for which $\sum_{k=1}^{n}(b_k - a_k) < \delta$. The function f is said to be singular if it is differentiable a.e. and if $f' = 0$ a.e. with respect to Lebesgue measure. The function f is lower semicontinuous at a point $x_0 \in \mathbb{R}$ if for each $\varepsilon > 0$ there exists an open set U such that $x_0 \in U$ and such that $f(x) > f(x_0) - \varepsilon$ for all $x \in U$. The function f is said to be lower semicontinuous if it is lower semicontinuous at each point in $\mathbb{R}$. The function f is upper semicontinuous if $-f$ is lower semicontinuous. The function f is said to be convex if for any $t \in [0, 1]$ it follows that $f(ta+(1-t)b) \leq tf(a)+(1-t)f(b)$ for all $a, b \in \mathbb{R}$. The function f is said to be midpoint convex if $f((a+b)/2) \leq [f(a)+f(b)]/2$ for all $a, b \in \mathbb{R}$. The function f is said to be Borel measurable if $f^{-1}(B)$ is a Borel set for each Borel set B. The function f is said to be Lebesgue measurable if $f^{-1}(B)$ is a Lebesgue measurable set for each Borel set B. The function f is said to be a Darboux function (or to have the intermediate value property) if given any real numbers x_1 and x_2 and any point y between $f(x_1)$ and $f(x_2)$ there exists some point x_3 between x_1 and x_2 such that $f(x_3) = y$. The function f is said to be a Lipschitz function if there exists some real number M such that $|f(x) - f(y)| \leq M|x - y|$ for all points x and y.

Example 2.1. *There exists a function f: $\mathbb{R} \to \mathbb{R}$ that is nowhere continuous yet is equal a.e. $[\lambda]$ to a continuous function.*

Proof: Let $f = I_{\mathbb{Q}}$ and note that f is nowhere continuous yet is equal a.e. $[\lambda]$ to the function that is identically zero. □

Example 2.2. *There exists a function f: $\mathbb{R} \to \mathbb{R}$ that is continuous at all but one point yet is not equal a.e. $[\lambda]$ to any continuous function.*

Proof: Let $f = I_{(0,\infty)}$. Note that f is continuous at all nonzero real numbers. Let g: $\mathbb{R} \to \mathbb{R}$ be any continuous function such that $\{0, 1\} \subset g(\mathbb{R})$. Then, since g is a Darboux function, the set $g^{-1}((0, 1))$ is nonempty, and it is open since g is continuous. Thus $\lambda(g^{-1}((0, 1))) > 0$ and the result follows. □

Example 2.3. *A sequence of Borel measurable functions on $\mathbb{R}$ can converge almost everywhere (with respect to Lebesgue measure) to a function that is not Borel measurable.*

Proof: Recall that the measure space given by $\mathbb{R}$, the Borel subsets of $\mathbb{R}$, and Lebesgue measure is not a complete measure space. Hence, there exists some Borel set B such that B has zero Lebesgue measure and such that there exists a subset A of B that is not a Borel subset of $\mathbb{R}$. An example of such a Borel set is the Cantor ternary set C given in Example 1.8. Define a real-valued function f on $\mathbb{R}$ via $f(x) = I_A(x)$. Now, consider the sequence $\{f_n\}_{n\in\mathbb{N}}$ of real-valued Borel measurable functions defined on $\mathbb{R}$ via $f_n(x) = 1/n$. Then, off the null set B, $f_n(x)$ converges to $f(x)$. Thus, in this situation, $f_n(x) \to f(x)$ almost everywhere, even though f is not Borel measurable. □

Example 2.4. *The composition of a Lebesgue measurable function mapping $\mathbb{R}$ to $\mathbb{R}$ with a continuous function mapping $\mathbb{R}$ to $\mathbb{R}$ need not be Lebesgue measurable.*

Proof: Let f be the Cantor–Lebesgue function defined on $I = [0, 1]$. (See Example 2.6.) Let g: $I \to I$ via $g(x) = [f(x) + x]/2$ and note that g is continuous and increasing. Let C denote the Cantor ternary set, let μ denote Lebesgue measure on the Lebesgue measurable subsets of I, and note that $\mu(g(C)) = 1/2$. Recall that any Lebesgue measurable set of positive Lebesgue measure has a subset that is not Lebesgue measurable. Let N be a non-Lebesgue measurable subset of $g(C)$. (See Example 1.23.) Let $h = g^{-1}$. Then $h(N) \subset C$, and hence $h(N)$ is a Lebesgue null set. Let $k = I_{h(N)}$. Note that h is continuous and k is Lebesgue measurable. However, $(k \circ h)(x) = I_{h(N)}(h(x))$. Hence, $(k \circ h)^{-1}(\{1\}) = N$, and thus $(k \circ h)$ is not Lebesgue measurable. □

Example 2.5. *A non-Lebesgue measurable function mapping the reals into the reals can be such that the image under this function of any Lebesgue measurable set is a Lebesgue measurable set.*

Proof: Let S be a non-Lebesgue measurable set and consider the function f defined via $f(x) = I_S(x)$. The result is obvious. □

Example 2.6. *There exists a continuous function f: $[0, 1] \to \mathbb{R}$ such that $f(0) = 0$ and $f(1) = 1$ and such that f is constant on each of the open intervals removed in the construction of the Cantor ternary set.*

Proof: Let C_n denote the union of the 2^n closed intervals that remain at the nth stage of the construction of the Cantor ternary set C and let $D_n = [0, 1] \setminus C_n$. Recall that $C = \bigcap_{n=1}^{\infty} C_n$. Let I_j^n for $1 \le j \le 2^n - 1$ denote the jth open interval (ordered left to right) that is removed during the first n stages of C's construction and note that $D_n = \bigcup_{j=1}^{2^n-1} I_j^n$. Let f_n: $[0, 1] \to \mathbb{R}$ be the continuous, piecewise affine function such that $f_n(0) = 0$, $f_n(1) = 1$, and such that $f_n(x) = j2^{-n}$ on I_j^n for $1 \le j \le 2^n - 1$. For each positive integer n, note that f_n is nondecreasing, that $f_{n+1} = f_n$ on I_j^n for $1 \le j \le 2^n - 1$, and that $|f_n - f_{n+1}| < 2^{-n}$. Thus, the f_n's converge uniformly on $[0, 1]$. Finally, let $f = \lim_{n\to\infty} f_n$ and note that $f(0) = 0$, $f(1) = 1$, f is nondecreasing and continuous on $[0, 1]$, and that f is constant on each interval removed during the construction of the Cantor ternary set C. The function f given here is known as the Cantor–Lebesgue function. □

Example 2.7. *There exists a strictly increasing, continuous, singular function g: $\mathbb{R} \to \mathbb{R}$.*

Proof: This example is from Freilich [1973]. Consider a function $f: \mathbb{R} \to \mathbb{R}$ such that $f(x) = 0$ if $x \leq 0$, $f(x) = 1$ if $x \geq 1$, and $f(x)$ is equal to the standard Cantor–Lebesgue function on $[0, 1]$. Let $\{q_1, q_2, \ldots\}$ be an enumeration of the rationals and define $g(x) = \sum_{n=1}^{\infty} f(2^n(x - q_n))/2^n$. Notice that since $|f(2^n(x - q_n))/2^n| \leq 1/2^n$ for all x, it follows that the sum defining $g(x)$ converges uniformly and hence that $g(x)$ is continuous. Next, let $x_2 < x_1$ and let $n \in \mathbb{N}$ be such that $x_2 < q_n < x_1$. Note that $f(2^n(x_1 - q_n)) > 0$ and that $f(2^n(x_2 - q_n)) = 0$. Further, if $m \neq n$, then $f(2^m(x_1 - q_m)) \geq f(2^m(x_2 - q_m))$ since f is nondecreasing and $x_1 > x_2$. Thus, $g(x)$ is strictly increasing. Finally, Fubini's theorem on the a.e. differentiation of a series with monotonic terms implies that $g'(x) = 0$ a.e. □

Example 2.8. *Lower semicontinuity of each element of the set $\{f_n: n \in \mathbb{N}\}$ of functions mapping $\mathbb{R}$ into $\mathbb{R}$ does not imply that $\lim_{n\to\infty} f_n$ is lower semicontinuous.*

Proof: Let $\{r_n: n \in \mathbb{N}\}$ be an enumeration of the rationals. For $n \in \mathbb{N}$, let $f_n: \mathbb{R} \to \mathbb{R}$ via $f_n(x) = I_{\mathbb{R}\setminus\{r_n\}}(x)$. Then for each $n \in \mathbb{N}$, f_n is lower semicontinuous, but $\lim_{n\to\infty} f_n = I_{\mathbb{R}\setminus\mathbb{Q}}$, which is not lower semicontinuous. □

Example 2.9. *Any real-valued function of three real variables is expressible as a combination of two real-valued functions of two real variables.*

Proof: This example, from Pólya and Szegö [1972], shows that any function $f(x, y, z): \mathbb{R}^3 \to \mathbb{R}$ may be expressed as $f(x, y, z) = g(h(x, y), z)$, where $g: \mathbb{R}^2 \to \mathbb{R}$ and $h: \mathbb{R}^2 \to \mathbb{R}$. Let $h(x, y)$ be any one-to-one function mapping $\mathbb{R}^2$ into $\mathbb{R}$ and let the subset A of $\mathbb{R}$ denote the range of h. Define $g(x, y)$ as follows. If $s \notin A$, then let $g(s, z)$ take on an arbitrary value, say $g(s, z) = 1$, for each $z \in \mathbb{R}$. If $s \in A$, then there exists a unique element (r, t) in $\mathbb{R}^2$ such that $h(r, t) = s$. For this case, define $g(s, z) = f(r, t, z)$ for each $z \in \mathbb{R}$. Then, it follows that for all real x, y, and z, $f(x, y, z) = g(h(x, y), z)$. □

Example 2.10. *There exists a real-valued function of a real variable that is continuous at every irrational point, discontinuous at every rational point, right continuous, strictly increasing, and that converges to zero at $-\infty$ and one at ∞.*

Proof: Let $\{r_n: n \in \mathbb{N}\}$ be an enumeration of the rationals and define a function $f: \mathbb{R} \to \mathbb{R}$ via

$$f(x) = \sum_{\{n:\ r_n \leq x\}} \frac{1}{2^n}.$$

It follows immediately that f is strictly increasing. Further, it is apparent that $f(x) \to 1$ as $x \to \infty$ and that $f(x) \to 0$ as $x \to -\infty$. Finally, it is clear that f possesses a jump discontinuity at each rational point. The remaining claims made above concerning f will now be proved.

Let y be an irrational number and let ε be a positive number. Further, let N_ε be a positive integer such that $\sum_{n \geq N_\varepsilon} 2^{-n} < \varepsilon$. Define a subset J

of $\mathbb{R}$ via

$$J = \left[\bigcap_{\{n \in \mathbb{N}:\ n < N_\varepsilon,\, r_n < y\}} (r_n,\, y] \right] \cup \left[\bigcap_{\{n \in \mathbb{N}:\ n < N_\varepsilon,\, r_n > y\}} [y,\, r_n) \right].$$

Notice that J is an open interval containing the point y. Further, every rational number r_n in J is such that $n \geq N_\varepsilon$. Let α and β represent the left and right endpoints, respectively, of J and let $\delta = \min\{\beta - y,\, y - \alpha\}$. Then, if $|x - y| < \delta$, it follows that $|f(x) - f(y)| < \varepsilon$. Hence, f is continuous at every irrational point.

Now, let q be some fixed rational number, let ε be a fixed positive number, and consider a sequence $\{x_j\}_{j \in \mathbb{N}}$ of real numbers such that $x_j \downarrow q$. Since f is strictly increasing, it is clear that $f(q) \leq f(x_j)$ for all j. Notice that

$$|f(x_j) - f(q)| = f(x_j) - f(q) = \sum_{\{n \in \mathbb{N}:\ q < r_n \leq x_j\}} \frac{1}{2^n}.$$

Let N_ε be a positive integer such that $\sum_{n \geq N_\varepsilon} (1/2^n) < \varepsilon$ and define a subset K of $\mathbb{R}$ via

$$K = \left[\bigcap_{\{n \in \mathbb{N}:\ n < N_\varepsilon,\, r_n > q\}} (q,\, r_n) \right].$$

Notice that K is an open interval of the form $(q,\, b)$ for some rational $b > q$. Further, every rational number in K is of the form r_n, where $n \geq N_\varepsilon$. Since $x_j \downarrow q$, there exists some integer M such that $x_j \in K$ for all $j > M$. Hence, $|f(x_j) - f(q)| < \varepsilon$ for all $j > M$, or, in other words, f is right continuous at q. □

Example 2.11. *If $f\colon \mathbb{R} \to \mathbb{R}$ is a pointwise limit of a sequence of continuous functions mapping $\mathbb{R}$ into $\mathbb{R}$, then f is continuous except at a set of points of the first category.*

Proof: This example is from Oxtoby [1980]. To begin we define the oscillation of f at x to be

$$\gamma(x) = \lim_{\delta \to 0} \left(\sup_{y \in (x-\delta,\, x+\delta)} f(y) - \inf_{y \in (x-\delta,\, x+\delta)} f(y) \right).$$

Note that for a given function f, γ is an extended real-valued function on $\mathbb{R}$. Further, note that for a given function f, $\gamma(x) = 0$ if and only if f is continuous at x. Finally, note that for a given function f, the set of points at which f is discontinuous is given by $\bigcup_{n=1}^{\infty} \{x \in \mathbb{R}\colon \gamma(x) \geq 1/n\}$.

Let $\varepsilon > 0$ be fixed. To show that f is continuous except on a set of first category, we will show that the set F of $x \in \mathbb{R}$ such that $\gamma(x) \geq \varepsilon$

is nowhere dense. Let $\{f_n\}_{n\in\mathbb{N}}$ be a sequence of continuous functions that converges to f and for each positive integer n let

$$D_n = \bigcap_{i=n}^{\infty}\bigcap_{j=n}^{\infty}\{x \in \mathbb{R}\colon |f_i(x) - f_j(x)| \le \varepsilon/5\}.$$

Note that for each positive integer n, D_n is closed and $D_n \subset D_{n+1}$. Note also that $\bigcup_{n=1}^{\infty} D_n = \mathbb{R}$. Consider a closed interval I and note that since $I = \bigcup_{n=1}^{\infty}(D_n \cap I)$ it follows that $D_n \cap I$ cannot be nowhere dense for each positive integer n. Thus, there exists a positive integer n_0 such that $D_{n_0} \cap I$ includes an open interval J. Note that $|f_i(x) - f_j(x)| \le \varepsilon/5$ for all $x \in J$ and for all positive integers i and j not less than n_0. Setting $j = n_0$ and letting $i \to \infty$, we see that $|f(x) - f_{n_0}(x)| \le \varepsilon/5$ for all $x \in J$. For any $x_0 \in J$ note that there exists an open subinterval J_{x_0} of J containing x_0 and such that $|f_{n_0}(x) - f_{n_0}(x_0)| \le \varepsilon/5$ for all $x \in J_{x_0}$. Thus we see that $|f(x) - f_{n_0}(x_0)| \le 2\varepsilon/5$ for all $x \in J_{x_0}$ and hence that $\gamma(x_0) \le 4\varepsilon/5$. Thus we see that $J \cap F = \emptyset$, which implies that for each closed interval I there exists an open interval J that is a subset of $I \setminus F$. Thus we conclude that F is nowhere dense, and the result follows. □

Example 2.12. *There exists a continuous function f: [0, 1] → [0, 1] such that it is zero on the Cantor ternary set and nonzero off the Cantor ternary set.*

Proof: Define f by letting $f(x)$ be the Euclidean distance from x to the Cantor ternary set. □

Example 2.13. *If $(\Omega, \mathcal{F}, \mu)$ is a measure space such that all singletons are measurable, then it need not follow that the function $f\colon \Omega \to \mathbb{R}$ via $f(x) = \mu(\{x\})$ is measurable.*

Proof: Let $\Omega = \mathbb{R}$, let $\mathcal{F}$ denote the σ-algebra of countable and co-countable subsets of $\mathbb{R}$, and let the measure μ be given by setting $\mu(A)$ equal to the counting measure [defined on $\mathbb{P}(\mathbb{R})$] of $A \cap (0, \infty)$. Then f is one on the positive reals and zero off it, and is thus not measurable. □

Example 2.14. *Given any three real numbers α, β, and γ that are linearly independent over $\mathbb{Q}$, there exist real-valued (not identically constant) functions f and g defined on $\mathbb{R}$ such that f is periodic with period α, g is periodic with period β, and $f + g$ is periodic with period γ.*

Proof: This example is from Golomb and Flanders [1957]. Let H be a Hamel basis for $\mathbb{R}$ over $\mathbb{Q}$ that contains the points α, β, and γ. Define f: $\mathbb{R} \to \mathbb{R}$ and g: $\mathbb{R} \to \mathbb{R}$ via

$$f(x) = \begin{cases} 0, & \text{if } x = \alpha \text{ or } 0 \\ 2, & \text{if } x = \gamma \\ 1, & \text{if } x \in H - \{\alpha, \gamma\} \\ \sum_{i=1}^{n} q_i f(h_i), & \text{otherwise, where } \sum_{i=1}^{n} q_i h_i \text{ is the Hamel basis} \\ & \quad \text{representation for } x \text{ using } H \end{cases}$$

and

$$g(x) = \begin{cases} 0, & \text{if } x = \beta \text{ or } 0 \\ -2, & \text{if } x = \gamma \\ 1, & \text{if } x \in H - \{\beta, \gamma\} \\ \sum_{i=1}^{n} q_i g(h_i), & \text{otherwise, where } \sum_{i=1}^{n} q_i h_i \text{ is the Hamel basis} \\ & \text{representation for } x \text{ using } H. \end{cases}$$

Note that f is periodic with period α and that g is periodic with period β. Further, $f + g$ is periodic with period γ. □

Example 2.15. *There does not exist a metric d on $C([0, 1])$ such that $\lim_{n\to\infty} d(f_n, f_0) = 0$ if and only if the sequence $\{f_n\}_{n\in\mathbb{N}}$ converges pointwise to f_0.*

Proof: This example is from Fort [1951]. First, we construct a double sequence $\{f_{m,n}\}_{m,n\in\mathbb{N}}$ of elements from $C([0, 1])$. By a normal subdivision of the closed interval $[a, b]$ of real numbers we will mean the sequence $x_1 = (a+b)/2$, $x_2 = (x_1+b)/2$, $x_3 = (x_2+b)/2$, etc. For a positive integer n and a positive real number ϵ, we will say that a function is of type (n, ϵ) on $[a, b]$ if the domain of f includes $[a, b]$ and if f restricted to $[a, b]$ consists of the broken line that joins successively the points with coordinates $(a, 0)$, $(x_n, 0)$, (x_{n+1}, ϵ), (x_{n+2}, ϵ), and $(b, 0)$.

Let S_1 be the normal subdivision on $[0, 1]$. If, for a positive integer n, a subdivision S_n of $[0, 1]$ has been defined, then let S_{n+1} be the refinement of S_n obtained by making a normal subdivision of each interval of S_n. Let T be the collection of all intervals J such that, for some n, J is an interval of the subdi''ision S_n. Note that T is countable, and hence we may let k be a one-to-one function whose domain is T and whose range is the set of all positive integers.

Let m and n be positive integers and define a function $f_{m,n}$ as follows. If J is an interval of the subdivision S_n then let $\epsilon_J = 1$ if $k(J) \leq m$ and $\epsilon_J = 1/k(J)$ if $k(J) > m$. Let $f_{m,n}$ be the function on $[0, 1]$ that is of type (m, ϵ_J) on each interval J of S_n. Note that, although the graph of $f_{m,n}$ has a hump on each interval of S_n, it is continuous since if $\delta > 0$, then its graph can contain at most a finite number of humps of height greater than δ. Let n be a positive integer and let $t \in [0, 1]$. It follows quickly that $f_{m,n}(t) = 0$ for all but at most three values of m. Thus it follows that $\lim_{m\to\infty} f_{m,n}(t) = 0$.

Now let f_0 be the identically zero function on $[0, 1]$ and assume that a metric d exists that defines convergence to be pointwise convergence. It follows that $\lim_{m\to\infty} d(f_{m,n}, f_0) = 0$ for each positive integer n. For each integer $n > 0$, choose an integer N_n so that if $j > N_n$, then $d(f_{j,n}, f_0) < 1/n$. It follows that if $\{m_n\}_{n\in\mathbb{N}}$ is any sequence such that $m_n > N_n$ for each n, then the sequence $\{f_{m_n,n}\}_{n\in\mathbb{N}}$ converges pointwise to f_0.

Now, for a contradiction, let J_1 be an interval of S_1 and choose m_1 to be any integer greater than $\max\{k(J_1), N_1\}$. It follows that $f_{m_1,1}$ is one on some interval J_2 of S_2 such that $J_2 \subset J_1$. Next, choose $m_2 > \max\{k(J_2), N_2\}$. It follows that $f_{m_2,2}$ is one on some interval J_3 of S_3

such that $J_3 \subset J_2$. Next choose $m_3 > \max\{k(J_3), N_3\}$ and continue this procedure to obtain a nested sequence $J_1 \supset J_2 \supset J_3 \supset \cdots$ of closed intervals and integers $m_n > N_n$ such that $f_{m_n,n}$ is one on J_{n+1}. Note that there exists a point p that belongs to each interval J_n. It follows that $\lim_{n\to\infty} f_{m_n,n}(p) = 1 \neq 0 = f_0(p)$. That is, $\{f_{m_n,n}\}_{n\in\mathbb{N}}$ does not converge pointwise to f_0 and hence no such metric d exists. □

Example 2.16. *There exists a periodic function $f\colon \mathbb{R} \to \mathbb{R}$ that has no chord of irrational length even though, for any $h \in \mathbb{R}$, $f(x+h) - f(x)$ is a constant function of x.*

Proof: This example is from Oxtoby [1972]. Let H be a Hamel basis for $\mathbb{R}$ over $\mathbb{Q}$ such that $1 \in H$. Let $x \in \mathbb{R} \setminus \{0\}$ and note that x may be expressed uniquely as $x = \alpha_x + \sum_{i=1}^{n} q_i h_i$, where $\alpha_x \in \mathbb{Q}$, $q_i \in \mathbb{Q} \setminus \{0\}$, and $h_i \in H - \{1\}$ for each $i \leq n$. For each such x, define $f\colon \mathbb{R} \to \mathbb{R}$ via $f(x) = x - \alpha_x$. Note that $f(x+y) = f(x) + f(y)$ since if $x = \alpha_x + \sum_{i=1}^{n} q_i h_i$ and $y = \alpha_y + \sum_{i=1}^{m} \hat{q}_i \hat{h}_i$, then the unique such representation for $x+y$ is given by $(\alpha_x + \alpha_y) + q_1 h_1 + \cdots + q_n h_n + \hat{q}_1 \hat{h}_1 + \cdots + \hat{q}_m \hat{h}_m$ with the obvious simplification if $h_i = \hat{h}_j$ for some i and j. Thus, if x, $h \in \mathbb{R}$, then $f(x+h) - f(x) = f(h)$. Further, f is periodic since $f(x+q) = f(x)$ for every $q \in \mathbb{Q}$. However, even though $f(x+h) - f(x)$ is constant for all x, $h \in \mathbb{R}$, f has no chord of irrational length. That is, there exists no positive irrational number h for which $f(x) = f(x+h)$ for some $x \in \mathbb{R}$. □

Example 2.17. *A real-valued midpoint convex function defined on the reals need not be continuous.*

Proof: Recall that a function $h\colon \mathbb{R} \to \mathbb{R}$ is said to be midpoint convex if, for any real numbers x and y, $h((x+y)/2) \leq [h(x) + h(y)]/2$. It follows easily that if $h\colon \mathbb{R} \to \mathbb{R}$ is both midpoint convex and continuous, then it is convex. The following example constructs a midpoint convex function that is not continuous and hence not convex.

Consider a function $f\colon \mathbb{R} \to \mathbb{R}$ for which $f(x+y) = f(x) + f(y)$ for all x, y in $\mathbb{R}$. It follows that $f(2x) = 2f(x)$ and hence that $2f((x+y)/2) = f(x+y) = f(x) + f(y)$. Thus, such a function f is midpoint convex.

Let B be a Hamel basis for $\mathbb{R}$ over $\mathbb{Q}$ and define a function $g\colon \mathbb{R} \to \mathbb{R}$ as follows. If $x \in B$ then let $g(x)$ take on an arbitrary value. For $x \in B^c$, let $g(x) = a_1 g(b_1) + \cdots + a_n g(b_n)$, where $x = a_1 b_1 + \cdots + a_n b_n$ with $n \in \mathbb{N}$, $a_i \in \mathbb{Q}$, and $b_i \in B$ for each i.

Next, for real numbers x and y, let $x = a_1 b_1 + \cdots + a_n b_n$ and $y = c_1 d_1 + \cdots + c_m d_m$, where n, $m \in \mathbb{N}$, a_i, $c_j \in \mathbb{Q}$, and b_i, $d_j \in B$ for each i and j. Notice that $x + y = a_1 b_1 + \cdots + a_n b_n + c_1 d_1 + \cdots + c_m d_m$, where the obvious simplification is made if $b_i = d_j$ for some i and j. If $b_i \neq d_j$ for all i and j, then it follows immediately that $g(x+y) = g(x) + g(y)$. For the case when $b_i = d_j$ for some i and j simply note that if $b \in B$ and r, $q \in \mathbb{Q}$ then $g((r+q)b) = (r+q)g(b) = g(rb) + g(qb)$. Thus, it follows that g is a midpoint convex function.

Notice, however, that since the values of $g(x)$ for $x \in B$ were chosen arbitrarily, g may be constructed so as to be discontinuous. In particular, let

$\varepsilon > 0$ be given and assume that there exists a $\delta > 0$ such that $|g(x)-g(y)| < \varepsilon$ whenever $|x-y| < \delta$. Let b_1 and b_2 be distinct elements of B, let $q \in \mathbb{Q}$ be such that $0 < qb_2 < \delta$, and note that $|b_1 - (b_1 + qb_2)| < \delta$. However, $g(b_2)$ may be chosen so that $|g(b_1) - g(b_1 + qb_2)| = |qg(b_2)| > \varepsilon$. Thus, g: $\mathbb{R} \to \mathbb{R}$ need not be continuous and hence need not be convex even though it is midpoint convex. □

Example 2.18. *Any real-valued function on* [0, 1] *may be expressed as the composition of a Lebesgue measurable function with a Borel measurable function.*

Proof: Let f be any real-valued function defined on [0, 1]. Define a real-valued function h on [0, 1] via $h(x) = \sum_{n=1}^{\infty} 2\alpha_n 3^{-n}$, where each point $x \in [0, 1]$ is expressed as $x = \sum_{n=1}^{\infty} \alpha_n 2^{-n}$, where $\alpha_n \in \{0, 1\}$ for each n and such that if $x \in (0, 1]$, then $\alpha_n = 1$ for infinitely many values of n. Note that h is strictly increasing on [0, 1] and that $h([0, 1])$ is a Lebesgue null set since it is a subset of the Cantor ternary set. Next, define g on [0, 1] via $g(x) = f(h^{-1}(x))$, when $x \in h([0, 1])$, and $g(x) = 0$ otherwise. Then, g is a Lebesgue measurable function and h is a Borel measurable function such that $f(x) = g(h(x))$ for $x \in [0, 1]$. □

Example 2.19. *There exists a nonnegative real-valued Lebesgue integrable function f defined on $\mathbb{R}$ such that for any $a < b$ and any $R > 0$ the set $(a, b) \cap \{x \in \mathbb{R}\colon f(x) \geq R\}$ has positive Lebesgue measure.*

Proof: Let $\{r_n\}_{n\in\mathbb{N}}$ be an enumeration of the rationals and for each $k \in \mathbb{N}$ define

$$f_k(x) = \begin{cases} 1/\sqrt{x - r_k}, & \text{if } x \in (r_k, r_k + 1) \\ 0, & \text{if } x \notin (r_k, r_k + 1). \end{cases}$$

Further, let $f(x) = \sum_{n=1}^{\infty} [f_n(x)/2^n]$ and note that

$$\int_{\mathbb{R}} f(x)\, dx = \sum_{n=1}^{\infty} \frac{1}{2^n} \int_{\mathbb{R}} f_n(x)\, dx = \sum_{n=1}^{\infty} \frac{2}{2^n} = 2.$$

However, if $a < b$ and $R > 0$, then there exists a rational number $q \in (a, b)$ and an open interval $(q, r) \subset (a, b)$ such that $f(x) > R$ for each $x \in (q, r)$. Thus, the set $(a, b) \cap \{x \in \mathbb{R}\colon f(x) \geq R\}$ has positive Lebesgue measure for any nonempty open interval (a, b). □

Example 2.20. *There exists a real-valued function defined on* [0, 1] *whose set of discontinuities has measure zero yet has an uncountable intersection with every nonempty open subinterval.*

Proof: Let C be the Cantor ternary subset of [0, 1] and define a sequence $\{A_n\}_{n\in\mathbb{N}}$ of subsets of [0, 1] as follows. Let $A_1 = C$. Let A_2 denote the union of Cantor ternary sets constructed on each of the open intervals removed during the construction of C. Let A_3 denote the union of Cantor ternary sets constructed on the open intervals removed during the construction of each of the Cantor sets whose union comprises A_2. In general, let A_n

denote the union of Cantor ternary sets constructed on the open intervals removed during the construction of A_{n-1}. Let $A = \bigcup_{n=1}^{\infty} A_n$ and note that A is a Lebesgue null set that has an uncountable intersection with each nonempty open subinterval of [0, 1]. Finally, let $f(x) = \sum_{n=1}^{\infty} 2^{-n} \mathrm{I}_{A_n}(x)$ and note that f is discontinuous on A. Note, however, that if $x \notin A$ and $\varepsilon > 0$, then, choosing $N \in \mathbb{N}$ so that $2^{-N} < \varepsilon$, there exists an open interval I containing x that is disjoint from $A_1, \ldots, A_N$ and hence is such that $f(x) < 2^{-N} < \varepsilon$ for all $x \in I$. Thus, f is continuous on $[0, 1] \setminus A$. □

Example 2.21. *Given any nonempty subset M of $\mathbb{R}$, there exists a function g: $\mathbb{R} \to \mathbb{R}$ such that g is constant a.e. with respect to Lebesgue measure and $g(\mathbb{R}) = M$.*

Proof: Let C denote the Cantor ternary subset of [0, 1], let $\alpha \in M$, and let f map C onto $\mathbb{R}$. Define g by letting $g(x) = \alpha$ if $x \notin C$, $g(x) = f(x)$ if $x \in C$ and $f(x) \in M$, and $g(x) = \alpha$ if $x \in C$ and $f(x) \notin M$. The function g is equal to α almost everywhere, yet $g(\mathbb{R}) = M$. □

Example 2.22. *There exists a function f: $\mathbb{R} \to \mathbb{R}$ such that $f(x) = 0$ holds for some value of x, such that f is everywhere discontinuous, such that f is unbounded and non-Lebesgue measurable on any nonempty open interval, yet such that, for any real number x_1 and the corresponding sequence $\{x_n\}_{n \in \mathbb{N}}$ defined via $x_{n+1} = f(x_n)$, $f(x_n) = 0$ for all $n \geq 3$.*

Proof: This example is from Antosiewicz and Hammersley [1953]. Let I denote the set of irrational real numbers that are not rational multiples of π and let S be a saturated non-Lebesgue measurable subset of I. Let $f(x) = 0$ if x is rational, $f(x) = 1/2$ if $x \in S$, $f(x) = q$ if $x = p\pi/q$, where p/q is a nonzero rational number expressed in its lowest terms, and $f(x) = 1$ otherwise. The claimed result follows immediately. Note that this result suggests that necessary conditions for convergence of such an iteration scheme to zero might be elusive. □

Example 2.23. *There exist two real-valued functions f and g of a real variable such that each is continuous almost everywhere with respect to Lebesgue measure and yet $f \circ g$ is nowhere continuous.*

Proof: This example is taken from Levine [1959]. Let

$$g(x) = \begin{cases} 0, & \text{if } x \text{ is irrational} \\ 1/p, & \text{if } x = p/q \text{ for } p \text{ and } q \text{ relatively prime integers and } p > 0. \end{cases}$$

Let

$$f(x) = \begin{cases} 1, & \text{if } x = 1/n \text{ where } n \text{ is a positive integer} \\ 0, & \text{otherwise.} \end{cases}$$

Note that g and f are each continuous almost everywhere. However, observe that $f \circ g$ is equal to the indicator function of the rationals, and hence is nowhere continuous. □

Example 2.24. *There exists a Lebesgue measurable function f: $[0, 1] \to \mathbb{R}$ such that, given any auteomorphism h on $[0, 1]$, there is no function of the first class of Baire that agrees with $f \circ h$ almost everywhere with respect to Lebesgue measure.*

Proof: This example is from Gorman [1966]. Let $I = [0, 1]$. Let μ be Lebesgue measure restricted to the Lebesgue measurable subsets of I. Let K_1 be a Cantor-like subset of I given by Example 1.8 with $\beta_n = 4^{-n}$ and note that $\lambda(K_1) = 1/2$. Let K_2 denote the subset of $I \setminus K_1$ that intersects each open interval of $I \setminus K_1$ in a translated and scaled replica of $K_1 \setminus \{0, 1\}$ so that the μ-measure of the intersection of K_2 with an open interval of $I \setminus K_1$ is one-half the length of the open interval. Continue in this manner, producing for each positive integer n the set K_n included in $I \setminus (K_1 \cup K_2 \cup \cdots \cup K_{n-1})$ and intersecting each open interval of the latter in a translated and scaled replica of $K_1 \setminus \{0, 1\}$ so that the μ-measure of the intersection of K_n with an open interval of $I \setminus (K_1 \cup K_2 \cup \cdots \cup K_{n-1})$ is one-half the length of the open interval. Finally, let $K = \bigcup_{n \in \mathbb{N}} K_n$. Note that $\mu(K) = 1$. Now, let B be a Bernstein set as in Example 1.22 and let $A_1 = (I \setminus K) \cap B$ and $A_2 = (I \setminus K) \cap (I \setminus B)$. Observe that A_1 and A_2 are μ-null sets. Now, let f: $I \to \mathbb{R}$ via

$$f(x) = -\mathrm{I}_{A_1}(x) - 2\mathrm{I}_{A_2}(x) + \sum_{n=1}^{\infty} n\mathrm{I}_{K_n}(x).$$

Note that f is clearly measurable.

Suppose that there exists an auteomorphism h on I such that $f \circ h$ agrees a.e. $[\mu]$ with some function g: $I \to \mathbb{R}$ of the first class of Baire. Then $f \circ h$ is μ-measurable. Note that $(f \circ h)^{-1}(\{-1\}) = h^{-1}(A_1)$. Since h^{-1} maps perfect sets to perfect sets and since A_1 includes no nonempty perfect set, we see that $h^{-1}(A_1)$ is a μ-null set. Similarly, $h^{-1}(A_2)$ is a μ-null set. Thus we see that $h^{-1}(K)$ must have μ-measure one. We see from the above construction of K that for any nonempty open subinterval of I, there are an infinite number of positive integers n for which $(f \circ h)^{-1}(\{n\})$, and therefore $g^{-1}(\{n\})$, intersects the subinterval in a set of positive measure. Thus, g has no points of continuity. From Example 2.11, we see that this is impossible for a function in the first class of Baire. □

Example 2.25. *There exists a nondecreasing continuous function f mapping $[0, 1]$ to $[0, 1]$ such that $|f'(x)| \leq 1$ off a Lebesgue null set and yet such that f is not a Lipschitz function.*

Proof: Let f be the Cantor–Lebesgue function presented in Example 2.6. For a positive integer n and for $x \in (3^{-n}, 2(3^{-n}))$, $f(x) = 2^{-n}$. Hence, we see that $f((3/2)3^{-n}) = 2^{-n}$. Thus,

$$\frac{f\left(\frac{3}{2}3^{-n}\right)}{\frac{3}{2}3^{-n}} = \left(\frac{3}{2}\right)^{n-1}.$$

Since this value can exceed any preassigned real number by choice of n, we see that f is not a Lipschitz function. □

Example 2.26. *There exists an absolutely continuous function f: $[0, 1] \to \mathbb{R}$ that is monotone in no nonempty open interval.*

Proof: Recall from Example 1.19 that there exists a Borel measurable subset A of $\mathbb{R}$ such that $A \cap I$ and $\mathbb{R} \setminus (A \cap I)$ both have positive Lebesgue measure for any nonempty open interval I. Let $B = A \cap [0, 1]$. Let g: $[0, 1] \to \mathbb{R}$ via $g(x) = \mathrm{I}_B(x) - \mathrm{I}_{[0,1]\setminus B}(x)$. Now, define f via $f(x) = \int_0^x g\, d\lambda$, where λ is Lebesgue measure. Note that f is absolutely continuous. Further, since for any nonempty open subinterval I of $[0, 1]$, $B \cap I$ and $[0, 1] \setminus (B \cap I)$ both have positive Lebesgue measure, it follows that f cannot be monotone on any such interval. □

Example 2.27. *There exists a net $\{f_t\}_{t\in\mathbb{R}}$ of Borel measurable real-valued functions defined on $(0, 1]$ such that the set of numbers $x \in (0, 1]$ for which $\lim_{t\to\infty} f_t(x) = 0$ is not a Borel set.*

Proof: This example is from Billingsley [1986]. Let H be a subset of $(0, 1]$ that is not a Borel set. For each real number t, and for each $x \in (0, 1]$, let $f_t(x)$ equal one if x is equal to the fractional part of t and is not in H and let $f_t(x)$ equal zero otherwise. Since $f_t(x)$ is zero except at possibly one point, we see that f_t is Borel measurable. However, note that $\lim_{t\to\infty} f_t(x) = 0$ if and only if $x \in H$. □

Example 2.28. *If f is a Lebesgue measurable real-valued function of a real variable, then the function of the real variable x given by $\sup_{t\in\mathbb{R}} |f(x+t) - f(x-t)|$ can be non-Lebesgue measurable.*

Proof: This example is taken from Rubel [1963]. Let A denote the subset of the reals constructed in Example 1.42. Choose a real number s such that $A \cap (A \oplus s) = \emptyset$, and let f: $\mathbb{R} \to \mathbb{R}$ via

$$f(x) = \begin{cases} 1, & \text{if } x \in A \\ -1, & \text{if } x \in (A \oplus s) \\ 0, & \text{otherwise.} \end{cases}$$

Now, $\sup_{t\in\mathbb{R}} |f(x+t) - f(x-t)| = 2$ precisely when there is a $t \in \mathbb{R}$ such that $(x+t) \in A$ and $(x-t) \in (A \oplus s)$. Thus, it follows that $\{x \in \mathbb{R}: \sup_{t\in\mathbb{R}} |f(x+t) - f(x-t)| = 2\} = (1/2)\{(A \oplus A) \oplus s\}$, which is a non-Lebesgue measurable set. Thus, $\sup_{t\in\mathbb{R}} |f(x+t) - f(x-t)|$ is not Lebesgue measurable. □

Example 2.29. *There exists a real-valued function defined on the reals that takes on every real value infinitely many times in every nonempty open interval yet is equal to zero a.e. with respect to Lebesgue measure.*

Proof: Let H be a Lebesgue measurable Hamel basis for $\mathbb{R}$ over $\mathbb{Q}$ (see Example 1.29 and note from Example 1.30 that H must have Lebesgue measure zero), let f: $H \to \mathbb{R}$ be surjective, and let $\{q_1, q_2, \ldots\}$ be an enumeration of $\mathbb{Q}$. Extend the function f to all of $\mathbb{R}$ by letting $f(qx) = f(x)$ if $q \in \mathbb{Q} \setminus \{0\}$ and $x \in H$ and by letting $f(x) = 0$ otherwise. Consider now the function f: $\mathbb{R} \to \mathbb{R}$, let (a, b) be a nonempty open interval, and let

$\lambda \in \mathbb{R}$. Let $h \in H$ be such that $f(h) = \lambda$ and let $N_0 \subset \mathbb{N}$ be such that $q_i h \in (a, b)$ for each $i \in N_0$. By construction, we see that $f(q_i h) = \lambda$ for each $i \in N_0$ and hence that f takes on every real value in every open interval infinitely many times. Further, since f is nonzero only on the set consisting of rational multiples of the Lebesgue null set H, it follows that f is zero a.e. □

Example 2.30. *There exists a function f: $\mathbb{R} \to \mathbb{R}$ such that f takes on every real value c times on every nonempty perfect subset of $\mathbb{R}$.*

Proof: This example is due to Halperin [1950]. Let $\{x_\alpha: \alpha < c\}$ be a well ordering of the reals. Note that since the reals are second countable, there are c open sets, and hence no more than c perfect sets. Further, there are clearly at least c perfect sets. Hence there are precisely c perfect subsets of $\mathbb{R}$. Let $\{P_\alpha\}_{\alpha<c}$ be a transfinite sequence of nonempty perfect sets such that each nonempty perfect set appears c times. Next, for each α less than c, define p_α by transfinite induction to be the first real number contained in P_α that has not been previously selected. Note that such a value will always exist since any nonempty perfect subset of $\mathbb{R}$ contains c elements. This procedure will select c elements from each nonempty perfect set since each such set appears c times in the transfinite sequence. Finally, arrange the selected elements in each nonempty perfect set as a double transfinite sequence $y_{\alpha,\beta}$, and define $f(y_{\alpha,\beta}) = x_\alpha$. For other real numbers, let f be arbitrarily defined. The advertised claim now follows immediately. Note that in contrast to the function obtained in Example 2.29, the function here cannot be Lebesgue measurable on any set of positive Lebesgue measure. □

Example 2.31. *For any function f: $\mathbb{R} \to \mathbb{R}$ there exists a Darboux function g: $\mathbb{R} \to \mathbb{R}$ such that the set $\{x \in \mathbb{R}: f(x) \neq g(x)\}$ is a set of Lebesgue measure zero.*

Proof: Let h: $\mathbb{R} \to \mathbb{R}$ denote the function constructed in Example 2.29 and let g be given by $g = f + h$. The desired result is then immediate. □

Example 2.32. *There exists a Darboux function mapping $\mathbb{R}$ into $\mathbb{R}$ that is non-Lebesgue measurable on every subset of $\mathbb{R}$ having positive Lebesgue measure.*

Proof: The function given in Example 2.30 is an example of this. □

Example 2.33. *If f is any real-valued function of a real variable, then there exist two Darboux functions g and h mapping the reals into the reals such that $f = g + h$.*

Proof: Let d be either the function given in Example 2.29 or in Example 2.30. Then simply let $g = (f/2) + d$ and let $h = (f/2) - d$. The functions g and h are clearly Darboux functions, and the advertised result follows. Thus the sum of Darboux functions need not be a Darboux function. □

Example 2.34. *There exists a sequence $\{f_n\}_{n\in\mathbb{N}}$ of Lebesgue measurable functions mapping the reals into the reals such that f_n converges pointwise* *to zero and such that for any positive integer n, f_n takes on each real value in any nonempty open set.*

Proof: Let H be a Lebesgue measurable Hamel basis for $\mathbb{R}$ over $\mathbb{Q}$ (see Example 1.31) and let h be a fixed element of H. Let Λ denote the collection of all real numbers whose Hamel basis representation does not use h. Let g map $H \setminus \{h\}$ onto $\mathbb{R}$. Further, let $\{q_n\colon n \in \mathbb{N}\}$ be an enumeration of the rationals. Extend the function g to all of Λ by letting $g(q_n x) = g(x)$ for all positive integers n and for all $x \in H \setminus \{h\}$ and by letting $g = 0$ otherwise. Now, extend g to $\mathbb{R}$ by defining it to be zero off Λ. Note that g is Lebesgue measurable since it is equal to zero off the Lebesgue null set given by the union of all rational multiples of $H \setminus \{h\}$. Similarly as in Example 2.29, it follows that g takes on each real value over any nonempty open subset of $\mathbb{R}$. Finally, define f_n mapping $\mathbb{R}$ into $\mathbb{R}$ via $f_n(x) = g(x+q_n h)$. The desired result follows. □

Example 2.35. *For any real-valued function f defined on the reals, there exists a sequence $\{g_n\}_{n\in\mathbb{N}}$ of measurable Darboux functions mapping the reals into the reals such that g_n converges pointwise to f.*

Proof: Consider the sequence $\{f_n\}_{n\in\mathbb{N}}$ of functions mapping the reals into the reals. Also, recall the set Λ from Example 2.34, and as in Example 1.35, let $\{P_n\colon n \in \mathbb{N}\}$ be a partition of $\mathbb{R}$ obtained by translating Λ over rational multiples of h. That is, let $P_n = \Lambda \oplus q_n h$ for each positive integer n. For positive integers n, let $R_n = \bigcup_{i=1}^{n} P_i$. Let $g_n = f\mathrm{I}_{R_n} + f_{n+1}$. The result follows. □

Example 2.36. *There exists a Darboux function $f\colon \mathbb{R} \to \mathbb{R}$ such that f, as a subset of $\mathbb{R}^2$, is not connected.*

Proof: Let h be the function given in Example 2.30, and define f via

$$f(x) = \begin{cases} 1, & \text{if } x = 0 \\ 0, & \text{if } x = h(x) \neq 0 \\ h(x), & \text{if } x \notin \{0, h(x)\}. \end{cases}$$

Then f takes on each real number (indeed, c times) on each nonempty open set and is thus a Darboux function. However, f, as a subset of $\mathbb{R}^2$, does not meet the line of unit slope through the origin and hence is not connected. □

Example 2.37. *If f is a Darboux function mapping the reals into the reals and if a is a real number, the function g mapping the reals into the reals via $g(x) = f(x) + ax$ need not be a Darboux function.*

Proof: Let f be the Darboux function given in Example 2.36, and let $a = -1$. Then for $g(x) = f(x) - x$, g, as a subset of $\mathbb{R}^2$, does not contain the horizontal line passing through the origin. Hence, g is not a Darboux function. □

Example 2.38. *For each $h \in [0, 1)$, there exists an almost everywhere continuous function $f_h\colon [0, 1) \to [0, 1]$ such that for each $x \in [0, 1)$, $f_h(x) \to 0$*

as $h \to 0$, yet this convergence is not uniform on any Lebesgue measurable set of positive Lebesgue measure.

Proof: This example is from Weston [1959]. Let V' be a Vitale set as given in Example 1.18 and let $V = V' \cap [0, 1)$. Recall that each point of $[0, 1)$ has a unique representation of the form $v \dotplus r$, where $v \in V$ and $r \in ([0, 1) \cap \mathbb{Q})$. For this example, when x and y are elements of $[0, 1)$, $x \dotplus y$ will equal $x+y$ if $x+y<1$, and it will equal $x+y-1$ if $x+y \geq 1$. Further, for a subset S of $[0, 1)$ and a point x in $[0, 1)$, $S \dotplus x = \{s \dotplus x \in [0, 1) \colon s \in S\}$. Also, let μ denote Lebesgue measure on the Lebesgue measurable subsets of $[0, 1)$.

First, we show that if A is a Lebesgue measurable subset of $[0, 1)$ that intersects only a finite number of the sets $\{V \dotplus r \colon r \in ([0, 1) \cap \mathbb{Q})\}$, then $\mu(A) = 0$. To show this, we first construct an infinite sequence $\{r_n\}_{n \in \mathbb{N}}$ of points in $[0, 1) \cap \mathbb{Q}$ such that the sets $A \dotplus r_n$ are disjoint from one another. Suppose that for a positive integer n, the first n terms of such a sequence are given. By hypothesis, there exists a positive integer m and points $a_1, a_2, \ldots, a_m$ in $[0, 1) \cap \mathbb{Q}$ such that $A \subset \bigcup_{i=1}^{m} (V \dotplus a_i)$. We can choose r_{n+1} in $[0, 1) \cap \mathbb{Q}$ such that $(r_{n+1} \dotplus a_j) \neq (a_i \dotplus r_k)$ whenever i and j are positive integers not greater than m and k is a positive integer not greater than n. For each positive integer k not greater than n, each point of $A \dotplus r_k$ is uniquely expressible as $r_k \dotplus a_i \dotplus v$, for some positive integer i not greater than m, with $v \in V$, and is therefore not expressible as $r_{n+1} \dotplus a_j \dotplus v$, for some positive integer j not greater than m, with $v \in V$. Hence, $A \dotplus r_k$ has no point in common with $A \dotplus r_{n+1}$. Thus, we obtain the desired infinite sequence by induction.

From the countable additivity and translation invariance of Lebesgue measure, it now follows that

$$\mu\left(\bigcup_{k=1}^{n} (A \dotplus r_k)\right) = \sum_{k=1}^{n} \mu(A \dotplus r_k) = n\mu(A).$$

Hence, $n\mu(A) \leq 1$ since A is a Lebesgue measurable subset of $[0, 1)$. Therefore, $\mu(A) = 0$.

Now, we construct the functions f_h. Let $\{q_n \colon n \in \mathbb{N}\}$ be an enumeration of $[0, 1) \cap \mathbb{Q}$, and, for each $n \in \mathbb{N}$, let $V_n = V \dotplus q_n$. Then each point x in $[0, 1)$ determines a unique positive integer $n(x)$ such that $x \in V_{n(x)}$. For a point x in $[0, 1)$, let $\hat{x}$ be defined as follows: if $0.\alpha_1\alpha_2\alpha_3\ldots$ is the decimal representation of $x/n(x)$, without an infinite number of consecutive recurring 9's, then $\hat{x}$ has the decimal representation $0.\beta_1\beta_2\beta_3\ldots$, where, for positive integers p, $\beta_{2p-1} = \mathrm{I}_{n(x)}(p)$, and $\beta_{2p} = \alpha_p$. Now, define $f_h(x)$, for each h and x in $[0, 1)$ in the following way: if $h = \hat{y}$ for some y in $[0, 1)$, then $f_h(x) = \mathrm{I}_{\{y\}}(x)$; otherwise, $f_h(x) = 0$ for all x in $[0, 1)$.

Now, it is clear that, for each value of h, there is at most one point x such that $f_h(x) = 1$, and for all other points x, $f_h(x) = 0$. Thus, for each $h \in [0, 1)$, f_h is almost everywhere continuous. Also, for each value of x, $f_h(x) = 0$ except for $h = \hat{x}$, so that $f_h(x) \to 0$ as $x \to 0$.

Now, let S be a subset of $[0, 1)$ that is Lebesgue measurable and has positive Lebesgue measure. The result mentioned above shows that S intersects infinitely many of the sets V_n, and so contains points x such that $n(x)$ can be made arbitrarily large, and thus $\hat{x}$ can be made arbitrarily small. Thus, there are arbitrarily small values of h such that $f_h(x) = 1$ for some x in S. The convergence of f_h to zero is thus not uniform on S.

Recall that Egoroff's theorem establishes the almost uniform convergence for a sequence of real-valued functions that converges almost everywhere on a measurable set of finite measure. The above example shows that if the index set is not countable, the result need not follow. With additional conditions (see Weston [1960]) on the family of functions, the countability assumption on the index set may be dropped. □

Example 2.39. *There exists a symmetric non-Lebesgue measurable, non-negative definite function mapping* $\mathbb{R}$ *to* $\mathbb{R}$.

Proof: This example is from Crum [1953]. Let $\{h_\alpha\}_{\alpha\in A}$ be a Hamel basis for $\mathbb{R}$ over $\mathbb{Q}$. Fix $\alpha \in A$ and let G_α denote the set of all real numbers that do not use h_α in their Hamel basis representation. If u and v are real numbers, then we will write $u \sim v$ if $u - v$ is in G_α. Note that G_α is an additive subgroup of $\mathbb{R}$ and that $\sim$ is an equivalence relation in $\mathbb{R}$. Thus, the relation $\sim$ partitions $\mathbb{R}$ into equivalence classes relative to $\sim$. For each $x \in \mathbb{R}$, let $[x]$ denote the equivalence class relative to $\sim$ that contains x. Note that $[x] = \{x + g\colon g \in G_\alpha\}$. Since the Hamel basis for any real number contains h_α with a (possibly zero) rational coefficient, it follows that there are only countably many different equivalence classes relative to $\sim$. Indeed, for any real number x we see that $[x] = [qh_\alpha]$ for some rational q. Let $C_q = [qh_\alpha]$ for each rational q.

Now, fix $K_\alpha \in (0, 1)$ and define a function $p_\alpha\colon \mathbb{R} \to \mathbb{R}$ via

$$p_\alpha(x) = \begin{cases} 1, & \text{if } x \in G_\alpha \\ K_\alpha, & \text{if } x \notin G_\alpha. \end{cases}$$

Let N be a positive integer, let $z_1, \ldots, z_N$ be complex numbers, let $x_1, \ldots, x_N$ be real numbers, and note that

$$\begin{aligned}
&\sum_{i=1}^{N}\sum_{j=1}^{N} p_\alpha(x_i - x_j) z_i z_j^* \\
&= \sum_{r\in\mathbb{Q}}\sum_{s\in\mathbb{Q}}\sum_{x_i\in C_r}\sum_{x_j\in C_s} p_\alpha(x_i - x_j) z_i z_j^* \\
&= \sum_{r\in\mathbb{Q}}\sum_{\substack{x_i\in C_r\\ x_j\in C_r}} p_\alpha(x_i - x_j) z_i z_j^* + \sum_{r\in\mathbb{Q}}\sum_{\substack{s\in\mathbb{Q}\\ s\neq r}}\sum_{x_i\in C_r}\sum_{x_j\in C_s} p_\alpha(x_i - x_j) z_i z_j^* \\
&= \sum_{r\in\mathbb{Q}}\sum_{\substack{x_i\in C_r\\ x_j\in C_r}} z_i z_j^* + K_\alpha \sum_{r\in\mathbb{Q}}\sum_{\substack{s\in\mathbb{Q}\\ s\neq r}}\sum_{x_i\in C_r}\sum_{x_j\in C_s} z_i z_j^* \\
&= \sum_{r\in\mathbb{Q}}\sum_{\substack{x_i\in C_r\\ x_j\in C_r}} z_i z_j^* + K_\alpha \sum_{r\in\mathbb{Q}}\sum_{s\in\mathbb{Q}}\sum_{x_i\in C_r}\sum_{x_j\in C_s} z_i z_j^* - K_\alpha \sum_{r\in\mathbb{Q}}\sum_{\substack{x_i\in C_r\\ x_j\in C_r}} z_i z_j^*
\end{aligned}$$

$$= (1 - K_\alpha) \sum_{r \in \mathbb{Q}} \sum_{\substack{x_i \in C_r \\ x_j \in C_r}} z_i z_j^* + K_\alpha \sum_{r \in \mathbb{Q}} \sum_{s \in \mathbb{Q}} \sum_{x_i \in C_r} \sum_{x_j \in C_s} z_i z_j^*$$

$$= (1 - K_\alpha) \sum_{r \in \mathbb{Q}} \sum_{\substack{x_i \in C_r \\ x_j \in C_r}} z_i z_j^* + K_\alpha \sum_{i=1}^{N} \sum_{j=1}^{N} z_i z_j^*$$

$$= (1 - K_\alpha) \sum_{r \in \mathbb{Q}} \left| \sum_{x_i \in C_r} z_i \right|^2 + K_\alpha \left| \sum_{i=1}^{N} z_i \right|^2 \geq 0.$$

Thus, p_α is nonnegative definite for each $\alpha \in A$. From Example 1.34 it follows that p_α is not Lebesgue measurable. Further, the function $p(x) = \prod_{\alpha \in A} p_\alpha(x)$ is nonnegative definite since it is the product of nonnegative definite functions. Further, $p(x) > 0$ for each x since if $x = \sum_{i=1}^{N} q_i h_{\alpha_i}$, then $p(x) = \prod_{i=1}^{N} K_{\alpha_i} > 0$. Finally, let Λ_α denote the set of all real numbers expressible in the form $\sum_{i=1}^{N} q_i h_{\alpha_i} - h_\alpha$, where $h_{\alpha_i} \neq h_\alpha$ for any i. Then, note that $p(x + h_\alpha) = K_\alpha p(x)$ on G_α, $p(x + h_\alpha) = p(x)$ on $G_\alpha^c \setminus \Lambda_\alpha$, and $p(x + h_\alpha) = p(x)/K_\alpha$ on Λ_α. Since none of these sets is Lebesgue measurable, it follows that p is not a Lebesgue measurable function. □

Example 2.40. *Assuming the continuum hypothesis, there exists a sequence of real-valued functions defined on the reals that converges on an uncountable subset but converges uniformly on none of its uncountable subsets.*

Proof: This example is due to Goffman [1953]. Let Λ denote an uncountable subset of $\mathbb{R}$ that has a countable intersection with any nowhere dense subset of $\mathbb{R}$. (Such a set is constructed under CH in Example 1.43.) Let $x \in \mathbb{R}$ and let U be an open subset of $\mathbb{R}$ containing x. Let N be an uncountable nowhere dense subset of U. (Simply construct a Cantor set in any nonempty open subinterval of U.) Since $\Lambda \cap N$ is countable, it follows that $\Lambda^c \cap U$ is not empty. Thus, Λ^c is dense in $\mathbb{R}$. Let $C = \{c_i\}_{i \in \mathbb{N}}$ be a countable subset of Λ^c that is also dense in $\mathbb{R}$. Define a function $f \colon \mathbb{R} \to \mathbb{R}$ via:

$$f(x) = \begin{cases} 1/n, & \text{if } x = c_n \\ 0, & \text{if } x \in C^c \end{cases}$$

and let $\{f_n\}_{n \in \mathbb{N}}$ be a sequence of continuous functions that converge to f.

Next, let T be an uncountable subset of Λ and assume that there exists no nonempty open interval I such that T is dense in I. Then, given any open interval J there exists an open subinterval of J that has no points in common with T. But, this implies that T is nowhere dense and uncountable which contradicts the definition of Λ, and hence we conclude that there exists some nonempty open interval on which T is dense. Assume that $\{f_n\}_{n \in \mathbb{N}}$ converges uniformly on some uncountable subset T of Λ. Then, from above, we see that $\{f_n\}_{n \in \mathbb{N}}$ converges uniformly on a set that is dense in some nonempty open subinterval K. Note, however, that if a sequence of functions converges uniformly on a dense subset of an interval, then it converges uniformly on the interval itself. Thus, f is continuous on K since it is a limit of a sequence of functions that converge uniformly on K. However, f is discontinuous at c_n for all n and since C is dense, there exists some element from C that is in K. The desired result follows immediately from this contradiction. □

References

Antosiewicz, H. A., and J. M. Hammersley, "The convergence of numerical iteration," *American Mathematical Monthly,* Vol. 60, pp. 604–607, November 1953.

Billingsley, P., *Probability and Measure,* Second edition, John Wiley: New York, 1986.

Crum, M. M., "On positive-definite functions," *Proceedings of the London Mathematical Society,* Third Series, Vol. 6, October 1953, pp. 548–560.

Fort, M. K., Jr., "A note on pointwise convergence," *Proceedings of the American Mathematical Society,* Vol. 2, February 1951, pp. 34–35.

Freilich, G., "Increasing, continuous singular functions," *American Mathematical Monthly,* October, 1973, pp. 918–919.

Goffman, C., *Real Functions,* Prindle, Weber, & Schmidt, Boston, 1953.

Golomb, S. W., and H. Flanders, "Problem 4706," *American Mathematical Monthly,* October 1957, pp. 598–599.

Gorman, W. J., "Lebesgue equivalence to functions of the first Baire class," *Proceedings of the American Mathematical Society,* Vol. 17, August 1966, pp. 831–834.

Halperin, I., "Discontinuous functions with the Darboux property," *American Mathematical Monthly,* Vol. 57, October 1950, pp. 539–540.

Levine, N., "A note on functions continuous almost everywhere," *American Mathematical Monthly,* Vol. 66, November 1959, pp. 791–792.

Oxtoby, J. C., "Horizontal chord theorems," *American Mathematical Monthly,* May, 1972, pp. 468–475.

Oxtoby, J. C., *Measure and Category,* Second edition, Springer-Verlag, New York, 1980.

Pólya, G., and G. Szegö, *Problems and Theorems in Analysis,* Volume 1, Springer-Verlag, New York, 1972.

Rubel, L. A., "A pathological Lebesgue-measurable function," *Journal of the London Mathematical Society,* Vol. 38, pp. 1–4, 1963.

Weston, J. D., "A counter-example concerning Egoroff's theorem," *Journal of the London Mathematical Society,* Vol. 34, pp. 139–140, 1959.

Weston, J. D., "Addendum to a note on Egoroff's theorem," *Journal of the London Mathematical Society,* Vol. 35, p. 366, 1960.

3. Differentiation

Consider a function f mapping the real line into the real line and consider a real number x_0. The upper right Dini derivative of f at x_0 is denoted by $D^+ f$ and is given by

$$D^+ f = \limsup_{h \downarrow 0} \frac{f(x_0 + h) - f(x_0)}{h}.$$

The upper left Dini derivative of f at x_0 is denoted by $D^- f$ and is given by

$$D^- f = \limsup_{h \uparrow 0} \frac{f(x_0 + h) - f(x_0)}{h}.$$

The lower right Dini derivative of f at x_0 is denoted by $D_+ f$ and is given by

$$D_+ f = \liminf_{h \downarrow 0} \frac{f(x_0 + h) - f(x_0)}{h}.$$

The lower left Dini derivative of f at x_0 is denoted by $D_- f$ and is given by

$$D_- f = \liminf_{h \uparrow 0} \frac{f(x_0 + h) - f(x_0)}{h}.$$

The function f is said to be differentiable at x_0 if the four Dini derivatives agree at x_0. If f is infinitely differentiable, then the power series $\sum_{n=0}^{\infty} [f^{(n)}(x_0)/n!](x - x_0)^n$ is called the Taylor's series about x_0 generated by f.

Example 3.1. *If f: $\mathbb{R} \to \mathbb{R}$ is a function having a Taylor's series at a real number a, then there exist c distinct functions mapping $\mathbb{R}$ into $\mathbb{R}$ having the same Taylor's series at the real number a.*

Proof: Let g: $\mathbb{R} \to \mathbb{R}$ via

$$g(x) = \begin{cases} e^{-(x-a)^{-2}}, & \text{if } x \neq a \\ 0, & \text{if } x = a. \end{cases}$$

Note that for any positive integer k, g has k derivatives at the point a and they are all equal to zero. Now consider the set of functions mapping $\mathbb{R}$ into $\mathbb{R}$ given by $\{f + bg: b \in \mathbb{R}\}$. For each positive integer k, the kth derivative at the point a of any function in this set is equal to the kth derivative at the point a of f. Thus, each function in this set of functions has the same Taylor's series at the real number a, and the cardinality of this set is c. □

Example 3.2. *There exists a function f: $\mathbb{R} \to \mathbb{R}$ that is everywhere differentiable yet whose derivative is not Lebesgue integrable.*

Proof: Let g and h map a nonempty interval of real numbers into $\mathbb{R}$. Further, assume that g is integrable and h is Lebesgue measurable. Then $g + h$ is integrable if and only if h is integrable. This follows immediately from noticing that $|h| - |g| \leq |h + g| \leq |h| + |g|$.

Now, define a function $f\colon \mathbb{R} \to \mathbb{R}$ via

$$f(x) = \begin{cases} x^2 \sin\left(\dfrac{1}{x^2}\right), & \text{if } x \neq 0 \\ 0, & \text{if } x = 0. \end{cases}$$

Clearly, f is differentiable away from zero. At $x = 0$, notice that for $h \neq 0$,

$$\left|\frac{f(x+h) - f(x)}{h}\right| = \left|\frac{f(h)}{h}\right| = \left|h \sin\left(\frac{1}{h^2}\right)\right| \leq |h|.$$

Letting $h \to 0$, it follows that $f'(0) = 0$. Thus, the derivative of f is given by

$$f'(x) = \begin{cases} 2x \sin\left(\dfrac{1}{x^2}\right) - \dfrac{2}{x} \cos\left(\dfrac{1}{x^2}\right), & \text{if } x \neq 0 \\ 0, & \text{if } x = 0. \end{cases}$$

Now, for any $\alpha > 0$, $f'(\cdot)$ is integrable over $[0, \alpha]$ if and only if the function mapping x to $(1/x)\cos(1/x^2)\, \mathrm{I}_{(0,\, \alpha)}(x)$ is integrable. Now, simply note that

$$\int_0^\alpha \left|\frac{1}{x} \cos\left(\frac{1}{x^2}\right)\right| dx = \int_{\alpha^{-2}}^\infty \left|\frac{\cos(y)}{2y}\right| dy \geq \sum_{n=N}^\infty \int_{(n-1)\pi}^{n\pi} \left|\frac{\cos(y)}{2y}\right| dy,$$

where N is a positive integer such that $(N-1)\pi \geq \alpha^{-2}$. Thus, we see that

$$\int_0^\alpha \left|\frac{1}{x} \cos\left(\frac{1}{x^2}\right)\right| dx \geq \sum_{n=N}^\infty \frac{K}{2n\pi} = \infty,$$

where $K = \int_0^\pi |\cos(u)|\, du > 0$. □

Example 3.3. *A differentiable function mapping the reals into the reals need not be absolutely continuous.*

Proof: This is a consequence of Example 3.2 since if the function were absolutely continuous, then its derivative would be locally integrable. □

Example 3.4. *An absolutely continuous function mapping the reals into the reals need not be differentiable.*

Proof: This is immediately revealed by considering the function f given by $f(x) = x + |x|$. □

Example 3.5. *There exists a function $f\colon [0, 1] \to \mathbb{R}$ that is differentiable with a bounded derivative but whose derivative is not Riemann integrable.*

Proof: This example is inspired by Volterra [1881]. Let K be a Cantor-like set of positive Lebesgue measure as constructed in Example 1.8. Further, let $\{\mathrm{I}_n\}_{n\in\mathbb{N}}$ denote the sequence of open intervals removed in the construction of K. Then, if $I_n = (a, b)$, define a real-valued function f_n on (a, b) via

$f_n(x) = (x-a)^2 \sin[1/(x-a)]$. Note that there exist points z in $(a, (a+b)/2)$ such that $f'(z) = 0$. Let y be such a point. Next, define a real-valued function g_n on (a, b) via

$$g_n(x) = \begin{cases} f_n(x), & \text{if } x \leq y \\ f_n(y), & \text{if } y \leq x \leq (a+b)/2 \\ f_n(a+b-x), & \text{if } (a+b)/2 \leq x. \end{cases}$$

Finally, define a real-valued function g on $[0, 1]$ via

$$g(x) = \begin{cases} g_n(x), & \text{if } x \in I_n \\ 0, & \text{if } x \in K. \end{cases}$$

Noticing that the real-valued function defined on the reals via

$$x \mapsto \begin{cases} x^2 \sin\left(\dfrac{1}{x}\right), & \text{if } x \neq 0 \\ 0, & \text{if } x = 0 \end{cases}$$

is everywhere differentiable but that the resulting derivative is discontinuous at zero, we see that although the function g is differentiable at all $x \in [0, 1]$, the resulting derivative is discontinuous at each point in K. Note also that g' is bounded. However, since K has positive Lebesgue measure, we see that g' is not Riemann integrable. □

Example 3.6. *There exists a continuously differentiable function defined on* $[0, 1]$ *whose set of critical values is uncountable.*

Proof: Recall that $x_0 \in \mathbb{R}$ is said to be a critical point of a function f: $\mathbb{R} \to \mathbb{R}$ if f is not differentiable at x_0 or if f is differentiable at x_0 and if $f'(x_0) = 0$. If x_0 is a critical point of f, then $f(x_0)$ is a critical value of the function f.

Let C denote the Cantor ternary subset of $[0, 1]$ and recall that C^c is expressible as a union of the form $\bigcup_{n \in \mathbb{N}} (\alpha_n, \beta_n)$, where, for each n, (α_n, β_n) is one of the open intervals removed from $[0, 1]$ during C's construction. Define f: $[0, 1] \to \mathbb{R}$ via

$$f(x) = \begin{cases} 0, & \text{if } x \in C \\ (x - \alpha_n)(\beta_n - x), & \text{if } x \in (\alpha_n, \beta_n) \end{cases}$$

and, define F: $[0, 1] \to \mathbb{R}$ via $F(x) = \int_0^x f(t)\, dt$. Since $F' = f$, it follows that F' is continuous. Further, the set $S = \{t\colon F'(t) = 0\}$ of critical points of F is the Cantor ternary set C, which is uncountable. Finally, since f is positive on the complement of a nowhere dense subset of $[0, 1]$, it follows that F is strictly increasing and hence that the set of critical values $\{F(t)$: $t \in S\}$ is also uncountable. □

Example 3.7. *A sequential limit of derivatives of real-valued functions of a real variable need not be the derivative of a real-valued function of a real variable.*

Proof: For each positive integer n, let $f_n: \mathbb{R} \to \mathbb{R}$ via $f_n(x) = \max\{0, 1 - n|x|\}$. Then f_n converges as $n \to \infty$ to $I_{\{0\}}$, which is not a Darboux function and is thus not a derivative. □

Example 3.8. *The product of the derivatives of two differentiable real-valued functions of a real variable need not be a derivative.*

Proof: This example is from Wilkosz [1921]. Let $F: \mathbb{R} \to \mathbb{R}$ be defined via

$$F(x) = \begin{cases} x^2 \sin(1/x) & \text{if } x \neq 0 \\ 0, & \text{if } x = 0. \end{cases}$$

Note that F is differentiable and has a derivative f given by

$$F'(x) = f(x) = \begin{cases} 2x \sin(1/x) - \cos(1/x), & \text{if } x \neq 0 \\ 0, & \text{if } x = 0. \end{cases}$$

Let $g: \mathbb{R} \to \mathbb{R}$ via

$$g(x) = \begin{cases} 2x \sin(1/x), & \text{if } x \neq 0 \\ 0, & \text{if } x = 0. \end{cases}$$

Further, note that since g is continuous, it is a derivative. Hence, $h = g - f$ is a derivative, and

$$h(x) = g(x) - f(x) = \begin{cases} \cos(1/x), & \text{if } x \neq 0 \\ 0, & \text{if } x = 0. \end{cases}$$

Assume that

$$h^2(x) = \begin{cases} [1 + \cos(2/x)]/2, & \text{if } x \neq 0 \\ 0, & \text{if } x = 0, \end{cases}$$

is a derivative. Then there exists a differentiable function $D: \mathbb{R} \to \mathbb{R}$ such that $D'(x) = h^2(x)$. Let $H: \mathbb{R} \to \mathbb{R}$ be a differentiable function such that $H' = h$. Then we have that

$$\frac{d}{dx} H\left(\frac{x}{2}\right) = \begin{cases} \frac{1}{2} \cos(2/x), & \text{if } x \neq 0 \\ 0, & \text{if } x = 0. \end{cases}$$

This implies that

$$\frac{d}{dx} D(x) = \begin{cases} \frac{1}{2} + \frac{d}{dx} H\left(\frac{x}{2}\right), & \text{if } x \neq 0 \\ \frac{d}{dx} H\left(\frac{x}{2}\right), & \text{if } x = 0, \end{cases}$$

and thus that

$$\frac{d}{dx}\left[D(x) - H\left(\frac{x}{2}\right)\right] = \begin{cases} \frac{1}{2}, & \text{if } x \neq 0 \\ 0, & \text{if } x = 0. \end{cases}$$

However, this is not possible since a derivative cannot have a discontinuity of the first kind. Thus we see that h^2 is not a derivative. □

Example 3.9. *For any Lebesgue null set A of $\mathbb{R}$, there exists a continuous nondecreasing function $f\colon \mathbb{R} \to \mathbb{R}$ such that each Dini derivative is equal to ∞ at each point of A.*

Proof: This example is due to Natanson [1955]. For each positive integer n, let U_n be an open subset of $\mathbb{R}$ that includes A and has Lebesgue measure less than 2^{-n}, and let $f_n\colon \mathbb{R} \to \mathbb{R}$ via $f_n(x) = \lambda(U_n \cap (-\infty, x])$. Note that for each positive integer n, f_n is nondecreasing, nonnegative, continuous, and is upper bounded by 2^{-n}. Thus, the function $f\colon \mathbb{R} \to \mathbb{R}$ given by

$$f(x) = \sum_{n=1}^{\infty} f_n(x)$$

is nondecreasing, nonnegative, and continuous. Now, for any positive integer n, if $x \in A$, there exists a positive real number ε such that $[x-\varepsilon, x+\varepsilon] \subset U_n$. For such ε we have

$$f_n(x+\varepsilon) = \lambda((U_n \cap (-\infty, x]) \cup (U_n \cap (x, x+\varepsilon])) = f_n(x) + \varepsilon.$$

Thus

$$\frac{f_n(x+\varepsilon) - f_n(x)}{\varepsilon} = 1.$$

Hence, it follows that for any positive integer K,

$$\frac{f(x+\varepsilon) - f(x)}{\varepsilon} \geq \sum_{n=1}^{K} \frac{f_n(x+\varepsilon) - f_n(x)}{\varepsilon} = K.$$

Thus it follows that each Dini derivative is ∞ at each point $x \in A$. □

Example 3.10. *The Dini derivatives of a continuous function $f\colon \mathbb{R} \to \mathbb{R}$ need not have the Darboux property.*

Proof: Letting f be defined via $f(x) = |x|$, the advertised result follows immediately. □

Example 3.11. *There exists a real-valued function defined on* $(0, 1)$ *whose derivative vanishes on a dense set and yet does not vanish everywhere.*

Proof: This example is due to Pompeiu [1906]. Recall that the function $x \mapsto \sqrt[3]{x}$ is differentiable at all nonzero x and at the origin has an extended derivative of infinity. Let $\{r_n\colon n \in \mathbb{N}\}$ be an enumeration of the rationals in $(0, 1)$. Define $f\colon (0, 1) \to \mathbb{R}$ via

$$f(x) = \sum_{n=1}^{\infty} \frac{1}{n^2} \sqrt[3]{x - r_n}.$$

Note that f is an increasing function, has a finite derivative at all points where the series

$$\sum_{n=1}^{\infty} \frac{1}{3n^2} \frac{1}{\sqrt[3]{(x - r_n)^2}}$$

converges [that is, at irrational numbers in $(0, 1)$], and an infinite extended derivative elsewhere [that is, at rational numbers in $(0, 1)$]. The inverse function f^{-1} is thus increasing, differentiable, and zero at each rational in $(0, 1)$. □

Example 3.12. *Let F_1 and F_2 be continuous, real-valued functions defined on an interval $[a, b]$ and assume that F_2 is a.e. differentiable. Given any $\varepsilon > 0$, there exists a continuous a.e. differentiable function G: $[a, b] \to \mathbb{R}$ such that $G' = F_2'$ a.e. and such that $|F_1(x) - G(x)| \leq \varepsilon$ for all $x \in [a, b]$.*

Proof: This example is suggested in Bruckner [1978]. Choose points $a = a_0 < a_1 < \cdots < a_n = b$ such that

$$\sup_{x,\, y \in [a_k,\, a_{k+1}]} \{|F_i(x) - F_i(y)|\} < \frac{\varepsilon}{2}$$

for $i \in \{1, 2\}$ and for each nonnegative integer $k < n$. Let f: $[0, 1] \to \mathbb{R}$ be the Cantor–Lebesgue function, and, for $0 \leq k < n$, define a function $H_k(x)$ on $[a_k, a_{k+1}]$ via

$$\begin{aligned} H_k(x) =& F_2(a_k) - F_1(a_k) \\ &+ [F_2(a_{k+1}) - F_1(a_{k+1}) - F_2(a_k) + F_1(a_k)]\, f\left(\frac{x - a_k}{a_{k+1} - a_k}\right). \end{aligned}$$

Further, define $H(x)$ on $[a, b]$ so that $H(x) = H_k(x)$ for $x \in [a_k, a_{k+1}]$. Finally, let $G(x) = F_2(x) - H(x)$ for $x \in [a, b]$. Note that G is continuous and that $G'(x) = F_2'(x)$ a.e. since $H'(x) = 0$ a.e. Finally, for each nonnegative integer $k < n$, let $\alpha_k = \min\{F_2(a_k) - F_1(a_k),\, F_2(a_{k+1}) - F_1(a_{k+1})\}$ and let $\beta_k = \max\{F_2(a_k) - F_1(a_k),\, F_2(a_{k+1}) - F_1(a_{k+1})\}$, and note that $\alpha_k \leq H(x) \leq \beta_k$ for $x \in [a_k, a_{k+1}]$. Hence, for $x \in [a_k, a_{k+1}]$,

$$\begin{aligned} |F_1(x) - G(x)| &= |F_1(x) - F_2(x) + H(x)| \\ &\leq \begin{cases} \beta_k - F_2(x) + F_1(x), & \text{if } F_2(x) - F_1(x) < H(x) \\ F_2(x) - F_1(x) - \alpha_k, & \text{if } F_2(x) - F_1(x) > H(x), \end{cases} \end{aligned}$$

which in either case implies that $|F_1(x) - G(x)| \leq \varepsilon$. Thus, $|F_1(x) - G(x)| \leq \varepsilon$ for all $x \in [a, b]$. □

Example 3.13. *Let g: $\mathbb{R} \to \mathbb{R}$ be Lebesgue integrable on $[a, b]$ and let $\varepsilon > 0$. There exists a continuous a.e. differentiable function G: $\mathbb{R} \to \mathbb{R}$ such that $G' = g$ a.e., $G(a) = G(b) = 0$, and $|G(x)| \leq \varepsilon$ for all $x \in [a, b]$.*

Proof: In Example 3.12, let $F_1(x) = 0$ and let $F_2(x) = \int_a^x g(y)\, dy$. The desired result then follows immediately. □

Example 3.14. *Let f be any real-valued function defined on $\mathbb{R}$ and let $\{h_n\}_{n \in \mathbb{N}}$ be any sequence of nonzero numbers converging to zero. There exists a function F: $\mathbb{R} \to \mathbb{R}$ such that $\lim_{n \to \infty}[F(x+h_n) - F(x)]/h_n = f(x)$ for all $x \in \mathbb{R}$.*

Proof: This example is from Bruckner [1978]. For real numbers x and y, we will write $x \sim y$ if $x - y$ may be written as a finite linear combination of elements in $\{h_n\colon n \in \mathbb{N}\}$ with integer coefficients. Notice that this relation is an equivalence relation and hence we may decompose $\mathbb{R}$ into a union of disjoint sets E^α, where the union is taken over all α from some index set A. Notice that E^α is countable for each $\alpha \in A$ since for any fixed real number x there exist only countably many values β such that $x \sim x + \beta$.

Fix $\alpha \in A$ and enumerate E^α as $\{x_1^\alpha, x_2^\alpha, \ldots\}$. For each $i \in \mathbb{N}$, define E_i^α to be the union of $\{x_i^\alpha\}$ with the set $\{x_i^\alpha + h_j \colon j \in \mathbb{N}\}$. Notice that E^α is simply the union of the E_i^α's over all $i \in \mathbb{N}$.

Define a real-valued function F on E_1^α by letting $F(x_1^\alpha) = 0$ and $F(x) = F(x_1^\alpha) + (x - x_1^\alpha)f(x_1^\alpha)$ for each $x \in E_1^\alpha - \{x_1^\alpha\}$. Proceeding inductively, assume for some integer $m > 1$ that F has been defined on $E_1^\alpha \cup \cdots \cup E_{m-1}^\alpha$ and let $R_m^\alpha = E_m^\alpha - (E_1^\alpha \cup \cdots \cup E_{m-1}^\alpha)$. (Define $R_1^\alpha = E_1^\alpha$.) If $x_m^\alpha \in R_m^\alpha$, then let $F(x_m^\alpha) = 0$ and for $x \in R_m^\alpha - \{x_m^\alpha\}$ let $F(x) = F(x_m^\alpha) + (x - x_m^\alpha)f(x_m^\alpha)$. This process defines F on E^α by induction. The function F is defined for all points in $\mathbb{R}$ by defining F on E^α in this way for each $\alpha \in A$.

Now let x be any real number. Since the E^α's partition $\mathbb{R}$, there exists some $\alpha \in A$ and some $m \in \mathbb{N}$ such that $x = x_m^\alpha$. If $m = 1$, then $x_m^\alpha + h_n \in R_m^\alpha$ for all $n \in \mathbb{N}$. If $m > 1$, then let I be an open interval such that $x_m^\alpha \in I$ but $x_i^\alpha \notin I$ for $i = 1, \ldots, m-1$. Since the h_n's converge to zero, there exists some integer N such that $x_i^\alpha + h_n \notin I$ and $x_m^\alpha + h_n \in I$ for $i = 1, \ldots, m-1$ and for all $n \geq N$ and hence $x_m^\alpha + h_n \in R_m^\alpha$ for all $n \geq N$. Thus, for any positive integer m, $x_m^\alpha + h_n \in R_m^\alpha$ for all n sufficiently large. Thus, for any $m \in \mathbb{N}$ and for all n sufficiently large, it follows that

$$\frac{F(x_m^\alpha + h_n) - F(x_m^\alpha)}{h_n} = \frac{F(x_m^\alpha) + (x_m^\alpha + h_n - x_m^\alpha)f(x_m^\alpha) - F(x_m^\alpha)}{h_n} = f(x_m^\alpha),$$

and hence the desired conclusion follows immediately. □

Example 3.15. *Given any closed subinterval $[a, b]$ of $\mathbb{R}$ with $a < b$ and any sequence $\{h_n\}_{n \in \mathbb{N}}$ of nonzero real numbers converging to zero, there exists a continuous function $F\colon [a, b] \to \mathbb{R}$ such that for any Lebesgue measurable function $f\colon [a, b] \to \mathbb{R}$ there exists a subsequence $\{h_{n_k}\}_{k \in \mathbb{N}}$ of $\{h_n\}_{n \in \mathbb{N}}$ such that*

$$\lim_{k \to \infty} \frac{F(x + h_{n_k}) - F(x)}{h_{n_k}} = f(x) \text{ a.e. on } [a, b].$$

Proof: This example, from Bruckner [1978], is due to Marcinkiewicz. Let $C([a, b])$ denote the normed linear space of all continuous real-valued functions defined on $[a, b]$ equipped with the sup norm and let $\{P_k\}_{k \in \mathbb{N}}$ be an enumeration of those elements in $C([a, b])$ that are polynomials with rational coefficients. It follows directly from the Weierstrass approximation theorem and a well-known approximation theorem due to Lusin that corresponding to any Lebesgue measurable function $f\colon [a, b] \to \mathbb{R}$ there exists a subsequence of $\{P_k\}_{k \in \mathbb{N}}$ that converges to f a.e. on $[a, b]$. Let $\mathcal{F}$ denote the subset of $C([a, b])$ consisting of all functions g from $C([a, b])$ such that for any positive integers n and k there exists an integer $m > n$ such that

$$\left|\frac{g(x+h_m)-g(x)}{h_m} - P_k(x)\right| < \frac{1}{n}$$

holds off a set having Lebesgue measure not greater than $1/n$. Note that if $\mathcal{F}$ is nonempty, then the desired result follows immediately. Indeed, any element in $\mathcal{F}$ may serve as the desired function.

Let $\mathcal{S}$ denote the set $C([a, b])\setminus\mathcal{F}$. We will show that $\mathcal{F}$ is nonempty by showing that $\mathcal{S}$ is a first category subset of $C([a, b])$. For positive integers n and k let $\mathcal{S}_{nk}$ denote the set of all functions g from $C([a, b])$ such that

$$\left|\frac{g(x+h_m)-g(x)}{h_m} - P_k(x)\right| \geq \frac{1}{n}$$

holds for each $m > n$ on some set having Lebesgue measure not less than $1/n$. Since $\mathcal{S} = \bigcup_{n=1}^{\infty}\bigcup_{k=1}^{\infty}\mathcal{S}_{nk}$, the desired result regarding $\mathcal{S}$ will follow if we show that $\mathcal{S}_{nk}$ is nowhere dense for all positive integers n and k. To accomplish this we will show that $\mathcal{S}_{nk}$ is closed and $C([a, b])\setminus\mathcal{S}_{nk}$ is dense in $C([a, b])$ for all positive integers n and k.

If $\mathcal{S}_{nk}$ is empty, the desired result follows immediately. Hence, assume $\mathcal{S}_{nk}$ is nonempty. Fix positive integers n and k and let $\{g_m\}_{m\in\mathbb{N}}$ be a sequence of functions in $\mathcal{S}_{nk}$ that converges (uniformly) to some function g. Let $\varepsilon > 0$ be given and choose $N \in \mathbb{N}$ so that $|g(x) - g_m(x)| < \varepsilon$ for all integers $m > N$ and all $x \in [a, b]$. Then, for $m > N$, we see that

$$\begin{aligned}&\left|\frac{g_m(x+h_j)-g_m(x)}{h_j} - \frac{g(x+h_j)-g(x)}{h_j}\right| \\ &\leq \left|\frac{g_m(x+h_j)-g(x+h_j)}{h_j}\right| + \left|\frac{g(x)-g_m(x)}{h_j}\right| \leq \frac{2\varepsilon}{|h_j|}\end{aligned}$$

for each positive integer j. Thus, since $g_m \in \mathcal{S}_{nk}$ for each positive integer m, it follows that

$$\begin{aligned}\frac{2\varepsilon}{|h_j|} &\geq \left|\frac{g_m(x+h_j)-g_m(x)}{h_j} - \frac{g(x+h_j)-g(x)}{h_j}\right| \\ &\geq \left|\frac{g_m(x+h_j)-g_m(x)}{h_j} - P_k(x)\right| - \left|\frac{g(x+h_j)-g(x)}{h_j} - P_k(x)\right| \\ &\geq \frac{1}{n} - \left|\frac{g(x+h_j)-g(x)}{h_j} - P_k(x)\right|\end{aligned}$$

for all integers $j > n$ on a set having Lebesgue measure not less than $1/n$ and hence we see that

$$\left|\frac{g(x+h_j)-g(x)}{h_j} - P_k(x)\right| \geq \frac{1}{n} - \frac{2\varepsilon}{|h_j|}$$

for all integers $j > n$ on a set having Lebesgue measure not less than $1/n$. Since ε may be any positive real number, we conclude that $g \in \mathcal{S}_{nk}$ and hence that $\mathcal{S}_{nk}$ is a closed subset of $C([a, b])$ for all positive integers n and k.

Next, fix positive integers n and k, let $\varepsilon > 0$ be given, let $g \in \mathcal{S}_{nk}$, and recall from Example 3.12 that there exists an a.e. differentiable function h in $C([a, b])$ such that $h' = P_k$ a.e. and $\|h - g\| < \varepsilon$. Since $h' = P_k$ off a Lebesgue null set it follows that h is not in $\mathcal{S}_{nk}$ and hence that $C([a, b]) \setminus \mathcal{S}_{nk}$ is dense in $C([a, b])$.

Finally, since $\mathcal{S}_{nk}$ is closed and has a dense complement, we see that $\mathcal{S}_{nk}$ is nowhere dense in $C([a, b])$ for all positive integers n and k. Thus, since $\mathcal{S}$ is a first category subset of $C([a, b])$, it follows that its complement $\mathcal{F}$ is not empty. Indeed, not only is $\mathcal{F}$ nonempty, but, as we have shown, "most" functions in $C([a, b])$ are elements of $\mathcal{F}$! □

References

Bruckner, A. M., *Differentiation of Real Functions,* Lecture Notes in Mathematics No. 659, Springer-Verlag: Berlin, 1978.

Natanson, I. P., *Theory of Functions of a Real Variable,* Ungar: New York, 1955.

Pompeiu, D., "Sur les fonctions dérivées," *Mathematische Annalen,* Vol. 63, 1906, pp. 326–332.

Volterra, V., "Sui principii del calcolo integrale," *Giornale di Battaglini,* Vol. 19, 1881, pp. 333–372.

Wilkosz, W., "Some properties of derivative functions," *Fundamenta Mathematicae,* Vol. 2, 1921, pp. 145–154.

4. Measures

A measurable space $(\Omega, \mathcal{F})$ is an ordered pair consisting of a set Ω and a σ-algebra $\mathcal{F}$ of subsets of Ω. Sets in $\mathcal{F}$ are known as measurable sets. Measures are nonnegative extended real-valued functions defined on σ-algebras that map the empty set to zero and that are countably additive. If $(\Omega, \mathcal{F})$ is a measurable space and μ is a measure defined on $\mathcal{F}$, then the ordered triple $(\Omega, \mathcal{F}, \mu)$ is called a measure space. If $(\Omega, \mathcal{F}, \mu)$ is a measure space and if $\mu(\Omega) < \infty$, the measure, as well as the measure space, is called finite. If Ω can be written as a union of a countable family of measurable sets each having finite measure, then the measure, as well as the measure space, is called σ-finite.

A signed measure on a σ-algebra is an extended real-valued function that assumes at most one of the values ∞ and $-\infty$, that maps the empty set to zero, and is such that the measure of a countable union of pairwise disjoint measurable sets is equal to the sum of the measures, where the sum converges to the same value under all possible rearrangements.

First, note that it is not true that on any measurable space, a normalized measure can be defined. For example, consider the empty set and the σ-algebra on the empty set consisting of the empty set; the only measure on this space is the zero measure. If, for example, $\mu: \mathcal{B}(\mathbb{R}) \to \overline{\mathbb{R}}$ via

$$\mu(B) = \sum_{n \in (\mathbb{N} \cap B)} \frac{(-1)^n}{n},$$

then μ is not a signed measure since under the appropriate rearrangement the summands for $\mu(\mathbb{R})$ could sum to any given element of $\overline{\mathbb{R}}$.

Given a measurable space $(\Omega, \mathcal{F})$, a measurable function $f: \Omega \to \mathbb{R}$ is any function such that $f^{-1}(\mathcal{B}(\mathbb{R})) \subset \mathcal{F}$.

Let k be a positive integer. Recall that Lebesgue measure on $\mathcal{B}(\mathbb{R}^k)$ is the unique measure on $\mathcal{B}(\mathbb{R}^k)$ such that the measure of a cube in $\mathbb{R}^k$ is the volume of the cube. Further, recall that the Lebesgue measurable subsets of $\mathbb{R}^k$ are the sets in the completion of $\mathcal{B}(\mathbb{R}^k)$ for Lebesgue measure.

If $(\Omega, \mathcal{F}, \mu)$ is a measure space, a measurable set has full measure if its complement has measure zero. If $(\Omega, \mathcal{F}, \mu)$ is a measure space such that Ω is a topological space and such that $\mathcal{F}$ includes the topology, then the support of the measure μ is the intersection of all closed sets having full measure.

If $(\Omega, \mathcal{F}, \mu)$ is a measure space, a measurable set A is called an atom if $\mu(A) > 0$ and if for any measurable set B, either $\mu(A \cap B) = 0$ or $\mu(A \setminus B) = 0$. The measure μ is said to be atomic if any measurable set of positive measure includes an atom. The measure μ is said to be nonatomic if μ has no atoms. The measure μ is said to be diffuse if singleton sets are measurable and have measure zero.

If $(\Omega, \mathcal{F})$ is a measurable space and μ and ν are measure on $\mathcal{F}$, then ν is absolutely continuous with respect to μ if for any $A \in \mathcal{F}$ such that $\mu(A) = 0$ it follows that $\nu(A) = 0$.

For a topological space, a neighborhood of a set is any superset that includes an open set that includes the set of interest. A neighborhood of a point is a neighborhood of the singleton set consisting of that point. Recall that a topological space is said to be Hausdorff if any two distinct points have disjoint neighborhoods. A topological space is said to be normal if any two disjoint closed sets have disjoint neighborhoods. Thus a normal space is not necessarily a Hausdorff space. For example, consider a topological space consisting of exactly two points with the indiscrete topology. A compact Hausdorff space is normal and metric spaces are normal. A topological space is said to be completely normal if any subspace is normal. Two sets are said to be separated if the closure of the first does not intersect the second and if the closure of the second does not intersect the first. Recall that a topological space is completely normal if and only if any two separated sets have disjoint neighborhoods.

A Borel measure on a Hausdorff space is a measure defined on the Borel sets, the σ-algebra generated by the topology, that is finite on compact sets. Consider a Borel measure μ on a Hausdorff space $(\Omega, \mathcal{T})$. The measure μ is said to be outer regular if for any Borel set B

$$\mu(B) = \inf\{\mu(U) \colon B \subset U \text{ and } U \in \mathcal{T}\}.$$

The measure μ is said to be inner regular if for any open set U

$$\mu(U) = \sup\{\mu(F) \colon F \subset U \text{ and } (\Omega \setminus F) \in \mathcal{T}\}.$$

The measure μ is said to be strongly inner regular if for any Borel set B

$$\mu(B) = \sup\{\mu(F) \colon F \subset B \text{ and } (\Omega \setminus F) \in \mathcal{T}\}.$$

Let $\mathcal{K}$ denote the compact subsets of Ω. Then the measure μ is said to be compact inner regular if for any open set U

$$\mu(U) = \sup\{\mu(K) \colon K \subset U \text{ and } K \in \mathcal{K}\}.$$

The measure μ is said to be strongly compact inner regular if for any Borel set B

$$\mu(B) = \sup\{\mu(K) \colon K \subset B \text{ and } K \in \mathcal{K}\}.$$

The measure μ is regular if it is outer regular and strongly inner regular, and it is compact regular if it is outer regular and strongly compact inner regular. Finally, the measure μ is said to be τ-smooth if

$$\mu\left(\bigcup_{\lambda \in \Lambda} U_\lambda\right) = \sup_{\lambda \in \Lambda} \mu(U_\lambda)$$

for every nondecreasing net $\{U_\lambda\}_{\lambda \in \Lambda}$ of open sets, with the partial order of set inclusion.

Recall that if Ω is any set, the family of all subsets that are either countable or co-countable (have a countable complement) is a σ-algebra on Ω. Also, note that a real-valued function on such a σ-algebra that maps countable sets to zero and sets having countable complements to one is a measure. Also, note that if $(\Omega, \mathcal{F})$ is a measurable space and if $\{\mu_\alpha: \alpha \in A\}$ is any nonempty family of measures on this measurable space, then the extended real-valued function μ on $\mathcal{F}$ defined via $\mu(B) = \sum_{\alpha \in A} \mu_\alpha(B)$ is a measure on $\mathcal{F}$.

If $(\Omega, \mathcal{F}, \mu)$ is a measure space, a subset A of Ω is said to be locally measurable if $A \cap F$ is measurable for all measurable sets F having finite measure. The family of all locally measurable sets is a σ-algebra. The measure μ is said to be saturated if all locally measurable sets are measurable.

An outer measure μ on a set Ω is a function mapping the power set of Ω into the nonnegative extended real numbers such that $\mu(\emptyset) = 0$, such that if $A \subset B \subset \Omega$ then $\mu(A) \leq \mu(B)$, and such that for any sequence $\{A_n\}_{n \in \mathbb{N}}$ of subsets of Ω, $\mu\left(\bigcup_{n \in \mathbb{N}} A_n\right) \leq \sum_{n=1}^{\infty} \mu(A_n)$. An outer measure μ on a set Ω is said to be regular if for any subset A of Ω there exists a measurable superset S of A such that $\mu(S) = \mu(A)$. If $\mathcal{A}$ is a σ-algebra on Ω, an outer measure μ on Ω is said to be $\mathcal{A}$ regular if for any subset A of Ω there exists an $\mathcal{A}$-measurable superset S of A such that $\mu(S) = \mu(A)$.

Example 4.1. *A σ-subalgebra of a countably generated σ-algebra need not be countably generated.*

Proof: Let $\Omega = [0, 1]$ and $\mathcal{F} = \mathcal{B}([0, 1])$. Further, let $\mathcal{G}$ be the σ-subalgebra of $\mathcal{F}$ given by the countable and co-countable subsets of $[0, 1]$. Since $\mathcal{F} = \sigma((a, b)\colon 0 \leq a < b \leq 1 \text{ and } a, b \in \mathbb{Q})$, it follows that $\mathcal{F}$ is countably generated. Assume now that $\mathcal{G}$ is also countably generated; that is, assume that $\mathcal{G} = \sigma(A_n\colon n \in \mathbb{N})$, where A_n is a subset of $[0, 1]$ for each n. Since $\mathcal{G}$ contains only countable and co-countable subsets of $[0, 1]$, A_n may be assumed without loss of generality to be countable for each n. Let $B = \bigcup_{n \in \mathbb{N}} A_n$ and note that B is also a countable subset of $[0, 1]$. Hence, there exists a real number x in $[0, 1]$ that is not an element of B. Notice also that if $\mathcal{D}$ is the family of all subsets of B and their complements, then $\mathcal{D}$ is a σ–subalgebra such that $\mathcal{G} \supset \mathcal{D} \supset \sigma(A_n\colon n \in \mathbb{N})$. But, $\mathcal{D} \neq \mathcal{G}$ since $\{x\}$ is in $\mathcal{G}$ but not in $\mathcal{D}$. This contradiction implies that $\mathcal{G}$ is not countably generated even though it is a σ-subalgebra of the countably generated σ-algebra $\mathcal{F}$. □

Example 4.2. *If $\mathcal{F}$ is a family of subsets of a set Ω and if A is an element of $\sigma(\mathcal{F})$, then there exists a countable subset $\mathcal{C}$ of $\mathcal{F}$ such that $A \in \sigma(\mathcal{C})$.*

Proof: Let $\mathcal{G}$ denote the family of sets $A \in \sigma(\mathcal{F})$ having the desired property. Note that $\mathcal{F}$ is a subset of $\mathcal{G}$ since for any $F \in \mathcal{F}$, $F \in \sigma(F)$. Thus we see that $\mathcal{F} \subset \mathcal{G} \subset \sigma(\mathcal{F})$. Now, if we show that $\mathcal{G}$ is a σ-algebra, then the desired result will follow due to the minimality of $\sigma(\mathcal{F})$. In this regard, note that the empty set is in $\mathcal{G}$ and that $\mathcal{G}$ is closed under complementation. Further, if $A_n \in \mathcal{G}$ for each positive integer n, then it follows that $\bigcup_{n \in \mathbb{N}} A_n \in \mathcal{G}$ since countable unions of countable sets are countable. Thus the result follows. □

Example 4.3. *If Ω is a set having cardinality $\aleph_1$, then there does not exist a diffuse probability measure μ defined on the power set of Ω.*

Proof: This example is from Oxtoby [1980] and is due to Ulam. Consider a well ordering of Ω denoted by $\leq$ and note that the set $\{x \in \Omega: x < y\}$ is countable for each $y \in \Omega$. For each $y \in \Omega$, let $f(x, y)$ be a bijective mapping of $\{x \in \Omega: x < y\}$ onto a subset of $\mathbb{N}$. Further, for each $x \in \Omega$ and each $n \in \mathbb{N}$, let F_x^n denote the subset of Ω given by $\{y \in \Omega: x < y$ and $f(x, y) = n\}$.

Consider points x and z from Ω with $x < z$ and assume that $y \in F_x^n \cap F_z^n$ for some positive integer n. Then $x < y$, $z < y$, and $f(x, y) = f(z, y) = n$. This implies, however, that $x = z$, since $f(s, u) \neq f(t, u)$ if $s < t < u$. Thus, for any positive integer n, the sets in $\{F_x^n: x \in \Omega\}$ are disjoint.

Next, note that if $x < y$, then $y \in F_x^n$ when $n = f(x, y)$. Further, if $y \leq x$ then y is in F_x^n for no positive integer n. Thus,

$$\Omega \setminus \bigcup_{n \in \mathbb{N}} F_x^n = \{y \in \Omega: y \leq x\}.$$

That is, for any $x \in \Omega$, the set $\bigcup_{n \in \mathbb{N}} F_x^n$ differs from Ω by only a countable set.

Now consider a finite measure μ defined on the power set of Ω and assume that μ assigns zero measure to each singleton set. We will show that $\mu(\Omega)$ cannot be positive. Since μ is finite and since, for any positive integer n, the sets in $\{F_x^n: x \in \Omega\}$ are disjoint, it follows that, for each $n \in \mathbb{N}$, $\mu(F_x^n)$ is positive for only countably many points x from Ω. Thus only countably many of the sets F_x^n with $n \in \mathbb{N}$ and $x \in \Omega$ have positive μ-measure. Therefore, since there are an uncountable number of points in Ω, there must exist some $z \in \Omega$ such that $\mu(F_z^n) = 0$ for each $n \in \mathbb{N}$, and hence, for such a choice of z, we see that

$$\mu\left(\bigcup_{n \in \mathbb{N}} F_z^n\right) = 0.$$

Further, since μ assigns zero measure to singletons, it follows that

$$\mu\left(\Omega \setminus \bigcup_{n \in \mathbb{N}} F_z^n\right) = 0.$$

Thus, we see that $\mu(\Omega) = 0$. □

Example 4.4. *Assuming the continuum hypothesis, there does not exist a diffuse probability measure on the power set of any interval of real numbers having a positive length.*

Proof: This is a direct consequence of Example 4.3. □

Example 4.5. *A σ-finite measure μ on $\mathcal{B}(\mathbb{R}^k)$ can be such that any Borel set with a nonempty interior has infinite measure.*

Proof: Recall that the set $\mathbb{Q}^k$ of all points in $\mathbb{R}^k$ that have rational coordinates is countable and dense. Let μ: $\mathcal{B}(\mathbb{R}^k) \to \overline{\mathbb{R}}$ via

$$\mu(B) = \sum_{x \in \mathbb{Q}^k} I_B(x).$$

Since any sum of measures is a measure, we see that μ is a measure. Further, since any Borel set with a nonempty interior contains infinitely many points in $\mathbb{Q}^k$, we see that the stated claim follows. □

Example 4.6. *A σ-finite measure on $\mathcal{B}(\mathbb{R}^k)$ need not be a Borel measure.*

Proof: The measure introduced in Example 4.5 is σ-finite, but it is not a Borel measure since any compact set with a nonempty interior is assigned infinite measure. □

Example 4.7. *A measure μ on $\mathcal{B}(\mathbb{R}^k)$ can be σ-finite but not locally finite.*

Proof: The measure space in Example 4.5 is σ-finite, but for any point x, any neighborhood of x has infinite measure. □

Example 4.8. *A Borel measure on a Hausdorff space need not be locally finite.*

Proof: Let Ω be the real line with the countable complement extension of the usual topology on the reals. That is, a set G is open in this topology if and only if $G = U \setminus K$, where U is an open set in the usual topology on the reals and K is a countable subset of the reals. Note that the resulting topological space is a Hausdorff space and that compact sets are precisely the finite subsets of the reals. Also, the Borel sets are the usual Borel subsets of the reals. Now define a measure μ on the Borel subsets of Ω given by $\mu(B) = 0$ if B is countable, and $\mu(B) = \infty$ otherwise. Since all compact sets are finite, this measure is a Borel measure. However, for any point in Ω, any neighborhood of it has uncountably many points. Thus, the Borel measure is not locally finite. □

Example 4.9. *A locally finite Borel measure on a locally compact Hausdorff space need not be σ-finite.*

Proof: Consider $\mathbb{R}$ with the discrete topology, and note that this is a locally compact Hausdorff space. Let the measure μ be defined on $\mathcal{B}(\mathbb{R}) = \mathbb{P}(\mathbb{R})$ by $\mu(B) = 0$ for all countable sets B and $\mu(B) = \infty$ for all uncountable sets B. Then μ is not σ-finite since any countable partition of $\mathbb{R}$ must contain at least one uncountable set, but since any set consisting of a single point is a neighborhood of that point, μ is locally finite. □

Example 4.10. *For a family $\mathcal{F}$ of open intervals of real numbers, the fact that the endpoints of the intervals in $\mathcal{F}$ are dense in $\mathbb{R}$ does not imply that $\sigma(\mathcal{F}) = \mathcal{B}(\mathbb{R})$.*

Proof: Recall that the positive rationals are countable, and let $\{r_n: n \in \mathbb{N}\}$ be an enumeration of $\mathbb{Q} \cap (0, \infty)$. Now, let $\mathcal{F} = \{(-r_n, r_n): n \in \mathbb{N}\}$. Note that the endpoints of the intervals in $\mathcal{F}$ are dense in $\mathbb{R}$. Obviously, $\sigma(\mathcal{F})$ contains only Borel subsets of $\mathbb{R}$ since $\sigma(\mathcal{F})$ is generated by a family of open sets. However, note that $\sigma(\mathcal{F})$ does not contain the Borel set $[0, 1]$. Thus, $\sigma(\mathcal{F})$ is a proper subset of $\mathcal{B}(\mathbb{R})$. □

Example 4.11. *The σ-algebra generated by the open balls of a metric space need not be the family of Borel subsets.*

Proof: Consider the metric space given by the reals with the discrete metric. Note that a Borel set is any set of real numbers; indeed, any set of real numbers is an open set in this metric space. Now, note that the family of open balls in this metric space consists of the whole space and all singleton sets. Hence, the σ-algebra generated by the open balls consists of all countable subsets and all subsets whose complements are countable sets. □

Example 4.12. *If H is a Hausdorff space and if B is a closed subset of H with the subspace topology, then the Borel sets of B need not be a subset of the Borel sets of H.*

Proof: Let H denote the Sorgenfrey plane; that is, H is $\mathbb{R}^2$ with the topology having a basis of all sets of the form $[a, b) \times [c, d)$ in the plane, where a, b, c, and d are real numbers. This topological space is obviously Hausdorff. Also, the Borel sets in H are the same as the Borel sets in $\mathbb{R}^2$ with the usual topology. Let B be the straight line in the plane that goes through the origin and has a slope of -1. It is clearly seen that B is a closed set since B contains all its limit points. For any real number x, note that the open set $[x, x+1) \times [-x, -x+1)$ intersects B at precisely the point $(x, -x) \in B$. Thus, each point of B is open, and hence the subspace topology on B is $\mathbb{P}(B)$. Now, simply choose an element from $\mathbb{P}(B)$ that is not in $\mathcal{B}(H)$, and the proof is complete. □

Example 4.13. *There does not exist any σ-algebra whose cardinality is $\aleph_0$.*

Proof: This example is taken from Billingsley [1986]. Assume that $\{A_n: n \in \mathbb{N}\}$ is a family of distinct sets in a σ-algebra $\mathcal{F}$. Let $\mathcal{G}$ consist of all nonempty sets that can be expressed as $\bigcap_{n\in\mathbb{N}} B_n$, where for each $n \in \mathbb{N}$, B_n is equal to either A_n or A_n^c. Note that for each $n \in \mathbb{N}$, A_n can be written as the union of all subsets of $\mathcal{G}$ that are included in A_n. Thus $\mathcal{G}$ is an infinite set. Consequently, there exist infinitely many disjoint sets in $\mathcal{G}$. Since there are uncountably many distinct countable unions of sets in $\mathcal{G}$, we see that $\mathcal{F}$ is uncountable. Thus, we see that a σ-algebra is either finite or uncountable. □

Example 4.14. *If $\{\mathcal{F}_n\}_{n\in\mathbb{N}}$ is a sequence of σ-algebras on a given set such that for each $n \in \mathbb{N}$, $\mathcal{F}_n$ is a proper subset of $\mathcal{F}_{n+1}$, then $\bigcup_{n\in\mathbb{N}} \mathcal{F}_n$ is not a σ-algebra on the given set.*

Proof: For each $n \in \mathbb{N}$, let A_n be a set contained in $\mathcal{F}_n$ that is not contained in $\mathcal{F}_{n+1}$. Then for each $n \in \mathbb{N}$, A_n is contained in $\bigcup_{n\in\mathbb{N}} \mathcal{F}_n$, and yet $\bigcup_{n\in\mathbb{N}} A_n$ is not contained in $\bigcup_{n\in\mathbb{N}} \mathcal{F}_n$, since there does not exist $k \in \mathbb{N}$ such that $\bigcup_{n\in\mathbb{N}} A_n$ is contained in $\mathcal{F}_k$. □

Example 4.15. *A sequence of decreasing measures on $\mathcal{B}(\mathbb{R}^k)$ may be such that the limiting set function is not a measure.*

Proof: Consider the measurable space $(\mathbb{R}, \mathcal{B}(\mathbb{R}))$ and Lebesgue measure μ on $\mathcal{B}(\mathbb{R})$. For each positive integer n define a measure μ_n via $\mu_n(A) = \mu(A \cap (n, \infty))$ for each Borel set A. Notice that $\{\mu_n\}$ is a family of decreasing measures. Let $\nu(A) = \lim_{n\to\infty} \mu_n(A)$ for $A \in \mathcal{B}(\mathbb{R})$. The set function ν is not a measure since $\nu([0, \infty)) = \lim_{n\to\infty} \mu_n([0, \infty)) = \lim_{n\to\infty} \infty = \infty$ is not equal to

$$\sum_{i\in\mathbb{N}} \nu([i-1, i)) = \sum_{i\in\mathbb{N}} \lim_{n\to\infty} \mu_n([i-1, i)) = \sum_{i\in\mathbb{N}} 0 = 0.$$

That is, even though the sequence $\{\mu_n\}$ of measures is decreasing, the set function given by the limit of the μ_n's is not a measure since it fails to be countably additive. □

Example 4.16. *The existence of a measure space $(\Omega, \mathcal{F}, \mu)$ and a sequence $\{A_n\}_{n\in\mathbb{N}}$ of measurable sets such that $A_n \supset A_{n+1}$ for each positive integer n and such that $\bigcap_{n\in\mathbb{N}} A_n = \emptyset$ does not imply that $\lim_{n\to\infty} \mu(A_n) = 0$.*

Proof: Let $(\Omega, \mathcal{F}, \mu)$ denote the real line, the Borel subsets, and Lebesgue measure. For $n \in \mathbb{N}$, let $A_n = [n, \infty)$. Then for $n \in \mathbb{N}$, $A_n \supset A_{n+1}$ and $\bigcap_{n\in\mathbb{N}} A_n = \emptyset$, and yet $\mu(A_n) = \infty$ for each $n \in \mathbb{N}$. □

Example 4.17. *A countably additive, real-valued function on an algebra $\mathcal{A}$ need not be extendable to a signed measure on $\sigma(\mathcal{A})$.*

Proof: The following example was suggested by Dudley [1989]. Consider the set of real numbers $\mathbb{R}$ and a partition of $\mathbb{R}$ given by $\{A, B\}$, where $A = (-\infty, 0]$ and $B = (0, \infty)$. Let $\mathcal{A}$ be the algebra consisting of all finite subsets of $\mathbb{R}$ and their complements and let λ be counting measure on $\mathcal{A}$. If G is a finite set in $\mathcal{A}$, then define $\mu(G) = \lambda(G \cap A) - \lambda(G \cap B)$. If G has a finite complement, then define $\mu(G) = -\mu(G^c)$.

Now, consider a sequence $\{C_n\}_{n\in\mathbb{N}}$ of disjoint elements from $\mathcal{A}$ and let $C = \bigcup_{n\in\mathbb{N}} C_n$. If C is finite, then C_n is finite for each n, which implies that all but a finite number of the C_n's are empty. Hence countable additivity holds if C is finite. If C has a finite complement, then there exists exactly one set C_m that also has a finite complement. [If C_m and C_k $(m \neq k)$ each had a finite complement, then C_m and C_k could not be disjoint. Conversely, if no such set C_m with a finite complement exists, then C must be countable and hence could not have a finite complement.] Assume without loss of generality that $m = 1$. Thus, for $n > 1$, the C_n's are each finite and hence all but a finite number are empty. (If infinitely many C_n's were nonempty, then, since they are disjoint, C_1 could not have a finite complement.) Notice that the finite set C_1^c may be represented as a disjoint union via $C_1^c = (C^c \cup C_2 \cup C_3 \cup \cdots)$. Thus, $-\mu(C_1) = \mu(C_1^c) = \mu(C^c) + [\mu(C_2) + \mu(C_3) + \cdots] = -\mu(C) + [\mu(C_2) + \mu(C_3) + \cdots]$, which implies that $\mu(C) = \mu(C_1) + \mu(C_2) + \cdots$. That is, countable additivity holds if C has a finite complement. Thus, this result and the previous result imply that μ is countably additive on $\mathcal{A}$.

Assume now that μ possesses an extension to a signed measure on $\sigma(\mathcal{A})$. The Hahn–Jordan decomposition theorem implies that there exist two measures μ^+ and μ^-, with at least one measure being finite, such that $\mu = \mu^+ - \mu^-$. This result implies that μ must be bounded either above or below depending on which of the two measures μ^+ or μ^- is finite. But in this case μ is not bounded above or below. Thus, this contradiction implies that μ does not possess an extension to a signed measure on $\sigma(\mathcal{A})$. □

Example 4.18. *There exists an algebra $\mathcal{A}$ on a set Ω and a σ-finite measure μ on $\sigma(\mathcal{A})$ such that μ is not σ-finite on $\mathcal{A}$.*

Proof: Let $\mathcal{A}$ be the algebra on the rationals $\mathbb{Q}$ given by the collection of all finite disjoint unions of sets of the form $\{q \in Q\colon a < q \leq b\}$, where $a \in \mathbb{Q}$ and $b \in \mathbb{Q}$, and of the form $\{q \in \mathbb{Q}\colon a < q < \infty\}$, where $a \in \mathbb{Q} \cup \{-\infty\}$. Since $\mathbb{Q}$ is countable, it follows that $\sigma(\mathcal{A}) = \mathbb{P}(\mathbb{Q})$.

Now, let μ be counting measure on $\mathcal{A}$ and on $\sigma(\mathcal{A})$. Since $\mathbb{Q}$ is a countable set, since $\mu(q)$ is finite for each q in $\mathbb{Q}$, and since $\sigma(\mathcal{A})$ contains all singletons, it follows immediately that μ is σ-finite on $\sigma(\mathcal{A})$. Note, however, that the empty set is the only finite set in $\mathcal{A}$ and μ is finite only for a finite set. Thus, μ is not σ-finite on the algebra $\mathcal{A}$ that generates $\sigma(\mathcal{A})$. □

Example 4.19. *There exists an algebra $\mathcal{A}$ on a set Ω and a σ-finite measure μ on $\sigma(\mathcal{A})$ such that there are sets in $\sigma(\mathcal{A})$ that cannot be well approximated by sets in the algebra $\mathcal{A}$.*

Proof: Consider a measure space $(\Omega, \sigma(\mathcal{A}), \mu)$, where $\mathcal{A}$ is an algebra on Ω. Recall that if μ is σ-finite on $\mathcal{A}$, then for any $\varepsilon > 0$ and any set A in $\sigma(\mathcal{A})$ for which $\mu(A) < \infty$, there exists a set B in $\mathcal{A}$ such that $\mu(A \triangle B) < \varepsilon$, where $A \triangle B$ denotes the symmetric difference between A and B. In this example it is shown that this approximation theorem can fail if σ-finiteness on the algebra $\mathcal{A}$ is replaced by σ-finiteness on the σ-algebra $\sigma(\mathcal{A})$.

Consider again the algebra $\mathcal{A}$ on the rationals $\mathbb{Q}$ that was presented in Example 4.18. Further, let μ be counting measure on $\mathcal{A}$ and on $\sigma(\mathcal{A})$ and recall that $\sigma(\mathcal{A}) = \mathbb{P}(\mathbb{Q})$. Let A be a finite nonempty subset of $\mathbb{Q}$ and let B be any element from $\mathcal{A}$. Then $\mu(A \triangle B) = \mu(A)$ if B is empty and $\mu(A \triangle B) = \infty$ otherwise. Hence, for any positive real value of $\varepsilon < \mu(A)$, there does not exist a set B in the algebra $\mathcal{A}$ for which $\mu(A \triangle B) < \varepsilon$, and thus the above approximation theorem fails. □

Example 4.20. *There exists an algebra $\mathcal{A}$ on a set Ω and an uncountable collection of distinct σ-finite measures on $\sigma(\mathcal{A})$ that each agree on the algebra $\mathcal{A}$, yet disagree on $\sigma(\mathcal{A})$.*

Proof: As in Example 4.19, let $\mathcal{A}$ be the algebra on the rationals $\mathbb{Q}$ given in Example 4.18 and let μ be counting measure on $\mathcal{A}$ and on $\sigma(\mathcal{A})$. Recall that μ is σ-finite on $\sigma(\mathcal{A})$. For $t \in (0, \infty)$, let ν_t be a measure on $\mathcal{A}$ and on $\sigma(\mathcal{A})$ given by $t\mu$. That is, $\nu_t(A) = t\mu(A)$ for any set A in $\sigma(\mathcal{A})$. Let M be a family of such measures given by $\{\nu_t\colon t \in (0, \infty); t \neq 1\}$. Note that on $\mathcal{A}$, the measure μ agrees with any measure from M since each assigns infinite measure to any nonempty set in $\mathcal{A}$. However, μ disagrees with every measure in M for any finite nonempty element in $\sigma(\mathcal{A})$. □

Example 4.21. *There exists a σ-finite measure on $\mathcal{B}(\mathbb{R})$ that is not a Lebesgue–Stieltjes measure.*

Proof: As in Example 4.5, define a σ-finite measure μ on $\mathcal{B}(\mathbb{R})$ via $\mu(B) = \sum_{n\in\mathbb{N}} I_B(r_n)$, where $\{r_n\colon n \in \mathbb{N}\}$ is an enumeration of $\mathbb{Q}$. Then for any real numbers a and b with $a < b$, $\mu((a, b]) = \infty$. Hence, there is no real valued nondecreasing function h mapping $\mathbb{R}$ into $\mathbb{R}$ such that $\mu((a, b]) = h(b) - h(a)$, and thus μ is not a Lebesgue–Stieltjes measure. □

Example 4.22. *Given a σ-subalgebra $\mathcal{A}_0$ of $\mathcal{B}(\mathbb{R})$, there need not exist a function $f\colon \mathbb{R} \to \mathbb{R}$ such that $f^{-1}(\{B \subset \mathbb{R}\colon f^{-1}(B) \in \mathcal{B}(\mathbb{R})\}) = \mathcal{A}_0$.*

Proof: This example is from Bahadur and Lehmann [1955]. Let $\mathcal{A}_0$ denote the σ-algebra on $\mathbb{R}$ consisting of all countable or co-countable subsets of $\mathbb{R}$, let f be a real-valued function defined on $\mathbb{R}$, and let $\mathcal{F}$ denote the collection of all subsets B of $\mathbb{R}$ such that $f^{-1}(B) \in \mathcal{B}(\mathbb{R})$. Assume that $f^{-1}(\mathcal{F}) = \mathcal{A}_0$. Let A be a nonempty real Borel set and let $x \in \mathbb{R}$ be such that $f(x) \in f(A)$. Then $f(x) = f(a)$ for some $a \in A$. Since $\{a\} \in \mathcal{A}_0$ and $\mathcal{A}_0 = f^{-1}(\mathcal{F})$, there exists some set $B \in \mathcal{F}$ such that $\{a\} = f^{-1}(B)$. But, since $f(x) = f(a)$ and $f(a) \in B$, we see that $x \in f^{-1}(B)$ and hence that $x = a$. Thus, if $f(x) \in f(A)$, then $x \in A$ and hence it follows that $f^{-1}(f(A)) \subset A$. Since the reverse set inclusion is always true, we conclude that $f^{-1}(f(A)) = A$. Further, since A is a real Borel set, we see that $f(A)$ must be in $\mathcal{F}$ by definition. Thus, by our assumption, we conclude that $f^{-1}(f(A))$ (which is equal to A) must be in $\mathcal{A}_0$ for any real Borel set A. This, however, is not possible since $\mathcal{A}_0$ is a proper subset of $\mathcal{B}(\mathbb{R})$, and thus we conclude that no such function f can exist. □

Example 4.23. *There exists a measure ν on $\mathcal{M}(\mathbb{R})$ that is absolutely continuous with respect to Lebesgue measure, and yet is such that $\nu(A) = \infty$ for any Lebesgue measurable set A with a nonempty interior.*

Proof: Let $f\colon \mathbb{R} \to \mathbb{R}$ via

$$f(x) = \begin{cases} 0, & \text{if } x = 0 \\ |x|^{-1}, & \text{if } x \neq 0. \end{cases}$$

Let $\{r_n\colon n \in \mathbb{N}\}$ be an enumeration of the rationals, and let $g\colon \mathbb{R} \to \mathbb{R}$ via $g(x) = \sum_{n=1}^{\infty} 2^{-n} f(x - r_n)$. Now, let ν be a measure on $\mathcal{M}(\mathbb{R})$ defined via $\nu(A) = \int_A g\, d\lambda$. Thus we see that $\nu \ll \lambda$ and yet for any Lebesgue measurable set A with a nonempty interior, $\nu(A) = \infty$. □

Example 4.24. *A σ-finite measure ν may be absolutely continuous with respect to a finite measure μ yet not be such that for every positive real number ε there exists a positive real number δ such that if $\mu(A) < \delta$ for some measurable set A, then $\nu(A) < \varepsilon$.*

Proof: This example was inspired by Billingsley [1986]. Consider the measurable space given by $\mathbb{N}$ and the power set $\mathbb{P}(\mathbb{N})$ of $\mathbb{N}$. Further, let ν

be counting measure on $\mathbb{P}(\mathbb{N})$ and define a measure μ on $\mathbb{P}(\mathbb{N})$ via $\mu(\{n\}) = 1/n^2$ for each n in $\mathbb{N}$. Notice that $\nu \ll \mu$; indeed, these measures are equivalent.

Now, let $\varepsilon \in (0, \infty)$ be given. Assume that there exists a positive real number δ such that if $\mu(A) < \delta$ for some subset A of $\mathbb{N}$, then $\nu(A) < \varepsilon$. Since $\sum_{n\in\mathbb{N}} 1/n^2$ is finite, there exists an integer N such that $\sum_{n=N}^{\infty} 1/n^2 < \delta$. Let $A = \{N, N+1, N+2, \ldots\}$. Then $\mu(A) < \delta$ even though $\nu(A) = \infty$. This contradiction implies that no such δ exists. □

Example 4.25. *There exist two measures μ and ν on $\mathcal{B}(\mathbb{R}^k)$ such that μ is nonatomic and ν is atomic, yet μ is absolutely continuous with respect to ν.*

Proof: Let μ be Lebesgue measure and let ν be counting measure, respectively, on $\mathcal{B}(\mathbb{R}^k)$. Then it is obvious that $\mu \ll \nu$, and yet μ is nonatomic and ν is atomic. □

Example 4.26. *For any positive integer k, Lebesgue measure restricted to a σ-subalgebra of the Lebesgue measurable subsets of $\mathbb{R}^k$ need not be σ-finite.*

Proof: Lebesgue measure restricted to the σ-subalgebra consisting of the countable and co-countable subsets of $\mathbb{R}^k$ is not σ-finite since the only sets in this σ-algebra having finite measure are the countable sets. □

Example 4.27. *If $(\Omega, \mathcal{F}, m)$ is a measure space, A_1 and A_2 are subsets of Ω, and C_1 and C_2 are measurable covers of A_1 and A_2, respectively, then $C_1 \cap C_2$ need not be a measurable cover of $A_1 \cap A_2$.*

Proof: Let $(\Omega, \mathcal{F}, m)$ be given by the real line, the Borel subsets of the real line, and Lebesgue measure restricted to the Borel subsets of the real line. Let A_1 denote a saturated non-m-measurable set and let $A_2 = \mathbb{R} \setminus A_1$. Then measurable covers are given by $C_1 = C_2 = \mathbb{R}$, and we see that $m^*(A_1 \cap A_2) = 0$ even though $m(C_1 \cap C_2) = \infty$. □

Example 4.28. *For a measure space $(\Omega, \mathcal{F}, \mu)$, the completion of $\mathcal{F}$ with respect to μ need not be the σ-algebra of sets measurable with respect to the outer measure on $\mathbb{P}(\Omega)$ generated by μ.*

Proof: This example is taken from Dudley [1989]. Let $\Omega = \{1, 2\}$, $\mathcal{F} = \{\emptyset, \Omega\}$, $\mu(\emptyset) = 0$, and $\mu(\Omega) = \infty$. Note that $(\Omega, \mathcal{F}, \mu)$ is complete. However, note that the outer measure μ^* generated by μ is zero on the empty set and infinity on any nonempty set. Thus, all subsets of Ω are μ^*-measurable. □

Example 4.29. *A measure induced by an outer measure need not be saturated.*

Proof: The example was suggested by Dotson [1970]. Define an outer measure μ^* on $\mathbb{P}(\mathbb{R})$ via $\mu^*(\emptyset) = 0$, $\mu^*(A) = 1$ if A is a nonempty countable subset of $\mathbb{R}$, and $\mu^*(A) = \infty$ for any uncountable subset A of $\mathbb{R}$. Let S be any nonempty proper subset of $\mathbb{R}$, and let $A = \{p, q\}$, where $p \in S$ and

$q \in (\mathbb{R} \setminus S)$. Then note that $\mu^*(A) = 1 < 1 + 1 = \mu^*(A \cap S) + \mu^*(A \setminus S)$. Thus, the only μ^*-measurable sets are $\mathbb{R}$ and $\emptyset$, and the measure μ induced by μ^* is given by $\mu = \mu^*|_{\{\mathbb{R}, \emptyset\}}$. Now, note that any subset C of $\mathbb{R}$ is locally measurable since the only μ^*-measurable subset of C of finite measure is $\emptyset$. Thus, μ is not saturated. □

Example 4.30. *If $(\Omega, \mathcal{F}, m)$ is a measure space and if W is any subset of Ω such that $m^*(W) < \infty$, then m^* is a finite measure on the σ-algebra given by $\{F \cap W\colon F \in \mathcal{F}\}$.*

Proof: This example is from Dudley [1989]. The fact that $\{F \cap W\colon F \in \mathcal{F}\}$ is a σ-algebra is immediate. Let $\{A_n\}_{n\in\mathbb{N}}$ be a sequence of sets from $\mathcal{F}$ such that the sets in $\{W \cap A_n\}_{n\in\mathbb{N}}$ are pairwise disjoint. Let C be a measurable cover of W, and note that $W \cap A_n = C \cap A_n$. Let $D_1 = A_1$ and, for integers $n > 1$, let $D_n = A_n \setminus \bigcup_{j=1}^{n-1} A_j$. Note that for positive integers n, $D_n \cap W = A_n \cap W$. Hence, without loss of generality, assume that the A_n's are disjoint subsets of C. We will show that

$$m^*\left(\bigcup_{n\in\mathbb{N}} W \cap A_n\right) = \sum_{n=1}^{\infty} m^*(W \cap A_n).$$

For each positive integer n, let F_n be a measurable cover of $A_n \cap W$ and be a subset of A_n. Note that $m(F_n) = m^*(A_n \cap W)$, and assume that $m(F_n) < m(A_n)$. Then $A_n \setminus F_n$ is a measurable set disjoint from W having positive measure. Thus the set $C \setminus (A_n \setminus F_n)$ is a measurable cover of W with measure less than that of C. This contradiction and the fact that $m^*(A_n \cap W) \le m(A_n)$ implies that $m^*(A_n \cap W) = m(A_n)$ for each positive integer n. Finally, an argument analogous to that above shows that

$$m^*\left(\bigcup_{n\in\mathbb{N}} A_n \cap W\right) = m\left(\bigcup_{n\in\mathbb{N}} A_n\right),$$

and hence it follows that m^* is countably additive. □

Example 4.31. *If $(\Omega, \mathcal{F}, \mu)$ is a measure space and if $\mathcal{G}$ is the σ-algebra of all locally measurable sets, there may exist different saturated extensions of μ to $\mathcal{G}$.*

Proof: Consider the measure μ on the measurable space $(\mathbb{R}, \{\mathbb{R}, \emptyset\})$ defined via $\mu(\mathbb{R}) = \infty$. As noted in Example 4.29, this measure is not saturated and the σ-algebra of locally measurable sets is $\mathbb{P}(\mathbb{R})$. Let μ_1 be a measure defined on $(\mathbb{R}, \mathbb{P}(\mathbb{R}))$ via $\mu_1(A) = 0$ if A is countable and $\mu_1(A) = \infty$ if A is uncountable. Let μ_2 be a measure defined on $(\mathbb{R}, \mathbb{P}(\mathbb{R}))$ via $\mu_2(A) = \infty$ if A is nonempty. Note that μ_1 and μ_2 are each saturated measures on $(\mathbb{R}, \mathbb{P}(\mathbb{R}))$ and that each is an extension of μ on $(\mathbb{R}, \{\mathbb{R}, \emptyset\})$. □

Example 4.32. *Let $(\Omega, \mathcal{F})$ be a measurable space and let μ and ν be σ-finite measures on $(\Omega, \mathcal{F})$. Let $\mathcal{P}$ be a collection of elements in $\mathcal{F}$ such* *that $\Omega \in \mathcal{P}$ and such that the intersection of any two sets in $\mathcal{P}$ is in $\mathcal{P}$. If $\mu(A) = \nu(A)$ for $A \in \mathcal{P}$, then $\mu(A)$ need not equal $\nu(A)$ for $A \in \sigma(\mathcal{P})$.*

Proof: Consider the measurable space $(\mathbb{R}, \mathcal{B}(\mathbb{R}))$, let μ denote Lebesgue measure, and let ν denote 2μ. Let $\mathcal{P} = \{(-\infty, a)\colon a \in (-\infty, \infty]\}$. Note that for $A \in \mathcal{P}$, $\mu(A) = \nu(A) = \infty$. However, note that $[-1, 1) \in \sigma(\mathcal{P})$ and that $\mu([-1, 1)) = 2$, whereas $\nu([-1, 1)) = 4$. □

Example 4.33. *A normalized regular Borel measure on a completely normal locally compact Hausdorff space can have an empty support.*

Proof: In this example we will develop some properties of the ordinal $\aleph_1 + 1$. Recall that we put the order topology on $\aleph_1 + 1$ and that $\aleph_1 + 1$ is a completely normal Hausdorff space. To show that $\aleph_1 + 1$ is compact, let $\mathcal{U}$ be any open cover of $\aleph_1 + 1$. Let α_1 be the least element of $\aleph_1 + 1$ such that $(\alpha_1, \aleph_1]$ is a subset of some set $U_1 \in \mathcal{U}$. If $\alpha_1 \neq \emptyset$, let α_2 be the least element of $\aleph_1 + 1$ such that $(\alpha_2, \alpha_1]$ is a subset of some set $U_2 \in \mathcal{U}$. If $\alpha_2 \neq \emptyset$, let α_3 be the least element of $\aleph_1 + 1$ such that $(\alpha_3, \alpha_2]$ is a subset of some set $U_3 \in \mathcal{U}$. Note that if we continue in this fashion, for some integer n we will have that $\alpha_n = \emptyset$. That is, if not, then there would exist a nonempty subset of $\aleph_1 + 1$ without a least element. However, this is impossible since $\aleph_1 + 1$ is a well ordered set. Now, select some element U_{n+1} from $\mathcal{U}$ that contains the point $\emptyset$. Then $\{U_1, U_2, \ldots, U_n, U_{n+1}\}$ is a finite subcover. Now, note that $\aleph_1$ is an open subset of $\aleph_1 + 1$. Thus, $\aleph_1$ is a locally compact completely normal Hausdorff space.

Next, notice that any countable subset of $\aleph_1$ has an upper bound in $\aleph_1$. That is, let A be a countable subset of $\aleph_1$. Then, for each $a \in A$, the set of ordinals S_a less than a is countable. Hence, the union $B = \bigcup_{a \in A} S_a$ is countable. Since $\aleph_1$ is uncountable, B is a proper subset of $\aleph_1$. Now, choose a point $\alpha \in \aleph_1$ such that $\alpha \notin B$. Note that if $\alpha < a$ for some $a \in A$, then $\alpha \in S_a$ and hence in B. Thus α is an upper bound of A.

Now, observe that any infinite subset of $\aleph_1$ has a limit point in $\aleph_1$. That is, let I be an infinite subset of $\aleph_1$, and let I_C be a countably infinite subset of I. Then there exists $\alpha \in \aleph_1 + 1$ such that α is an upper bound for I_C. Hence, I_C is a subset of the closed set $[\alpha_0, \alpha]$. Note that since $[\alpha_0, \alpha]$ is a closed subset of $\aleph_1 + 1$, it is compact. Thus, I_C has a limit point in $[\alpha_0, \alpha]$. This point in $\aleph_1$ is also a limit point of I.

Let $\{x_n\}_{n \in \mathbb{N}}$ and $\{y_n\}_{n \in \mathbb{N}}$ be sequences in $\aleph_1$ such that for each $n \in \mathbb{N}$, $x_n \leq y_n \leq x_{n+1}$. From this and the above, it follows that each of these sequences converges to the same limit in $\aleph_1$.

Let $\mathcal{C}$ be the set of all closed uncountable subsets of $\aleph_1$. Observe that a subset S of $\aleph_1$ is uncountable if and only if for any $\beta \in \aleph_1$ there is an $\alpha \in S$ such that $\beta < \alpha$. Let A and B be elements of $\mathcal{C}$. Note that if $\{x_n\}_{n \in \mathbb{N}}$ and $\{y_n\}_{n \in \mathbb{N}}$ are sequences such that $\{x_n\colon n \in \mathbb{N}\} \subset A$, $\{y_n\colon n \in \mathbb{N}\} \subset B$, and such that for each $n \in \mathbb{N}$, $x_n \leq y_n \leq x_{n+1}$, then, by the above reasoning, the sequences $\{x_n\}_{n \in \mathbb{N}}$ and $\{y_n\}_{n \in \mathbb{N}}$ each converge to an element of $\aleph_1$, and since A and B are closed, this element is in each of these sets. Note that for any $\alpha \in \aleph_1$, such a sequence can be chosen such that $\alpha < x_1$. Further, note that $\aleph_1 = \bigcup_{\alpha \in \aleph_1} \{\beta \in \aleph_1\colon \beta < \alpha\}$. Now, for each $\alpha \in \aleph_1$, choose a sequence taking values in A and a sequence taking values in B such that each of the sequences converges to a common point in $\aleph_1$ greater than α. Thus, we see that $A \cap B$ is a closed uncountable set. Hence, finite intersections of sets in $\mathcal{C}$ are in $\mathcal{C}$.

Now, consider a countable family of sets $\{F_n: n \in \mathbb{N}\} \subset \mathcal{C}$, and let F denote the intersection of this family. Now, for each $\alpha \in \aleph_1$, choose $\alpha_n \in \bigcap_{k=1}^n F_k$ such that $\alpha < \alpha_1 < \alpha_2 < \alpha_3 < \cdots$. As noted above, this sequence must converge to some ordinal $\gamma < \aleph_1$. For each fixed $n \in \mathbb{N}$, $\{a_{n-1+m}: m \in \mathbb{N}\} \subset \bigcap_{k=1}^n F_k$, and since $\bigcap_{k=1}^n F_k$ is closed, $\gamma \in \bigcap_{k=1}^n F_k$. Thus, $\gamma \in F$. Further, recall that there are only countably many ordinals less than γ. Now, for each $\alpha \in \aleph_1$, find such a point $\gamma \in F$. Thus, $F \in \mathcal{C}$.

Let $\mathcal{M}$ consist of all subsets M of $\aleph_1 + 1$ such that there exists a set $C \in \mathcal{C}$ so that either M or $(\aleph_1 + 1) \setminus M$ is a superset of C. Note that $\emptyset$ and $\aleph_1 + 1$ each are elements of $\mathcal{M}$. Further, if $M \in \mathcal{M}$ and if $C \in \mathcal{C}$ is a subset of M, then C is a subset of $(\aleph_1 + 1) \setminus ((\aleph_1 + 1) \setminus M) = M$, and, if $C \in \mathcal{C}$ is a subset of $(\aleph_1 + 1) \setminus M$, then M is a subset of $(\aleph_1 + 1) \setminus ((\aleph_1 + 1) \setminus ((\aleph_1 + 1) \setminus C)) = (\aleph_1 + 1) \setminus C$. Thus, $\mathcal{M}$ is an algebra of subsets of $\aleph_1 + 1$. Now, let $\{M_n\}_{n \in \mathbb{N}}$ be a sequence of sets in $\mathcal{M}$. Then, for each $n \in \mathbb{N}$, there exists a $C_n \in \mathcal{C}$ such that C_n is either a subset of M_n or of $(\aleph_1 + 1) \setminus M_n$. If for any one $n \in \mathbb{N}$, C_n is a subset of M_n, then C_n is a subset of $\bigcup_{j \in \mathbb{N}} M_j$. Alternatively, if for no $n \in \mathbb{N}$, C_n is a subset of M_n, then $\bigcup_{j \in \mathbb{N}} M_j$ is a subset of $(\aleph_1 + 1) \setminus \bigcap_{j \in \mathbb{N}} C_j$, and it follows from the above reasoning that $\bigcap_{j \in \mathbb{N}} C_j \in \mathcal{C}$, and hence $\bigcup_{j \in \mathbb{N}} M_j \in \mathcal{M}$. Thus, $\mathcal{M}$ is a σ-algebra. Note that the complement of any neighborhood of $\aleph_1$ is countable and thus $\mathcal{M}$ contains all closed subsets of $\aleph_1 + 1$. That is, if a closed set does not contain $\{\aleph_1\}$ or if it does contain $\{\aleph_1\}$ as an isolated point, then it must be countable and a set of the form $[\beta, \aleph_1)$ for some ordinal $\beta < \aleph_1$, is an element of $\mathcal{C}$ that is a subset of the complement of the closed set of interest. Further, if a closed set does contain $\{\aleph_1\}$ and if $\{\aleph_1\}$ is not an isolated point, then a set of the form $[\beta, \aleph_1)$ for some ordinal $\beta < \aleph_1$ is an element of $\mathcal{C}$ that is a subset of it. Therefore, $\mathcal{M}$ is a σ-superalgebra of $\mathcal{B}(\aleph_1 + 1)$.

We will now show that $\mathcal{M} = \mathcal{B}(\aleph_1 + 1)$, following a method given by Rao and Rao [1971]. Note that we only need show that $\mathcal{M} \subset \mathcal{B}(\aleph_1 + 1)$. Let A be any closed uncountable subset of $\aleph_1$. Note that it suffices to show that every subset of $(\aleph_1 + 1) \setminus A$ is in $\mathcal{B}(\aleph_1 + 1)$. Let B be such a subset, and assume without loss of generality that $\aleph_1 \notin B$ and $\emptyset \in A$. Now, for each $\alpha \in A$, let α' denote the first succeeding ordinal in A. Observe that α' exists for any point in A since A is uncountable. Let C be that subset of A consisting of all ordinals $\alpha \in A$ such that α' is not the first succeeding ordinal in $\aleph_1 + 1$. If C is empty, we are through. Thus, assume that $C \neq \emptyset$. Let $f: C \to (\aleph_1 + 1) \setminus A$ such that $\alpha < f(\alpha) < \alpha'$ for every $\alpha \in C$. Thus, for a given f, for each $\alpha \in C$, we can write $f(\alpha)$ as the element in a singleton set given by $\bigcap_{i \in \mathbb{N}} A_{i\alpha}$, where $A_{i\alpha} \subset (\alpha, \alpha')$ for each $i \in \mathbb{N}$. Now,

$$f(C) = \bigcup_{\alpha \in C} \bigcap_{i \in \mathbb{N}} A_{i\alpha} = \bigcap_{i \in \mathbb{N}} \bigcup_{\alpha \in C} A_{i\alpha}.$$

Thus, $f(C)$ is a G_δ, and is hence a Borel set.

Now, since A is closed and uncountable, for any $\beta \in ((\aleph_1+1)\setminus A)\setminus\{\aleph_1\}$, there exists $\alpha \in A$ such that $\alpha < \beta < \alpha'$. Now define a function g: $A \to \mathbb{P}(B)$ via $g(\alpha) = \{\beta \in B\colon \alpha < \beta < \alpha'\}$. Note that $g(A) = B$. Note also that for each $\alpha \in A$, $g(\alpha)$ is countable. Thus, for each $\alpha \in A$, fix an enumeration of $g(\alpha)$ whenever $g(\alpha) \neq \emptyset$. The sets $A_n = \bigcup_{\alpha \in A} \{n$th element in the enumeration of $g(\alpha)$ whenever there is an nth element in the enumeration of $g(\alpha)\}$, for $n \in \mathbb{N}$, are Borel sets by the reasoning in the preceding paragraph. Thus, $g(A) = B = \bigcup_{n\in\mathbb{N}} A_n$ is a Borel set. Hence, we conclude that $\mathcal{M} = \mathcal{B}(\aleph_1 + 1)$. Thus, a Borel subset of $\aleph_1 + 1$ is any subset of $\aleph_1 + 1$ such that either it includes an element of $\mathcal{C}$ as a subset or its complement in $\aleph_1 + 1$ includes an element of $\mathcal{C}$ as a subset.

Define μ as a $\{0, 1\}$-valued function on $\mathcal{B}(\aleph_1+1)$ via $\mu(M) = 1$ if there is a set in $\mathcal{C}$ that is a subset of M, and $\mu(M) = 0$ otherwise. Note that $\mu(\emptyset) = 0$ and that $\mu(\aleph_1 + 1) = 1$. Let M_1 and M_2 be disjoint Borel sets. Then they cannot each include elements from $\mathcal{C}$, because if they did they could not be disjoint due to the earlier reasoning. Thus $M_1 \cup M_2$ includes an element of $\mathcal{C}$ if and only if precisely one of M_1 and M_2 includes an element from $\mathcal{C}$. Thus, $\mu(M_1 \cup M_2) = \mu(M_1) + \mu(M_2)$. Therefore, we see that μ is finitely additive. Now, let $\{B_n\}_{n\in\mathbb{N}}$ be any sequence of Borel sets such that for each $n \in \mathbb{N}$, $B_n \supset B_{n+1}$ and such that $\bigcap_{n\in\mathbb{N}} B_n = \emptyset$. Then, there must exist an $m \in \mathbb{N}$ such that for $n > m$, B_n does not include an element of $\mathcal{C}$. Hence, as $n \to \infty$, $\mu(B_n) \downarrow 0$, and thus we see that μ is countably additive. This resulting measure is known as the Dieudonné measure. Observe that any measurable neighborhood of $\{\aleph_1\}$ has measure 1, and any measurable set not including a neighborhood of $\{\aleph_1\}$ has measure 0. Thus, the support of μ is the singleton set $\{\aleph_1\}$.

Finally, consider the measure space given by $\aleph_1$, the σ-algebra $\mathcal{B}(\aleph_1)$, and the measure ν obtained by restricting μ to $\mathcal{B}(\aleph_1)$. Since every closed set of the form $[\alpha, \aleph_1)$ has full measure, it immediately follows that the support of this measure ν is empty.

Now, for the measure space $(\aleph_1, \mathcal{B}(\aleph_1), \nu)$, if a measurable set B has zero measure, then any closed subset of B must be countable, and hence also has zero measure. Further, the complement of B must include a closed uncountable set F, and hence, the complement of F is an open set having zero measure and including B. Further, if the measurable set B has unit measure, then by definition of the measure, the set B must include a closed uncountable set which also has unit measure. Further, any open superset of B also includes a closed uncountable set, and thus such an open superset has unit measure. Hence this measure is regular. □

Example 4.34. *A measure can be diffuse and atomic.*

Proof: Consider the measure space given by the reals, the σ-algebra consisting of the sets that are either countable or have countable complements, and the measure μ that maps countable sets to zero and sets with countable complements to one. Since singleton sets are measurable and have measure zero, μ is diffuse. However, given any measurable set B with positive mea-

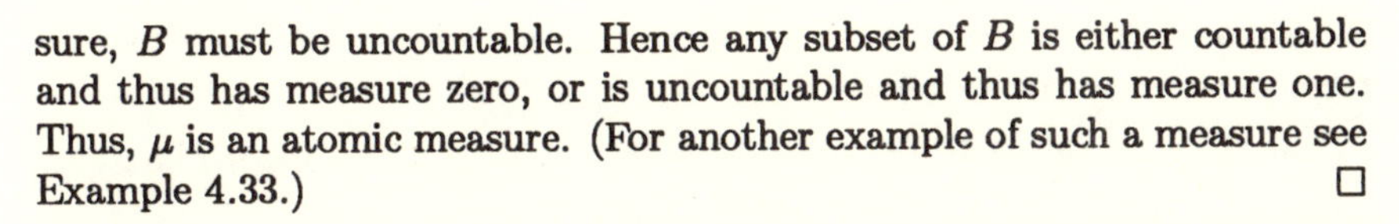

sure, B must be uncountable. Hence any subset of B is either countable and thus has measure zero, or is uncountable and thus has measure one. Thus, μ is an atomic measure. (For another example of such a measure see Example 4.33.) □

Example 4.35. *For a measure space $(\Omega, \mathcal{F}, \mu)$ such that there exists an atom A, a measurable subset B of A may be such that neither B nor $A \setminus B$ is a null set.*

Proof: Consider the set $\{1, 2\}$, the power set thereof, and the measure μ equal to infinity on any nonempty set. Then $\mu(\{1, 2\}) = \infty$, $\mu(\{1\}) = \infty$, and $\mu(\{2\}) = \infty$. □

Example 4.36. *The sum of an atomic measure and a nonatomic measure, each defined on a given measurable space, can be atomic.*

Proof: Consider the measurable space $(\mathbb{R}, \mathcal{B}(\mathbb{R}))$, and let μ be Lebesgue measure and let ν be counting measure. Note that μ is nonatomic and that ν is atomic. However, the measure m defined by $m(B) = \mu(B) + \nu(B)$ is atomic since each singleton is an atom. □

Example 4.37. *A normalized Borel measure on a completely normal compact Hausdorff space can have a nonempty support that has zero measure.*

Proof: Consider the measure space $(\aleph_1 + 1, \mathcal{B}(\aleph_1 + 1), \mu)$, where μ is Dieudonné measure, that was constructed in Example 4.33. As noted in that example, the support of μ is $\{\aleph_1\}$. However, $\mu(\{\aleph_1\}) = 0$. □

Example 4.38. *A normalized Borel measure ν on a compact completely normal Hausdorff space can have a nonempty support S such that the support of ν restricted to S is a nonempty proper subset of S.*

Proof: This example is taken from Okada [1979]. Consider the measure space $(\aleph_1 + 1, \mathcal{B}(\aleph_1 + 1), \mu)$, where μ is Dieudonné measure, that was constructed in Example 4.33. Let $\alpha \in \aleph_1 + 1$ be such that $\alpha < \aleph_1 + 1$. Consider the measure ν on $\mathcal{B}(\aleph_1 + 1)$ given by $\nu(M) = \mathrm{I}_{\{\alpha\}}(M)/2 + \mu(M)/2$. Then the support of ν is $\{\alpha, \{\aleph_1\}\}$. However, ν restricted to $\{\alpha, \{\aleph_1\}\}$ has support given by $\{\alpha\}$. □

Example 4.39. *A Borel measure on a compact completely normal Hausdorff space can fail to be outer regular or strongly inner regular.*

Proof: Consider the measure space $(\aleph_1 + 1, \mathcal{B}(\aleph_1 + 1), \mu)$, where μ is Dieudonné measure, that was constructed in Example 4.33 and let $\mathcal{U}$ denote the topology on $\aleph_1 + 1$. This is an example of a Borel measure on a compact completely normal Hausdorff space. Notice that any open set containing the point $\{\aleph_1\}$ includes a closed uncountable set. Thus, $\mu(\{\aleph_1\}) = 0$, and yet for any open set U containing $\{\aleph_1\}$, $\mu(U) = 1$. So μ is not outer regular. Also, for the Borel set $[\alpha_0, \aleph_1)$, no closed subset is uncountable, and thus $\sup\{\mu(F) \colon F \subset [\alpha_0, \aleph_1) \text{ and } (\aleph_1 + 1) \setminus F \in \mathcal{U}\} = 0$, even though $\mu([\alpha_0, \aleph_1)) = 1$. Hence, μ is not strongly inner regular. □

Example 4.40. *A normalized Borel measure on a Hausdorff space can be τ-smooth but not outer regular.*

Proof: Let $\Omega = [0, 1]$ and let H be a subset of $[0, 1]$ whose inner Lebesgue measure is zero and whose outer Lebesgue measure is one. Let Ω be equipped with the topology $\mathcal{U}$ generated by H and the usual topology $\mathcal{U}_0$ on $[0, 1]$, that is, $\mathcal{U} = \{U_1 \cup (H \cap U_2)$: U_1 and U_2 are in $\mathcal{U}_0\}$. Then Ω is a second countable Hausdorff space. Further, $\mathcal{B}(\Omega) = \{(B_1 \cap H) \cup ((\Omega \setminus H) \cap B_2)$: B_1 and B_2 are Borel sets with respect to $\mathcal{U}_0\}$. For any two $\mathcal{U}_0$ Borel sets B_1 and B_2, let

$$\mu((B_1 \cap H) \cup ((\Omega \setminus H) \cap B_2)) = \frac{1}{2}(\lambda(B_1) + \lambda(B_2)),$$

where λ denotes Lebesgue measure restricted to the $\mathcal{U}_0$ Borel sets. That μ is well defined follows from the properties of the set H. Thus, $(\Omega, \mathcal{B}(\Omega), \mu)$ is a measure space. Now, $\mu(\Omega \setminus H) = 1/2$, but $\inf\{\mu(U)$: $\Omega \setminus H \subset U$ and $U \in \mathcal{U}\} = \inf\{\mu(U)$: $\Omega \setminus H \subset U$ and $U \in \mathcal{U}_0\} = \inf\{\lambda(U)$: $\Omega \setminus H \subset U$ and $U \in \mathcal{U}_0\}$, which equals the outer Lebesgue measure of $\Omega \setminus H$, which is given by one minus the inner Lebesgue measure of H, which equals one. Thus, μ is not outer regular.

Now, since the measure is finite and the topological space is second countable, μ is τ-smooth since for any nondecreasing net $\{U_\lambda\}_{\lambda \in \Lambda}$ of open sets, $\bigcup_{\lambda \in \Lambda} U_\lambda$ can be written as a countable union of open sets. □

Example 4.41. *A normalized regular Borel measure on a locally compact completely normal Hausdorff space can fail to be τ-smooth.*

Proof: In this example, we draw upon the development in Example 4.33. Consider the measure space $(\aleph_1, \mathcal{B}(\aleph_1), \nu)$, where ν is Dieudonné measure restricted to $\mathcal{B}(\aleph_1)$. Recall that $\aleph_1$ is a locally compact completely normal Hausdorff space. Further, as shown in Example 4.33, ν is regular.

For each ordinal $\alpha < \aleph_1$, let S_α denote the set of all ordinals less than α. Note that for each such α, S_α is open and countable. Thus, $\mu(S_\alpha) = 0$ for each such ordinal $\alpha < \aleph_1$. However, note that $\{S_\alpha\}_{\alpha < \aleph_1}$ is a nondecreasing net of open sets, and $\bigcup_{\alpha < \aleph_1} S_\alpha = \aleph_1$. However, $\mu(\aleph_1) = 1 \neq 0 = \sup_{\alpha < \aleph_1} \mu(S_\alpha)$. Thus, μ is not τ-smooth. □

Example 4.42. *A Borel measure on a Hausdorff space can be inner regular, but fail to be compact inner regular, or outer regular.*

Proof: Consider the measure space given in Example 4.8, and recall that the measure is a Borel measure. First, notice that any open set is of the form $U \setminus K$, where U is open in the usual topology and K is countable. If U is nonempty, then it has positive Lebesgue measure. Now, since K is countable, we can find a subset C closed in the usual topology that includes K as a subset and that has positive Lebesgue measure less than the Lebesgue measure of U. Thus, the set $U \setminus C$ is open in the usual topology and has positive Lebesgue measure. Hence, we can find a small closed interval of positive length that is a subset of $U \setminus C$, and hence of $U \setminus K$. Thus the measure is inner regular.

Next, notice that the measure of any uncountable open set is ∞, but the measure of any compact set is zero. Thus, the measure is not compact inner regular. Finally, notice that the measure of any singleton set is zero, but the measure of any neighborhood of a singleton set is ∞. Thus, it is not outer regular. $\square$

Example 4.43. *A regular Borel measure on a locally compact metric space may be neither compact inner regular nor σ-finite.*

Proof: Consider the real numbers with the discrete metric, and note that for any point, the singleton set containing that point is a compact neighborhood of the point. Define the measure μ on the power set of the reals via $\mu(B)$ is zero if B is countable and $\mu(B)$ is infinity if B is not countable. Since all sets of real numbers are both closed and open, this measure is obviously regular. However, all compact sets are finite, so μ is not inner compact regular. Also, since any countable cover of the reals must contain an uncountable set, the measure cannot be σ-finite. $\square$

Example 4.44. *A Borel measure on a completely normal Hausdorff space might not be outer regular, compact inner regular, or inner regular.*

Proof: Consider the interval $[0, 1]$ and define a set to be open if and only if its complement is either countable or contains the point 0. First, we show that this topological space is a completely normal Hausdorff space. Let A and B be separated subsets. If neither contains 0, then they are each open and thus they are separated by disjoint neighborhoods. If $0 \in A$, then A is closed and B is open. But $([0, 1] \setminus B)$ is also open since $0 \in ([0, 1] \setminus B)$. In this case, $A \subset ([0, 1] \setminus B)$, which is open, and since B is a neighborhood of itself, A and B are separated by disjoint neighborhoods. Thus, this topological space is completely normal. Since singleton sets are, by definition, closed, this topological space is also Hausdorff. Let Ω denote this topological space.

Note that every singleton set except for $\{0\}$ is both open and closed. Further, note that $\{0\}$ is closed but not open. Indeed, any neighborhood of $\{0\}$ contains uncountably many points. Further, observe that the only compact sets are finite sets.

Now, define the measure μ on $\mathcal{B}(\Omega)$ via $\mu(B) = 0$ if B is countable and $\mu(B) = \infty$ if B is not countable. Note that for compact sets K, $\mu(K) = 0$. Thus, μ is a Borel measure. Now, note that $\mu(\{0\}) = 0$, but for any open neighborhood U of 0, $\mu(U) = \infty$. Thus, μ is not outer regular. Next, note that $(1/2, 1]$ is a Borel set such that any closed subset of it is countable and any compact subset of it is finite, even though $\mu((1/2, 1]) = \infty$. Thus, μ is not compact inner regular or inner regular. $\square$

Example 4.45. *A Borel measure on a locally compact metric space may be compact inner regular but not be outer regular and not be σ-finite.*

Proof: This example is taken from Gruenhage and Pfeffer [1978]. Let Y be an uncountable set with the discrete metric, let $\mathbb{R}$ have the usual metric, and let $\Omega = \mathbb{R} \times Y$. Put the product topology on Ω. Since $\mathbb{R}$ and Y are each locally compact metric spaces, Ω is also. For $B \in \mathcal{B}(\Omega)$, let $B^y = \{x \in \mathbb{R}: (x, y) \in B\}$. Note that for each $y \in Y$, $B^y \in \mathcal{B}(\mathbb{R})$. Let λ denote Lebesgue measure on $\mathcal{B}(\mathbb{R})$. For each $B \in \mathcal{B}(\Omega)$, let $\mu(B) = \sum_{y \in Y} \lambda(B^y)$. Now, for any open subset G of Ω, by appropriately choosing a finite subset S_G of Y and compact subsets C^y of the corresponding B^y's, we can construct a compact subset of Ω, $C = \bigcup_{y \in S_G} C^y$ such that $\mu(C)$ approximates $\mu(G)$. In this way, we observe that μ is inner compact regular. Now, note that $\mu(\{0\} \times Y) = 0$, but that any open superset U of $(\{0\} \times Y)$ is such that $\mu(U) = \infty$. Thus, μ is not outer regular.

Consider any covering of Ω by countably many measurable sets. Since Y is uncountable, there must exist at least one of these sets B and an uncountable set S_B of $y \in Y$, such that $\lambda(B_y) > 0$ for all $y \in S_B$, and thus $\mu(B) = \infty$. Thus, μ is not σ-finite. □

Example 4.46. *If $(\Omega, \mathcal{F})$ is a measurable space and μ and ν are finite measures on $(\Omega, \mathcal{F})$, it need not be true that the smallest measure not less than μ or ν is* $\max\{\mu(A), \nu(A)\}$.

Proof: This example is from Kelley and Srinivasan [1988]. Let $\Omega = \{0, 1\}$ and $\mathcal{F} = \mathbb{P}(\Omega)$. Let μ be Dirac measure at the point 0 and let ν be Dirac measure at the point 1. Then $\max\{\mu(\{0\}), \nu(\{0\})\} = \max\{\mu(\{0\}), \nu(\{1\})\} = \max\{\mu(\{1\}), \nu(\{0\})\} = \max\{\mu(\{1\}), \nu(\{1\})\} = 1$. However, the smallest measure not less than μ or ν is $\mu + \nu$, which is 2 at $\{0, 1\}$. □

Example 4.47. *There exists a compact subset K of the reals that is either null or else non-σ-finite for any translation-invariant measure defined on the real Borel sets.*

Proof: This example is taken from Davies [1971b]. For this example, we will write $A \leq B$ to mean that A and B are compact subsets of $\mathbb{R}$ such that A is the union of a finite number of disjoint subsets each of which is a translate of a certain compact set A_0 and B is a superset of $2^{\aleph_0}$ disjoint translates of A_0. Observe that in this case, if μ is a translation-invariant measure defined on $\mathcal{B}(\mathbb{R})$ such that $\mu(A) > 0$, then B is non-σ-finite. Suppose we can construct a sequence $\{K_n\}_{n \in \mathbb{N}}$ of compact subsets of $\mathbb{R}$ such that $\sum_{n=1}^{\infty} \operatorname{diam}(K_n) < \infty$ and $K_n \leq K_{n+1}$ for each $n \in \mathbb{N}$, where $\operatorname{diam}(K_n)$ denotes the diameter of K_n. If we place translates of these respective sets in successive intervals and let K be their union with the limit point adjoined, then this set K is our required example.

Now we construct the sets K_n. For each positive integer n, let N_n denote the set of all positive integers divisible by 2^n, and let K_n denote the set of all real numbers x of the form

$$x = \frac{c_n}{8^n} + \frac{c_{n+1}}{8^{n+1}} + \frac{c_{n+2}}{8^{n+2}} + \cdots,$$

with each c_k a nonnegative integer less than 8 and with c_k equal to 1 or 5 for $k \in N_n$. Note that this set K_n is a nowhere dense perfect set, and $\operatorname{diam}(K_n) < 8^{-n+1}$. Moreover, K_n is the union of two or eight, depending, respectively, on whether $n \in N_n$ or not, translates of the set $K_{n,0}$ of all real numbers x of the form

$$x = \frac{c_{n+1}}{8^{n+1}} + \frac{c_{n+2}}{8^{n+2}} + \cdots,$$

with the c_k's satisfying the same conditions as before. Finally, K_{n+1} includes $2^{\aleph_0}$ disjoint translates of $K_{n,0}$, namely $K_{n,0} \oplus y$, where y is a real number of the form

$$y = \frac{d_{n+1}}{8^{n+1}} + \frac{d_{n+2}}{8^{n+2}} + \cdots,$$

with d_k equal 2 or 0 if $k \in N_n \setminus N_{n+1}$, and otherwise $d_k = 0$. To see that these sets $K_{n,0} \oplus y$ are disjoint, note that two different y's must have in some position 0 and 2, respectively. All elements of the corresponding set $K_{n,0} \oplus y$ have 1 or 5 in that position for the first y, and 3 or 7 for the second y. Thus, $K_n \leq K_{n+1}$ for each $n \in \mathbb{N}$, and the proof is complete. □

Example 4.48. *There exists a compact Hausdorff space H and a probability measure P on $\mathcal{B}(H)$ and a $\mathcal{B}(H)$-measurable real-valued function f such that for any continuous real-valued function g, $P(\{x \in H\colon f(x) \neq g(x)\}) \geq 1/2$.*

Proof: Let H denote the compact Hausdorff space $\aleph_1 + 1$. (See Example 4.33.) Let D be the measure on $\mathcal{B}(H)$ given by the Dirac measure at the point $\{\aleph_1\}$, and let μ be the Dieudonné measure on $\mathcal{B}(H)$. Let the probability measure P on $\mathcal{B}(H)$ be defined via $P(B) = [\mu(B) + D(B)]/2$. Let the real-valued $\mathcal{B}(H)$-measurable function f be defined on H via $f(x) = I_{\{\aleph_1\}}(x)$. Recall from Example 5.32 that any real-valued continuous function g defined on H is such that there exists an ordinal α less than $\aleph_1$ such that g is constant on all ordinals no less than α and no greater than $\aleph_1$. From this, the result follows immediately. □

Example 4.49. *There exists a compact metric space Ω and a probability measure P on $\mathcal{B}(\Omega)$ such that for any countable set C of disjoint closed balls of radius not exceeding unity, $\sum_{B \in C} P(B) < 1/2$.*

Proof: This example is from Davies [1971a].. Let C be a circle with unit radius, and for θ and ϕ in C, let $|\theta - \phi|$ denote the length of the shorter arc joining them. Let Ω denote the set of all sequences $\theta = \{\theta_n\}_{n \in \mathbb{N}}$ of points in C. Let $\{\epsilon_n\}_{n \in \mathbb{N}}$ be a decreasing sequence of positive real numbers converging to zero so slowly that $\sum_{m=1}^{\infty} 3/\text{card}(R_m) < 1/2$, where

$$R_m = \left\{ n \in \mathbb{N}\colon \frac{1}{\pi}\left(\frac{3}{4}\right)^{m-2} < \epsilon_n \leq \frac{2}{\pi}\left(\frac{3}{4}\right)^{m} \right\}.$$

For θ and ϕ in Ω, define $\rho(\theta, \phi) = \sup_{n \in \mathbb{N}} \epsilon_n |\theta_n - \phi_n|$ and note that (Ω, ρ) is a compact metric space. Let μ denote the product probability measure on $\mathcal{B}(\Omega)$ such that for Borel sets $A_1, A_2, \ldots$ in C we have

$$\mu\left(\bigtimes_{n=1}^{\infty} A_n \right) = \prod_{n=1}^{\infty} \frac{\Lambda(A_n)}{2\pi},$$

where Λ denotes normalized arc-length measure on the Borel subsets of C.

Given any countable collection $\mathcal{S}$ of (nondegenerate, as we may suppose) disjoint closed balls in Ω, each of radius not exceeding unity, let $\mathcal{S}_m$ denote the subfamily of $\mathcal{S}$ consisting of balls S whose radius $r(S)$ satisfies

$$\left(\frac{3}{4}\right)^m < r(S) < \left(\frac{3}{4}\right)^{m-1}$$

and observe that $\mathcal{S} = \bigcup_{m=1}^{\infty} \mathcal{S}_m$. We shall show how to associate with each ball $S \in \mathcal{S}_m$ a "fringe" F_S consisting of card(R_m) mutually disjoint "near-copies" $F_S(n)$, $n \in R_m$ of S, each of measure $\mu(F_S(n)) > \mu(S)/3$. Thus, $\mu(F_S) > \text{card}(R_m)\mu(S)/3$. Moreover, this will be done in such a way that the fringes corresponding to different balls $S \in \mathcal{S}_m$ are disjoint. Hence, we shall have

$$\sum_{S \in \mathcal{S}_m} \mu(S) < \frac{3}{\text{card}(R_m)} \sum_{S \in \mathcal{S}_m} \mu(F_S) \leq \frac{3}{\text{card}(R_m)},$$

and consequently,

$$\sum_{S \in \mathcal{S}} \mu(S) < \sum_{m=1}^{\infty} \frac{3}{\text{card}(R_m)} \leq \frac{1}{2},$$

as required.

Now, given any ball $S \in \mathcal{S}_m$, with center ϕ, and any integer $n \in R_m$, denote by ϕ_n^* the point on C diametrically opposite to ϕ_n. Let $F_S(n)$ denote the Cartesian product $I_1 \times I_2 \times \cdots$, where I_k is the arc $\pi_k(S)$ if $k \neq n$ (π_k denotes projection on the kth coordinate circle), and I_n is the closed arc on C of length $\pi/2$ with center at ϕ_n^*. Since $\pi_n(S)$ is an arc of length $2r(S)/\varepsilon_n$, where by the above $2r(S)/\varepsilon_n < 3\pi/2$, and since $S = \bigtimes_{k=1}^{\infty} \pi_k(S)$, we have

$$\frac{\mu(F_S(n))}{\mu(S)} = \frac{\Lambda(I_n)}{\Lambda(\pi_n(S))} = \frac{\pi/2}{2r(S)/\varepsilon_n} > \frac{1}{3},$$

which is as given above.

Now we will show that the sets $F_S(n)$ are disjoint. First we consider two distinct sets $F_S(n)$ and $F_S(n')$ with n and n' in R_m and $n \neq n'$ associated with the same ball $S \in \mathcal{S}_m$. If $F_S(n)$ and $F_S(n')$ had a nonempty intersection, then so also would $\pi_n(F_S(n))$ and $\pi_n(F_S(n'))$. But these are arcs of lengths $\pi/2$ and $2r(S)/\varepsilon_{n'}$ with centers at the diametrically opposite points ϕ_n^* and ϕ_n and so they would cover C, which is impossible since, as above, $2r(S)/\varepsilon_n < 3\pi/2$.

Finally, we must show that the sets $F_S(n)$ and $F_{S'}(n')$ with n and n' in R_m associated with distinct balls S and S' in $\mathcal{S}_m$ are disjoint. Now since $S \cap S' = \emptyset$, and $S = \bigtimes_{k=1}^{\infty} \pi_k(S)$ and $S' = \bigtimes_{k=1}^{\infty} \pi_k(S')$, we have that $\pi_\nu(S) \cap \pi_\nu(S') = \emptyset$ for some index ν. Let N denote the least such index. We cannot have $N \in R_m$, because otherwise since

$$R_m = \left\{ n \in \mathbb{N}: \frac{1}{\pi}\left(\frac{3}{4}\right)^{m-2} < \varepsilon_n \leq \frac{2}{\pi}\left(\frac{3}{4}\right)^m \right\}$$

and

$$\left(\frac{3}{4}\right)^m < r(S) < \left(\frac{3}{4}\right)^{m-1}$$

we should have the contradiction $2\pi = \Lambda(C) \geq \Lambda(\pi_N(S)) + \Lambda(\pi_N(S')) = 2[r(S) + r(S')]/\varepsilon_N > 2\pi$. Hence $\pi_N(F_S(n)) \cap \pi_N(F_{S'}(n')) = \pi_N(S) \cap \pi_N(S') = \emptyset$, and it follows that $F_S(n) \cap F_{S'}(n') = \emptyset$, which completes the proof. □

Example 4.50. *If B is a Bernstein subset of $\mathbb{R}$, then B is not measurable for any nonatomic, $\mathcal{B}(\mathbb{R})$ regular outer measure μ that is not identically zero.*

Proof: This example is taken from Ostaszewski [1975]. Let B be a Bernstein set. Since μ is nonatomic and not identically zero, there exists a real Borel set A with $0 < \mu(A) < \infty$. Let A be such a set. Assume that B is measurable. Then $\mu(A) = \mu(A \cap B) + \mu(A \setminus B)$. Now, recalling that μ is $\mathcal{B}(\mathbb{R})$ regular, choose a Borel superset S of $A \cap B$ such that $\mu(S) = \mu(A \cap B)$. Now B does not intersect $A \setminus S$, and hence $A \setminus S$ is countable, otherwise it would include an uncountable closed set that would intersect B. Thus $\mu(A \setminus S) = 0$, since μ is nonatomic. Consequently, $\mu(A) = \mu(A \cap S) \leq \mu(S) = \mu(A \cap B)$. Similarly, $\mu(A) \leq \mu(A \setminus B)$, and thus we have the contradiction $\mu(A) \geq 2\mu(A)$. □

Example 4.51. *There exists a compact Hausdorff space that is not the support of any compact inner regular Borel measure.*

Proof: This example was inspired by Dudley [1989]. Consider the set $H = [0, 1]$ with the topology defined via all positive singleton sets are open and any set containing zero and all but a finite number of other points is open. Any two distinct points can be separated by open sets. Further, observe that H is compact. Now, if μ is a compact inner regular measure and its support is H, then note that $\mu(H) = \infty$ since any open subset of H must have positive measure. Thus, there does not exists any compact inner regular Borel measure whose support is H. □

References

Bahadur, R. R., and E. L. Lehmann, "Two comments on 'Sufficiency and statistical decision functions,'" *The Annals of Mathematical Statistics,* Vol. 26, 1955, pp. 286–293.

Billingsley, P., *Probability and Measure,* Second edition, New York: Wiley, 1986.

Davies, R. O., "Measures not approximable or not specifiable by means of balls," *Mathematika,* Vol. 18, pp. 157–160, 1971a.

Davies, R. O., "Sets which are null or non-sigma-finite for every translation-invariant measure," *Mathematika,* Vol. 18, pp. 161–162, 1971b.

Dotson, W. G., "Saturation of measures," *American Mathematical Monthly,* Vol. 77, pp. 179–180, February 1970.

Dudley, R. M., *Real Analysis and Probability,* Pacific Grove, California: Wadsworth & Brooks/Cole, 1989.

Gruenhage, G., and W. F. Pfeffer, "When inner regularity of Borel measures implies regularity," *Journal of the London Mathematical Society (2),* Vol. 17, pp. 165–171, 1978.

Kelley, J. L., and T. P. Srinivasan, *Measure and Integral,* Vol. 1, Springer-Verlag, New York, 1988.

Okada, S., "Supports of Borel measures," *Journal of the Australian Mathematical Society (Series A),* Vol. 27, pp. 221–231, 1979.

Ostaszewski, A. J., "Absolutely non-measurable and singular co-analytic sets," *Mathematika,* Vol. 22, pp. 161–163, 1975.

Oxtoby, J. C., *Measure and Category,* Second Edition, Springer-Verlag, New York, 1980.

Rao, M. Bhaskara, and K. P. S. Bhaskara Rao, "Borel σ-algebra on $[0, \Omega]$," *Manuscripta Mathematica,* Vol. 5, pp. 195–198, 1971.

5. Integration

Although students are often first introduced to Riemann integrals and improper Riemann integrals, the integral that is the cornerstone of integration theory is the Lebesgue integral. If $(\Omega, \mathcal{F}, \mu)$ is a measure space and if f is an extended real-valued measurable function, then f is said to be integrable if $\int_\Omega |f|\, d\mu < \infty$. Note in particular that a bounded real-valued function f such that $\int_\Omega |f|\, d\mu < \infty$ need not be integrable; indeed, just consider a situation in which there exists a nonmeasurable set A included within a measurable set B with finite measure and let f be 1 on A and -1 on $B \setminus A$.

A family $\mathcal{F}$ of extended real-valued integrable functions is said to be uniformly integrable if and only if for each positive real ε there corresponds a positive real δ such that the modulus of the integral of f over M is less than ε whenever $f \in \mathcal{F}$ and the measure of M is less than δ.

If $(\Omega, \mathcal{F}, \mu)$ is a measure space, and $p \in [0, \infty)$, $\mathbf{L}_p(\Omega, \mathcal{F}, \mu)$ will denote the set of all μ-equivalence classes of measurable functions f mapping Ω into $\mathbb{R}$ such that $|f|^p$ is integrable.

If $(\Omega, \mathcal{F})$ is a measurable space and if μ and ν are measures on $(\Omega, \mathcal{F})$ such that $\nu \ll \mu$, Radon–Nikodym theory is concerned with conditions under which there exists a function f in $\mathbf{L}_1(\Omega, \mathcal{F}, \mu)$ such that $\nu(A) = \int_A f\, d\mu$. Such results comprise a fundamental part of analysis and also play a prominent role in probability theory. Indeed, for a given measure, results such as this permits us to represent other measures defined on $\mathcal{F}$ in terms of extended real-valued functions defined on Ω. Also, in probability theory, conditional expectations and condition probabilities are defined in terms of Radon–Nikodym theory.

Example 5.1. *A nondecreasing sequence of functions mapping* $[0, 1]$ *into* $[0, 1]$ *may be such that each term in the sequence is Riemann integrable and such that the limit of the resulting sequence of Riemann integrals exists, but the limit of the sequence of functions is not Riemann integrable.*

Proof: Let $\{r_n\colon n \in \mathbb{N}\}$ be an enumeration of the rationals in $[0, 1]$. For each $n \in \mathbb{N}$, let f_n map $[0, 1]$ into $[0, 1]$ via $f_n(x) = \sum_{k=1}^{n} \mathrm{I}_{\{r_k\}}(x)$. Then, for each n, f_n is Riemann integrable over $[0, 1]$, and the resulting Riemann integral is zero. However, as $n \to \infty$, $f_n \to \mathrm{I}_{(\mathbb{Q}\cap[0,\,1])}$, which is not Riemann integrable. □

Example 5.2. *A function* $f\colon \mathbb{R} \to \mathbb{R}$ *that has a finite improper Riemann integral need not be bounded.*

Proof: Let $T\colon \mathbb{R} \to \mathbb{R}$ via

$$T(x) = \begin{cases} 1 - |x|, & \text{if } |x| \leq 1 \\ 0, & \text{otherwise.} \end{cases}$$

For positive integers n, let f_n map the reals into the reals via $f_n(x) = nT(n^3x - n)$. Finally, let f map the reals into the reals via $f(x) = \sum_{n=1}^{\infty} f_n(x)$. This function f is not bounded and yet has a finite improper Riemann integral. □

Example 5.3. *If f and g map [0, 1] into [0, 1] and f is continuous and g is Riemann integrable, $g \circ f$ need not be Riemann integrable.*

Proof: Let C be a Cantor subset of [0, 1] having positive Lebesgue measure. Let f: $[0, 1] \to \mathbb{R}$ via $f(x) = d(x, C)$, where d denotes the Euclidean metric on $\mathbb{R}$, and let g: $[0, 1] \to \mathbb{R}$ via $g(x) = 1 - I_{\{0\}}(x)$. Then f is continuous and g is Riemann integrable, and yet $g \circ f = I_C$, which is not Riemann integrable since the set of discontinuities of I_C is C and C has positive Lebesgue measure. □

Example 5.4. *There exists a bounded infinitely differentiable function that is not Lebesgue integrable yet whose improper Riemann integral exists and is finite.*

Proof: Consider the function f: $[0, \infty) \to \mathbb{R}$ defined via $f(x) = \sin(x)/x$ for $x > 0$ and $f(0) = 1$. First we will show that the improper Riemann integral of this function exists and is finite.

Note that $\int_0^x [|\sin(y)|/y]\, dy$ exists since the integrand is continuous in $[0, x]$ for any positive real number x. Recall that $\int_0^\infty \exp(-yz)\, dz = 1/y$. Hence, Tonelli's theorem implies that

$$\int_0^x \frac{\sin(y)}{y}\, dy = \int_0^\infty \int_0^x \exp(-yz)\sin(y)\, dy\, dz.$$

Integrating by parts twice implies that

$$\int_0^x \exp(-yz)\sin(y)\, dy = \frac{1}{1+z^2} - \exp(-xz)\frac{\cos(x) + z\sin(x)}{1+z^2}.$$

Notice that if $x \geq 1$ and $z \geq 0$, then $\exp(xz) \geq 1 + z$, and hence,

$$\begin{aligned}\left|\exp(-xz)\frac{\cos(x) + z\sin(x)}{1+z^2}\right| &\leq |\exp(-xz)|\frac{|\cos(x) + z\sin(x)|}{1+z^2}\\ &\leq \left|\frac{1}{1+z}\right|\frac{|1+z|}{1+z^2} = \frac{1}{1+z^2}.\end{aligned}$$

Thus, it follows that

$$\begin{aligned}\lim_{x\to\infty}\int_0^x \frac{\sin(y)}{y}\, dy &= \lim_{x\to\infty}\int_0^\infty \int_0^x \exp(-yz)\sin(y)\, dy\, dz\\ &= \lim_{x\to\infty}\int_0^\infty \frac{1}{1+z^2} - \exp(-xz)\frac{\cos(x) + z\sin(x)}{1+z^2}\, dz\\ &= \int_0^\infty \left(\frac{1}{1+z^2} - \lim_{x\to\infty}\exp(-xz)\frac{\cos(x) + z\sin(x)}{1+z^2}\right) dz\\ &\qquad\text{(via dominated convergence)}\\ &= \int_0^\infty \frac{1}{1+z^2}\, dz = \frac{\pi}{2}.\end{aligned}$$

Thus, the improper Riemann integral of f exists and is finite.

It will now be shown that f is not Lebesgue integrable. Note that for each positive integer n

$$\int_{(n-1)\pi}^{n\pi} \left|\frac{\sin(x)}{x}\right| dx \geq \frac{1}{n\pi}\int_{(n-1)\pi}^{n\pi} |\sin(x)|\, dx = \frac{2}{n\pi}.$$

Hence,

$$\int_{0}^{n\pi} \left|\frac{\sin(x)}{x}\right| dx \geq \frac{2}{\pi}\left(1+\frac{1}{2}+\frac{1}{3}+\cdots+\frac{1}{n}\right),$$

which is not bounded. Thus, the bounded infinitely differentiable function f is not Lebesgue integrable even though its improper Riemann integral exists and is finite. □

Example 5.5. *There exists a bounded Lebesgue integrable function f: $[0, 1] \to \mathbb{R}$ such that there does not exist a Riemann integrable function g: $[0, 1] \to \mathbb{R}$ that is equal to f almost everywhere with respect to Lebesgue measure.*

Proof: This example was inspired by Wilansky [1953]. Let A be an open subset of $[0, 1]$ that contains each of the rationals in $[0, 1]$ and that has Lebesgue measure less than one. (See Example 1.4.) Let the bounded Lebesgue integrable function f be given by the indicator function of A. Let g: $[0, 1] \to \mathbb{R}$ be any function that is equal to f almost everywhere with respect to Lebesgue measure. Then there exists a Lebesgue null subset N of $[0, 1]$ such that $g(x) = 1$ for $x \in A \setminus N$, and $g(x) = 0$ for $x \notin A \cup N$. Since A is an open set that includes the rationals, $A \setminus N$ is dense. Thus $g = 1$ on a dense set and is hence discontinuous at any point x such that $g(x) \neq 1$. In particular, g is discontinuous off the set $A \cup N$, which has positive Lebesgue measure. Hence, g is not Riemann integrable. □

Example 5.6. *Integration by parts may fail for Stieltjes integrals.*

Proof: Consider a subset $[a, b]$ of $\mathbb{R}$ and let F and G be two real-valued, nondecreasing, right continuous functions defined on $[a, b]$. The following counterexample shows that the commonly used integration by parts formula does not hold in general. That is, in general it is not true that

$$\int_{(a,\, b]} G(x)\, dF(x) = F(b)G(b) - F(a)G(a) - \int_{(a,\, b]} F(x)\, dG(x).$$

In particular, consider the case in which $a = -1$, $b = 1$, and $F(x) = G(x) = \mathrm{I}_{[0,\, 1]}(x)$. Then

$$\int_{(-1,\, 1]} G(x)\, dF(x) = \int_{(-1,\, 1]} F(x)\, dG(x) = F(1)G(1) = 1,$$

yet $F(-1)G(-1) = 0$. Hence,

$$\int_{(-1,\, 1]} G(x)\, dF(x) \neq F(1)G(1) - F(-1)G(-1) - \int_{(-1,\, 1]} F(x)\, dG(x). \quad \square$$

Example 5.7. *There exist monotone nondecreasing functions f and g mapping $[-1, 1]$ into $[-1, 1]$ such that the Riemann–Stieltjes integrals*

$$\int_{-1}^{0} f(x)\, dg(x)$$

and

$$\int_{0}^{1} f(x)\, dg(x)$$

exist but the Riemann–Stieltjes integral

$$\int_{-1}^{1} f(x)\, dg(x)$$

does not exist.

Proof: This example is from Natanson [1955]. Let f and g be defined by $f(x) = I_{(0,\,1]}$ and $g(x) = I_{[0,\,1]}$. Then note that

$$\int_{-1}^{0} f(x)\, dg(x) = 0,$$

since all of the approximating sums are zero. Similarly,

$$\int_{0}^{1} f(x)\, dg(x) = 0.$$

Now, let $\{x_1, x_2, \dots, x_n\}$ be a set of points in $[-1, 1]$ such that $x_1 = -1$, $x_n = 1$, $x_k < x_{k+1}$ for each positive integer $k \le n - 1$, and $x_k \neq 0$ for each positive integer $k \le n$. Now, notice that if i is the index such that $x_i < 0 < x_{i+1}$, then only the ith term remains in the approximating sum and it is equal to $f(y_i)[g(x_{i+1}) - g(x_i)] = f(y_i)$, where y_i is some element of $[x_i, x_{i+1}]$. Now, depending upon whether y_i is positive or negative, we see that the approximating sum is either one or zero, respectively. Hence the approximating sum has no limit. □

Example 5.8. *If f maps $[0, 1]$ into $\mathbb{R}$ and is continuous and nondecreasing, then $\int_{[0,\,1]} df$ need not equal $\int_{[0,\,1]} f'\, dx$.*

Proof: Let f denote the Cantor–Lebesgue function, and recall that $f(0) = 0$, $f(1) = 1$, and f has a zero derivative almost everywhere with respect to Lebesgue measure. Thus, $\int_{[0,\,1]} f'\, dx = 0$, but $\int_{[0,\,1]} df = 1$. □

Example 5.9. *If f and g map $[0, 1]$ into $\mathbb{R}$, f is nondecreasing and continuous, and g is a polynomial, then $\int_{[0,\,1]} fg'\, dx$ need not equal $f(1)g(1) - f(0)g(0) - \int_{[0,\,1]} gf'\, dx$.*

Proof: Let f denote the Cantor–Lebesgue function. Recall that f is nondecreasing and continuous and that $f(0) = 0$ and $f(1) = 1$. Further, f has a zero derivative almost everywhere with respect to Lebesgue measure. Let $g\colon [0, 1] \to \mathbb{R}$ via $g(x) = 1 - x$. Note that $f(1)g(1) - f(0)g(0) = 0$. Note that $\int_{[0,\,1]} gf'\, dx = 0$. However, $\int_{[0,\,1]} fg'\, dx = -\int_{[0,\,1]} f\, dx < 0$. □

Example 5.10. *If f and g map $\mathbb{R}$ into $\mathbb{R}$, f is strictly increasing, and g is continuously differentiable with compact support, then $\int_{\mathbb{R}} fg'\,dx$ need not equal $-\int_{\mathbb{R}} gf'\,dx$.*

Proof: Let f be the strictly increasing function constructed in Example 2.7, and then we see that $\int_{\mathbb{R}} gf'dx = 0$ since f has a zero derivative almost everywhere with respect to Lebesgue measure. Now, continuously differentiable functions g having compact support such that $\int_{\mathbb{R}} fg'\,dx \neq 0$ exist in profusion. For example, if g has a derivative given by

$$g'(x) = \begin{cases} x, & \text{if } x \in [0,\, 1] \\ 2-x, & \text{if } x \in [1,\, 3] \\ x-4, & \text{if } x \in [3,\, 4] \\ 0, & \text{otherwise,} \end{cases}$$

then the desired result will follow. □

Example 5.11. *If $f\colon \mathbb{R} \to \mathbb{R}$ is such that $\int_{[-1,0)} f\,d\lambda$ exists and is such that $\int_{[0,\,1)} f\,d\lambda$ exists, then the integral of f over $[-1,\, 1)$ need not exist.*

Proof: Choose f such that $\int_{[-1,0)} f\,d\lambda = -\infty$ and such that $\int_{[0,\,1)} f\,d\lambda = \infty$, and the result follows. As an example, let $f(x) = (1/x)\mathrm{I}_{[-1,\,1)\setminus\{0\}}(x) + \mathrm{I}_{\{0\}}(x)$. □

Example 5.12. *The limit of a uniformly convergent sequence of integrable functions mapping $\mathbb{R}$ into $\mathbb{R}$ need not be integrable.*

Proof: Consider the real line, the Lebesgue measurable subsets of $\mathbb{R}$, and Lebesgue measure. Let f be any real-valued, bounded, Lebesgue measurable function that is not integrable and that converges to zero at infinity and at minus infinity. For positive integers n, let $f_n(x) = \mathrm{I}_{[-n,\,n]}(x)f(x)$. Then, for each positive integer n, f_n is integrable, and as $n \to \infty$, $f_n(x)$ converges uniformly to $f(x)$, since $f(x)$ converges to zero at infinity and at minus infinity. This uniform limit, however, is not integrable. □

Example 5.13. *There exists a sequence $\{f_n\}_{n\in\mathbb{N}}$ of continuous real-valued functions defined on $[0,\, 1]$ such that $f_1(x) \geq f_2(x) \geq \cdots \geq 0$ for all $x \in [0,\, 1]$, such that the only continuous function f for which $f_n(x) \geq f(x) \geq 0$ for all $x \in [0,\, 1]$ and all $n \in \mathbb{N}$ is the identically zero function, and such that $\int_0^1 f_n(x)\,dx \not\to 0$ as $n \to \infty$.*

Proof: This example is from Langford and Kestelman [1973]. To begin, cover the points in $\mathbb{Q} \cap [0,\, 1]$ by a sequence of open intervals such that the sum of the lengths of the intervals is $1/2$. Let K denote the closed, nowhere dense subset given by the relative complement in $[0,\, 1]$ of the union of these intervals. For each nonnegative integer n, let $f_n(x) = [1-d(x,\, K)]^n$, where $d(x,\, K) = \inf_{y\in K} |x-y|$. Note that, for each n, f_n is continuous and that the sequence $\{f_n\}_{n\in\mathbb{N}}$ is nonincreasing. Further, since K has Lebesgue measure not less than $1/2$ and $f_n(x) = 1$ for each n when $x \in K$, it follows that $\int_0^1 f_n(x)\,dx \geq 1/2$ for each n. Finally, assume that f is nonnegative and continuous and that $f_n(x) \geq f(x)$ for each $n \in \mathbb{N}$ and each $x \in \mathbb{Q} \cap [0, 1]$. Then, $f(x)$ must equal zero for each $x \in \mathbb{Q} \cap [0, 1]$ since $\lim_{n\to\infty} f_n(x) = 0$ when x is rational. [This follows since $d(x, K)$ is positive when x is rational.] Thus, since f is continuous, it must be identically zero. □

Example 5.14. *The existence of a sequence of nonnegative Borel measurable functions* $\{f_n\}_{n\in\mathbb{N}}$ *defined on the same measure space* $(\Omega, \mathcal{F}, \mu)$ *such that for* $n \in \mathbb{N}$ *and some positive real constant* K, $K \geq f_n \geq f_{n+1} \geq 0$, *does not imply that* $\lim_{n\to\infty} \int_\Omega f_n(\omega)\, d\mu = \int_\Omega \lim_{n\to\infty} f_n(\omega)\, d\mu$.

Proof: Consider the measure space given by the reals, the real Borel sets, and Lebesgue measure λ on the real Borel sets. Define $f_n(x) = I_{[n,\infty)}(x)$. Then $\lim_{n\to\infty} f_n(x) = 0$ for each x in $\mathbb{R}$, which implies that $\int_\Omega \lim_{n\to\infty} f_n(x)\, d\lambda = 0$. However, $\int_\Omega f_n(x)\, d\lambda = \lambda([n, \infty)) = \infty$ for each n in $\mathbb{N}$. Thus, $\lim_{n\to\infty} \int_\Omega f_n(x)\, d\lambda \neq \int_\Omega \lim_{n\to\infty} f_n(x)\, d\lambda$ for this example, in which the f_n's are nonincreasing, bounded above by one, and bounded below by zero. □

Example 5.15. *If* p *is a real number not less than one,* I *is any interval of real numbers having positive length,* μ *is Lebesgue measure restricted to* $\mathcal{M}(I)$, *and* g *is any element of* $\mathbf{L}_p(I, \mathcal{M}(I), \mu)$, *then the subset* S *of* $\mathbf{L}_p(I, \mathcal{M}(I), \mu)$ *consisting of all points* f *that are almost everywhere not greater than* g *on any subinterval of positive length is a closed nowhere dense subset of* $\mathbf{L}_p(I, \mathcal{M}(I), \mu)$.

Proof: This example was inspired by Oxtoby [1937]. Select any sequence from the set S that converges in $\mathbf{L}_p(I, \mathcal{M}(I), \mu)$ to f. Then there exists a subsequence that converges almost everywhere $[\mu]$ to f, and hence $f \leq g$ a.e. $[\mu]$. Thus we see that S is closed. Now, to show that S is nowhere dense, we will show that S is equal to its boundary. Choose a point f in S and let $\tilde{I}$ be a subinterval of I having positive length. Then by redefining this point on a small subinterval of $\tilde{I}$ to equal $g + 1$ on that subinterval and to equal f off that subinterval, we can obtain a new element $\tilde{f}$ of $\mathbf{L}_p(I, \mathcal{M}(I), \mu)$ that is not in S, but is such that $\|f - \tilde{f}\| < \varepsilon$ for any preassigned $\varepsilon > 0$.

Note that via a similar argument we see that the subset of $\mathbf{L}_p(I, \mathcal{M}(I), \mu)$ consisting of all points f that are almost everywhere not less than g on any subinterval $\tilde{I}$ of positive length is a closed nowhere dense subset of $\mathbf{L}_p(I, \mathcal{M}(I), \mu)$.

Note also that similar results may be obtained if $\mathcal{M}(I)$ is replaced with $\mathcal{B}(I)$ and μ is replaced by $\mu|_{\mathcal{B}(I)}$. □

Example 5.16. *The set of all nonnegative Lebesgue integrable functions defined up to almost everywhere equivalence is a nowhere dense closed subset of* $\mathbf{L}_1(\mathbb{R}, \mathcal{M}(\mathbb{R}), \lambda)$.

Proof: This is a direct consequence of Example 5.15. □

Example 5.17. *A sequence of real-valued functions defined on* $[0, 1]$ *exists that converges in* $\mathbf{L}_p([0, 1], \mathcal{M}([0, 1]), \lambda|_{\mathcal{M}([0,1])})$ *to a function in* $\mathbf{L}_p([0, 1], \mathcal{M}([0, 1]), \lambda|_{\mathcal{M}([0,1])})$ *for any* $p > 0$, *yet converges pointwise at no point.*

Proof: Consider the sequence of integrable, real-valued functions $\{f_n\}_{n\in\mathbb{N}}$ defined by $f_n(x) = I_{[(n-k-1)2^{-k},\,(n-k)2^{-k}]}(x)$, where k is a nonnegative integer such that $2k \le n < 2k+1$. Let p be a positive real number. Then $\int_0^1 |f_n|^p\,dx \to 0$ as $n \to \infty$. However, the set $\{x \in [0, 1]\colon f_n(x) \to 0$ as $n \to \infty\}$ is empty. □

Example 5.18. *There exists a sequence of bounded measurable functions defined on* $[0, 1]$ *that converges pointwise, yet for any* $p > 0$, *does not converge in* $\mathbf{L}_p([0, 1], \mathcal{M}([0, 1]), \lambda|_{\mathcal{M}([0,1])})$.

Proof: Consider the sequence of integrable real-valued functions $\{f_n\}_{n\in\mathbb{N}}$ defined by $f_n(x) = n^{1/p} I_{(0,\,1/n]}(x)$. Notice that as $n \to \infty$, $f_n(x) \to 0$ for all $x \in [0, 1]$. Thus the sequence of bounded measurable functions $\{f_n\}_{n\in\mathbb{N}}$ converges to 0 pointwise. However, $\int_0^1 |f_n|^p\,dx = 1$ for all n. □

Example 5.19. *There exists a sequence* $\{f_n\}_{n\in\mathbb{N}}$ *of bounded Lebesgue integrable functions mapping* $\mathbb{R}$ *into* $\mathbb{R}$ *such that* $\liminf_{n\to\infty} \int_{\mathbb{R}} f_n\,d\lambda < \int_{\mathbb{R}} \liminf_{n\to\infty} f_n\,d\lambda$.

Proof: For positive integers n, let $f_n(x) = -I_{[n,\,n+1]}(x)$. Then

$$\liminf_{n\to\infty} f_n(x) = 0.$$

However, $\int_{\mathbb{R}} f_n\,d\lambda = -1$. □

Example 5.20. *If* $(\Omega, \mathcal{F}, \mu)$ *is a finite measure space and if* $\{f_\alpha\}_{\alpha\in A}$ *is an increasing net of integrable functions taking values in* $[0, 1]$, *then*

$$\lim_\alpha \int_\Omega f_\alpha\,d\mu$$

need not equal

$$\int_\Omega \lim_\alpha f_\alpha\,d\mu.$$

Proof: Consider the measure space $(\aleph_1, \mathcal{B}(\aleph_1), \mu)$, where μ is Dieudonné measure restricted to $\mathcal{B}(\aleph_1)$. Let $A = \aleph_1$, and note that A is a directed set. For each $\alpha \in A$, let $f_\alpha(x)$ denote the indicator function of all ordinals less than α. Then note that for $\alpha \in A$,

$$\int_\Omega f_\alpha\,d\mu = 0,$$

and yet

$$\int_\Omega \lim_\alpha f_\alpha\,d\mu = 1.$$

□

Example 5.21. *If* $(\Omega, \mathcal{F}, \mu)$ *is a finite measure space and if* $\{f_\alpha\}_{\alpha\in A}$ *is a net of integrable functions taking values in* $[0, 1]$ *for which there exists an integrable function* g *such that* $|f_\alpha| \le g$, *then*

$$\lim_\alpha \int_\Omega f_\alpha\,d\mu$$

need not equal

$$\int_\Omega \lim_\alpha f_\alpha\,d\mu.$$

Proof: Consider Example 5.20, let $g = 1$, and the result follows. □

Example 5.22. *If $(\Omega, \mathcal{F}, \mu)$ is a finite measure space and if $\{f_\alpha\}_{\alpha\in A}$ is a net of integrable functions taking values in [0, 1], then*

$$\liminf_{\alpha} \int_{\Omega} f_\alpha \, d\mu$$

need not be lower bounded by

$$\int_{\Omega} \liminf_{\alpha} f_\alpha \, d\mu.$$

Proof: Consider the measure space given in Example 5.20, and for each $\alpha \in A$, let f_α denote the negative of the indicator function of the set of ordinals not less than α and yet less than the first uncountable ordinal. Then for each $\alpha \in A$,

$$\int_{\Omega} f_\alpha \, d\mu = -1,$$

and yet

$$\int_{\Omega} \liminf_{\alpha} f_\alpha \, d\mu = 0.$$

□

Example 5.23. *There exists an integrable function mapping $\mathbb{R}$ into $\mathbb{R}$ that is not square integrable on (a, b) for any real numbers a and b such that $a < b$.*

Proof: To begin, define g: $\mathbb{R} \to \mathbb{R}$ via $g(x) = I_{(0,1)}(x)/\sqrt{x}$ and let $\{r_k$: $k \in \mathbb{N}\}$ be an enumeration of the rationals. Further, let $\{\alpha_k$: $k \in \mathbb{N}\}$ be a summable sequence of positive real numbers and define f: $\mathbb{R} \to \mathbb{R}$ via $f(x) = \sum_{k\in\mathbb{N}} \alpha_k g(x - r_k)$. Since $\int_{\mathbb{R}} g(x)\, dx = 2$ and $\{\alpha_k$: $k \in \mathbb{N}\}$ is absolutely summable, it is clear that f is integrable. Consider now an open interval (a, b) where $a < b$. Further, let $q = r_m$ be a rational number such that $a < q < b$. Notice that

$$\begin{aligned} \int_a^b f^2(x)\, dx &> \int_q^b f^2(x)\, dx \\ &> \int_q^b \alpha_m^2 [g(x-q)]^2\, dx = \alpha_m^2 \int_0^{b-q} \frac{1}{x}\, dx = \infty. \end{aligned}$$

Hence, even though f is integrable, f is not square integrable on (a, b) for any real numbers a and b such that $a < b$. □

Example 5.24. *There exists a real-valued Lebesgue integrable function that is not bounded over any neighborhood of any point.*

Proof: Example 5.23 presented a function f: $\mathbb{R} \to \mathbb{R}$ that is Lebesgue integrable yet not square integrable on (a, b) for any $a < b$. Such a function f is not bounded over any neighborhood of any point. For a simpler example, let $\{r_k$: $k \in \mathbb{N}\}$ be an enumeration of the rationals and define h: $\mathbb{R} \to \mathbb{R}$ via $h(x) = k$ if $x = r_k$ and zero otherwise. Then, even though it is clear that h is unbounded over any neighborhood of any point, h is a nonnegative Lebesgue integrable function whose Lebesgue integral is equal to zero. □

Example 5.25. *A bounded continuous function on* $(0, \infty)$ *may vanish at infinity yet not be in* $\mathbf{L}_p((0, \infty), \mathcal{B}((0, \infty)), \lambda|_{\mathcal{B}((0, \infty))})$ *for any* $p > 0$.

Proof: Let f: $(0, \infty) \to (0, \infty)$ via $f(x) = 1/[1 + \ln(x + 1)]$. Then f is a continuous, positive function upper bounded by one. Also, $f(x) \to 0$ as $x \to \infty$. However, f is not integrable, and for $p \in (0, 1)$, $f(x) < (f(x))^p$. Thus, f is not in $\mathbf{L}_p((0, \infty), \mathcal{B}((0, \infty)), \lambda|_{\mathcal{B}((0, \infty))})$. For p real and greater than one, there exists a positive real number M such that for $x > M$, $(f(x))^p > 1/(1 + x)$, and thus f is not in $\mathbf{L}_p((0, \infty), \mathcal{B}((0, \infty)), \lambda|_{\mathcal{B}((0, \infty))})$. □

Example 5.26. *For any positive integer* k, *there exists a* σ*-finite measure* μ *on* $\mathcal{B}(\mathbb{R}^k)$ *such that there exists no continuous nonidentically zero real-valued function defined on* $\mathbb{R}^k$ *that is integrable.*

Proof: Consider the σ-finite measure μ on $\mathcal{B}(\mathbb{R}^k)$ given in Example 4.5. For any continuous real-valued function f: $\mathbb{R}^k \to \mathbb{R}$ that is not identically zero, $\int_{\mathbb{R}^k} |f|\, d\mu = \infty$, since $|f|$ is bounded away from zero on a nonempty open set. □

Example 5.27. *For any real number* $p > 1$, *there exists a measure space* $(\Omega, \mathcal{F}, \mu)$ *and a measurable, nonnegative, real-valued function* f *such that* fg *is in* $\mathbf{L}_1(\Omega, \mathcal{F}, \mu)$ *for all* g *in* $\mathbf{L}_p(\Omega, \mathcal{F}, \mu)$, *and yet* f *is not in* $\mathbf{L}_q(\Omega, \mathcal{F}, \mu)$ *for any real positive* q.

Proof: Let $\Omega = \{1, 2\}$, $\mathcal{F} = \mathbb{P}(\Omega)$, and define the measure μ on $\mathcal{F}$ via $\mu(\{1\}) = 1$ and $\mu(\{2\}) = \infty$. Let f be the constant function equal to 1. Then any g in $\mathbf{L}_p(\Omega, \mathcal{F}, \mu)$ is equal to zero at 2 and assumes a real value at 1. Hence, $gf = g \in \mathbf{L}_p(\Omega, \mathcal{F}, \mu)$ for all $g \in \mathbf{L}_p(\Omega, \mathcal{F}, \mu)$, yet $f \notin \mathbf{L}_q(\Omega, \mathcal{F}, \mu)$ for any $q \in (0, \infty)$. □

Example 5.28. *For any real number* $p > 1$, *there exists a measure space* $(\Omega, \mathcal{F}, \mu)$ *such that for any two functions* f *and* g *in* $\mathbf{L}_p(\Omega, \mathcal{F}, \mu)$ *for which* $\|f\|_{\mathbf{L}_p} \neq 0$ *and* $\|g\|_{\mathbf{L}_p} \neq 0$,

$$2^{-|2-p|/p} \leq \frac{\|f+g\|^2_{\mathbf{L}_p} + \|f-g\|^2_{\mathbf{L}_p}}{2\left(\|f\|^2_{\mathbf{L}_p} + \|g\|^2_{\mathbf{L}_p}\right)} \leq 2^{|2-p|/p},$$

and such that for $p \neq 2$, *these bounds are not achievable by any two functions* f *and* g *in* $\mathbf{L}_p(\Omega, \mathcal{F}, \mu)$ *for which* $\|f\|_{\mathbf{L}_p} \neq 0$ *and* $\|g\|_{\mathbf{L}_p} \neq 0$.

Proof: Consider the measure space $(\mathbb{R}, \mathcal{F}, \mu)$, where $\mathcal{F}$ is the σ-algebra consisting of the subsets of $\mathbb{R}$ that are either countable or have countable complements, and $\mu(A)$ equals zero if A is countable and equals one if the complement of A is countable. Note that any function in $\mathbf{L}_p(\mathbb{R}, \mathcal{F}, \mu)$ is constant almost everywhere with respect to μ. Thus, for any function h in $\mathbf{L}_p(\mathbb{R}, \mathcal{F}, \mu)$, $\|h\|_{\mathbf{L}_p} = |h|$. Therefore, for any f and g in $\mathbf{L}_p(\mathbb{R}, \mathcal{F}, \mu)$, such that $\|f\|_{\mathbf{L}_p} \neq 0$ and $\|g\|_{\mathbf{L}_p} \neq 0$,

$$\frac{\|f+g\|^2_{\mathbf{L}_p} + \|f-g\|^2_{\mathbf{L}_p}}{2\left(\|f\|^2_{\mathbf{L}_p} + \|g\|^2_{\mathbf{L}_p}\right)} = 1,$$

and $2^{|2-p|/p} > 1$ and $2^{-|2-p|/p} < 1$. This example is a counterexample to a misunderstanding of Clarkson's condition. □

Example 5.29. *There exists a measure space* $(\Omega, \mathcal{F}, \mu)$ *such that* $\mathbf{L}_1(\Omega, \mathcal{F}, \mu) = \mathbf{L}_1^*(\Omega, \mathcal{F}, \mu) \neq \mathbf{L}_\infty(\Omega, \mathcal{F}, \mu)$.

Proof: This example is a minor modification of one suggested in McShane [1950], where credit is also given to T. A. Botts and V. L. Klee. Let $\Omega = \mathbb{R}$, let $\mathcal{F} = \mathbb{P}(\mathbb{R})$, and let μ be defined via $\mu(B) = 0$ if B is countable and does not contain the origin, $\mu(B) = 1$ if B is countable and does contain the origin, and $\mu(B) = \infty$ if B is not countable. Notice that $\mathbf{L}_1(\Omega, \mathcal{F}, \mu)$ consists of all extended real-valued functions that are zero off a countable set and are finite at the origin. Further, for any $f \in \mathbf{L}_1(\Omega, \mathcal{F}, \mu)$, $\int_\Omega f\,d\mu = f(0)$.

Note that $\mathbf{L}_\infty(\Omega, \mathcal{F}, \mu)$ is not the topological dual of $\mathbf{L}_1(\Omega, \mathcal{F}, \mu)$ since there exist distinct functions h_1 and h_2 in $\mathbf{L}_\infty(\Omega, \mathcal{F}, \mu)$ such that $\int_\Omega f h_1\,d\mu = \int_\Omega f h_2\,d\mu$ for all $f \in \mathbf{L}_1(\Omega, \mathcal{F}, \mu)$. For example, let h_1 be the constant function equal to one, and let $h_2(0) = 1$ and for $x \neq 0$, let $h_2(x) = 2$. Then h_1 and h_2 are distinct elements of $\mathbf{L}_\infty(\Omega, \mathcal{F}, \mu)$ such that $\int_\Omega f h_1\,d\mu = \int_\Omega f h_2\,d\mu$ for all $f \in \mathbf{L}_1(\Omega, \mathcal{F}, \mu)$.

Next, note that for all real $p \geq 1$, $\mathbf{L}_1(\Omega, \mathcal{F}, \mu) = \mathbf{L}_p(\Omega, \mathcal{F}, \mu)$. Now, since the topological dual of $\mathbf{L}_2(\Omega, \mathcal{F}, \mu)$ is given by $\mathbf{L}_2(\Omega, \mathcal{F}, \mu)$, it follows that $\mathbf{L}_1^*(\Omega, \mathcal{F}, \mu) = \mathbf{L}_1(\Omega, \mathcal{F}, \mu)$. □

Example 5.30. *There exists a measure space* $(\Omega, \mathcal{F}, \mu)$ *and an element* $L \in \mathbf{L}_1^*(\Omega, \mathcal{F}, \mu)$ *such that there exists no* $g \in \mathbf{L}_\infty(\Omega, \mathcal{F}, \mu)$ *for which* $L(f) = \int_\Omega fg\,d\mu$.

Proof: This example is from Hewitt and Stromberg [1965]. Let m denote Lebesgue measure on $\mathcal{B}([0, 1])$. Let $\Omega = [0, 1] \times [0, 1]$ and let $\mathcal{F}$ denote the Borel subsets in the usual product topology. Define the measures ν and μ on $\mathcal{F}$ via

$$\nu(B) = \int_{[0,1]} m(\{y \in [0, 1]: (x, y) \in B\})\,dC$$

and

$$\mu(B) = \nu(B) + \int_{[0,1]} m(\{x \in [0, 1]: (x, y) \in B\})\,dC,$$

where C is counting measure. Note that $\nu \ll \mu$. Let L: $\mathbf{L}_1(\Omega, \mathcal{F}, \mu) \to \mathbb{R}$ via $L(f) = \int_\Omega f\,d\nu$. Observe that $L \in \mathbf{L}_1^*(\Omega, \mathcal{F}, \mu)$.

Now, assume that there exists $g \in \mathbf{L}_\infty(\Omega, \mathcal{F}, \mu)$ such that $L(f) = \int_\Omega fg\,d\mu$. For $y \in [0, 1]$, let $H_y = \{(x, y) \in \Omega: x \in [0, 1]\}$. Choose $f = I_{H_y}\text{sgn}(g)$, where $\text{sgn}(z)$ is 1 for z positive, 0 for z equal to zero, and -1 for z negative. Then f is in $\mathbf{L}_1(\Omega, \mathcal{F}, \mu)$, and we have that

$$\int_\Omega fg\,d\mu = \int_{[0,1]} |g(x, y)|\,dm(x) = L(f) = \int_\Omega f\,d\nu = 0.$$

Thus, for each $y \in [0, 1]$, $g(\cdot, y) = 0$ a.e. $[m]$. Now it follows from Fubini's theorem that for almost all $[m]$ x, $g(x, \cdot) = 0$ a.e. $[m]$. Select such an x, and call it x_0. Let $V = \{(x_0, y) \in \Omega: y \in [0, 1]\}$ and let $f = I_V$. Then f is in $\mathbf{L}_1(\Omega, \mathcal{F}, \mu)$, and we have that

$$\nu(V) = 1 = \int_\Omega f\, d\nu = L(f) = \int_\Omega fg\, d\mu$$
$$= \int_V g\, d\mu = \int_{[0,1]} g(x_0, y)\, dm(y) = 0.$$

Hence, we see that no such g exists. □

Example 5.31. *If $(\Omega, \mathcal{F}, \mu)$ is a normalized measure space and S is a subset of $\mathbf{L}_1(\Omega, \mathcal{F}, \mu)$ that is uniformly integrable, there need not exist $h \in \mathbf{L}_1(\Omega, \mathcal{F}, \mu)$ such that $|f| \leq |h|$ for all $f \in S$.*

Proof: This example is from Biler and Witkowski [1990]. Let $\Omega = \mathbb{N}$, $\mathcal{F} = \mathbb{P}(\mathbb{N})$, and $\mu(n) = 2^{-n}$ for $n \in \mathbb{N}$. Note that $\mu(\Omega) = 1$. Let S denote the family of functions $f_n(x) = (2^n/n)I_{\{n\}}(x)$, where $n \in \mathbb{N}$. Then $\int_\Omega |f_n|\, d\mu = 1/n$ and hence it straightforwardly follows that S is uniformly integrable. However, if h is such that $|h| \geq |f_n|$ for each $n \in \mathbb{N}$, then $h \notin \mathbf{L}_1(\Omega, \mathcal{F}, \mu)$. □

Example 5.32. *There exists a compact, completely normal Hausdorff space Ω and a nonnegative linear function L defined on the continuous real-valued functions on Ω such that there are two distinct normalized measures on $\mathcal{B}(\Omega)$, μ and ν, such that $L(f) = \int_\Omega f\, d\mu = \int_\Omega f\, d\nu$ for all continuous real-valued functions f defined on Ω and such that μ and ν each have the same nonempty support S, and yet $\mu(S) = 0$ and $\nu(S) = 1$.*

Proof: Recall the Dieudonné measure space $(\aleph_1 + 1, \mathcal{B}(\aleph_1 + 1), \mu)$ constructed in Example 4.33, where μ is Dieudonné measure. Consider any continuous real-valued function g defined on $\aleph_1$. We will show that there must exist an ordinal $\alpha < \aleph_1$ such that g is constant on ordinals β for which $\alpha < \beta < \aleph_1$. For the purpose of contradiction, assume that there does not exist a sequence $\{\alpha_n\}_{n\in\mathbb{N}}$ taking values in $\aleph_1$ such that $|g(\beta) - g(\alpha_n)| < 1/n$ whenever $\beta > \alpha_n$. Then, if no such sequence existed, there would be a positive real number ε such that we could construct an increasing sequence $\{\gamma_n\}_{n\in\mathbb{N}}$ taking values in $\aleph_1$ such that $|g(\gamma_n) - g(\gamma_{n-1})| > \varepsilon$ for each $n \in \mathbb{N}$. However, the sequence $\{\gamma_n\}_{n\in\mathbb{N}}$ converges to its least upper bound γ in $\aleph_1$, and the real numbers $g(\gamma_n)$ cannot converge. Since g is continuous, this is impossible, and hence we see that there exists a sequence $\{\alpha_n\}_{n\in\mathbb{N}}$ taking values in $\aleph_1$ such that $|g(\beta) - g(\alpha_n)| < 1/n$ whenever $\beta > \alpha_n$. Now since such a sequence $\{\alpha_n\}_{n\in\mathbb{N}}$ has a least upper bound α in $\aleph_1$, we see that that g is constant on ordinals β such that $\alpha < \beta < \aleph_1$.

Now, if f is any real-valued function defined on $\aleph_1 + 1$, then its restriction to $\aleph_1$ must be continuous. Hence we see that there is some ordinal α such that f is equal to a constant, say f_0, on ordinals β for which $\alpha < \beta < \aleph_1$. Now, if $f(\aleph_1) \neq f_0$, then f would not be continuous. Hence, any real-valued continuous function f defined on $\aleph_1 + 1$ is such that there is some ordinal α such that f is equal to a constant on ordinals β for which $\alpha < \beta \leq \aleph_1$.

Now, note that for any real-valued continuous function f defined on $\aleph_1 + 1$, $\int_{\aleph_1+1} f\,d\mu = f(\aleph_1)$. Further, observe that this integral defines a nonnegative linear function L defined on the continuous functions on $\aleph_1+1$. Thus, letting ν denote the Dirac measure on $\mathcal{B}(\aleph_1+1)$ at the point $\{\aleph_1\}$, we see that

$$L(f) = \int_{\aleph_1+1} f\,d\mu = \int_{\aleph_1+1} f\,d\nu = f(\aleph_1),$$

and further, note that $\mu(\aleph_1) = 1$ and $\nu(\aleph_1) = 0$. Finally, note that the support of μ and of ν is each equal to $\{\aleph_1\}$, and yet $\mu(\{\aleph_1\}) = 0$ and $\nu(\{\aleph_1\}) = 1$. □

Example 5.33. *If H is a compact Hausdorff space and m is a normalized measure on $\mathcal{B}(H)$, $C(H)$ need not be dense in $\mathbf{L}_1(H, \mathcal{B}(H), m)$.*

Proof: Let $H = \aleph_1 + 1$ and recall from Example 4.33 that H is a compact Hausdorff space. Recall from Example 4.33 that if μ is the Dieudonné measure on $\mathcal{B}(H)$ and $f \in C(H)$, then $\int_H f\,d\mu = f(\{\aleph_1\})$. Let ν be Dirac measure at the point $\{\aleph_1\}$, and let m be defined on $\mathcal{B}(H)$ via $m = (\mu+\nu)/2$. Let the real-valued function g be defined on H via $g(x) = 0$ if $x \neq \aleph_1$ and $g(\aleph_1) = 1$. Then $g \in \mathbf{L}_1(H, \mathcal{B}(H), m)$ and $\int_H g\,dm = 1/2$. If $f \in C(H)$, then $\int_H f\,dm = f(\{\aleph_1\})$. Now, since $\big|\,\|f\|_{\mathbf{L}_1} - \|g\|_{\mathbf{L}_1}\big| \leq \|f-g\|_{\mathbf{L}_1}$, we see that for any $f \in C(H)$, $\|f-g\|_{\mathbf{L}_1} \geq |f(\{\aleph_1\}) - 1/2|$. However, for any $f \in C(H)$, if $f(\{\aleph_1\}) = 1/2$, then $\|f-g\|_{\mathbf{L}_1} = 1/2$. Thus, $C(H)$ is not dense in $\mathbf{L}_1(H, \mathcal{B}(H), m)$. □

Example 5.34. *There exists a compact Hausdorff space H and a finite nonidentically zero signed measure m on $\mathcal{B}(H)$ such that for any $f \in C(H)$, the integral over H of f with respect to m is zero.*

Proof: As in Example 5.33, let $H = \aleph_1 + 1$ and let μ denote Dieudonné measure on $\mathcal{B}(H)$. Let ν denote Dirac measure at the point $\{\aleph_1\}$, and define the signed measure m on $\mathcal{B}(H)$ via $m = \mu - \nu$. Recall from Example 5.32 that if $f \in C(H)$, then the integral of f over H with respect to μ is $f(\{\aleph_1\})$. Hence, we see that for any $f \in C(H)$, $\int_H f\,dm = 0$. □

Example 5.35. *A Radon–Nikodym derivative may exist yet not be real-valued.*

Proof: Let $\mathcal{S}$ denote the set of all subsets of $\mathbb{R}$ that are either countable or co-countable and consider the measurable space $(\mathbb{R}, \mathcal{S})$. Let ν denote the restriction of Lebesgue measure to $\mathcal{S}$. Further, define a measure μ on $\mathcal{S}$ via $\mu(A) = 0$ if A is a countable element of $\mathcal{S}$ and $\mu(A) = 1$ if A is a co-countable element of $\mathcal{S}$. Note that $\mu(\mathbb{R}) = 1$ and $\nu(\mathbb{R}) = \infty$. In addition, note that $(\mathbb{R}, \mathcal{S}, \mu)$ is a finite measure space, $(\mathbb{R}, \mathcal{S}, \nu)$ is not a σ-finite measure space, and $\nu \ll \mu$. Hence, it is clear that μ being finite and $\nu \ll \mu$ does not imply that ν is finite or even σ-finite.

Applying the Radon–Nikodym theorem to this situation, it follows that for any set A in $\mathcal{S}$, $\nu(A) = \int_A \infty\,d\mu$, where ∞ is the a.e. $[\mu]$ unique such function. Thus, in this situation, even though the Radon–Nikodym derivative exists, it is real-valued almost nowhere $[\mu]$. □

Example 5.36. *The existence of two measures on a measurable space, the first absolutely continuous with respect to the second, does not imply the existence of a Radon–Nikodym derivative of the first with respect to the second.*

Proof: Consider the measurable space given by the interval [0, 1] and the Borel subsets of [0, 1]. Let ν denote Lebesgue measure on $\mathcal{B}([0, 1])$ and let μ be counting measure on $\mathcal{B}([0, 1])$. Notice that ν is absolutely continuous with respect to μ. Assume that there exists an extended real-valued Borel measurable function f such that $\nu(A) = \int_A f\,d\mu$ for every A in $\mathcal{B}([0, 1])$. Since $\nu(\{x\}) = 0$ and $\mu(\{x\}) = 1$ for every x in [0, 1], it is clear that $f(x) = 0$ for every x in [0, 1]. This observation, however, implies that $\nu(A) = 0$ for every A in $\mathcal{B}([0, 1])$. This contradiction implies that no such function f exists, and hence the Radon–Nikodym theorem fails to hold even though $\nu \ll \mu$. □

Example 5.37. *The result of the Radon–Nikodym Theorem may hold even when neither measure of interest is σ-finite.*

Proof: The following example shows that, in contrast to Example 5.36, the result of the Radon–Nikodym Theorem may hold even when neither measure of interest is σ-finite. In particular, consider the measurable space given by the interval [0, 1] and the Borel subsets of [0, 1]. Let μ_1 denote counting measure on $\mathcal{B}([0, 1])$ and let $\mu_2(A)$ equal infinity for every nonempty subset A in $\mathcal{B}([0, 1])$ and equal zero when $A = \emptyset$. Notice that μ_2 is absolutely continuous with respect to μ_1 and that neither μ_1 nor μ_2 is σ-finite. However, $f = \infty$ is an extended real-valued Borel measurable function for which $\mu_2(A) = \int_A f\,d\mu_1$ for every A in $\mathcal{B}([0, 1])$. Hence, f is a Radon–Nikodym derivative of μ_2 with respect to μ_1. □

Example 5.38. *There exists a measurable space $(\Omega, \mathcal{F})$ and two equivalent measures μ_1 and μ_2 defined on $(\Omega, \mathcal{F})$ such that the Radon–Nikodym derivative of μ_2 with respect to μ_1 exists, but a Radon–Nikodym derivative of μ_1 with respect to μ_2 does not exist.*

Proof: Consider the situation described in Example 5.37. Recall that a Radon–Nikodym derivative of μ_2 with respect to μ_1 exists and notice that μ_1 and μ_2 are equivalent measures. Even so, it is clear that a Radon–Nikodym derivative of μ_1 with respect to μ_2 does not exist. □

Example 5.39. *An integrable, square-integrable, band-limited function may be everywhere discontinuous.*

Proof: Consider a function $g: \mathbb{R} \to \mathbb{R}$ defined by

$$g(x) = \begin{cases} \dfrac{\sin(x)}{\pi x}, & \text{if } x \neq 0 \\ \dfrac{1}{\pi}, & \text{if } x = 0 \end{cases}$$

and define $f(x) = g^2(x)I_{\mathbb{R}-\mathbb{Q}}(x) + I_{\mathbb{Z}}(x/\pi)$. Then f is integrable and square integrable, since $\int_{\mathbb{R}} f(x)\,dx = 1/\pi$ and $\int_{\mathbb{R}} f^2(x)\,dx = 2/(3\pi^3)$. Further, f

is band-limited since its Fourier transform is zero off the interval $[-2, 2]$. Nevertheless, it follows easily that f is nowhere continuous. □

Example 5.40. *The Fourier transform of a Lebesgue integrable function $f\colon \mathbb{R} \to [0, \infty)$ need not be integrable.*

Proof: See the situation in Example 7.10. □

Example 5.41. *If $f \in \mathbf{L}_1(\mathbb{R}, \mathcal{M}(\mathbb{R}), \lambda)$ has a Fourier transform F such that $F(\omega) = 0$ for $\omega > W$, where W is a positive real number, then f need not be equal to*

$$\sum_{n=1}^{\infty} f\left(\frac{n\pi}{w}\right) \frac{\sin(n\pi - wt)}{wt},$$

where w is a real number greater than W.

Proof: Let f be given by

$$f(t) = \begin{cases} \left(\dfrac{\sin(t)}{t}\right)^2, & \text{if } t \text{ is not a rational multiple of } \pi \\ 0, & \text{otherwise.} \end{cases}$$

Then, choosing $w = 10$, we see that

$$\sum_{n=1}^{\infty} f\left(\frac{n\pi}{w}\right) \frac{\sin(n\pi - wt)}{wt} = 0. \qquad \square$$

Example 5.42. *Multidimensional convolution need not be associative.*

Proof: This example is suggested in Rudin [1988]. Recall that the convolution of two integrable real-valued functions f and g defined on $\mathbb{R}^k$ is denoted by $f * g$ and is defined via $(f * g)(x) = \int_{\mathbb{R}^k} f(x - y)\, g(y)\, dy$ provided that this integral exists for all $x \in \mathbb{R}^k$. Consider first the special case when $k = 1$. For t and x real, define $p(t) = [1 - \cos(t)]I_{[0,\, 2\pi]}(t)$, and let $f(x) = 1$, $g(x) = p'(x)$, and $h(x) = \int_{-\infty}^{x} p(t)\, dt$. Note that $(f * g)(x) = \int_{\mathbb{R}} f(x - t)\, g(t)\, dt = \int_{\mathbb{R}} p'(t)\, dt = p(2\pi) - p(0) = 0$. Further, $(g * h)(x) = \int_{\mathbb{R}} g(x-t)\, h(t)\, dt = \int_{\mathbb{R}} p'(x-t) \int_{-\infty}^{t} p(s)\, ds\, dt = (p*p)(x)$ via integration by parts. Note that $(p * p)(x) = \int_0^{2\pi} [1 - \cos(x - y)][1 - \cos(y)]I_{[x-2\pi,\, x]}(y)\, dy$. Hence, $(g * h)(x)$ is positive on $(0, 4\pi)$ and zero elsewhere. Finally, even though $(f * g) * h = 0$, it follows that $f * (g * h)$ is a positive constant.

Now, let k be an integer greater than 1. With f, g, and h defined as in the preceding paragraph, let $\tilde{f}$, $\tilde{g}$, and $\tilde{h}$ map $\mathbb{R}^k$ into $\mathbb{R}$ via $\tilde{f}(x) = 1$, $\tilde{g}(x) = g(x_1)g(x_2)\cdots g(x_k)$, and $\tilde{h}(x) = h(x_1)h(x_2)\cdots h(x_k)$. It follows immediately that $(\tilde{f} * \tilde{g}) = 0$ and $(\tilde{g} * \tilde{h})$ is positive on $(0, 4\pi)^k$ and zero elsewhere. Hence, $(\tilde{f} * \tilde{g}) * \tilde{h} = 0$ but $\tilde{f} * (\tilde{g} * \tilde{h})$ is a positive constant. □

Example 5.43. *For any positive integer k, the multidimensional convolution of two real-valued, bounded, integrable, nowhere zero functions defined on $\mathbb{R}^k$ can be identically equal to zero.*

Proof: This example is from Hall and Wise [1990]. Let k be a positive integer and let $\mathbf{L}_1(\mathbb{R}^k)$ denote the set of all extended real-valued Lebesgue integrable functions modulo almost everywhere equivalence defined on $\mathbb{R}^k$ equipped with the norm given by the integral of the absolute value of an element of $\mathbf{L}_1(\mathbb{R}^k)$. By a k-sequence of real numbers we will mean any function mapping $\mathbb{Z}^k$ into $\mathbb{R}$, and we will denote the value of such a function α at the point x via α_x. A k-sequence will be called absolutely summable if it is integrable with respect to counting measure on the power set of $\mathbb{Z}^k$. Further, we will occasionally denote points x in $\mathbb{R}^k$ as $x = (x_1, \ldots, x_k)$, where the x_i's are real numbers. Further, for two points x and y in $\mathbb{R}^k$, we will denote the Euclidean inner product via $\langle x, y\rangle = \sum_{j=1}^k x_j y_j$.

For an absolutely summable k-sequence of real numbers α, define a bounded linear operator on $\mathbf{L}_1(\mathbb{R}^k)$ to $\mathbf{L}_1(\mathbb{R}^k)$ via

$$(T_\alpha(f))(x) = \int_{\mathbb{Z}^k} \alpha_y f(x-y)\, dC(y),$$

where C denotes counting measure on the power set of $\mathbb{Z}^k$, for any element f from $\mathbf{L}_1(\mathbb{R}^k)$. For any two absolutely summable k-sequences of real numbers α and β, it follows that

$$\begin{aligned}((T_\alpha \circ T_\beta)(f))(x) &= T_\alpha\left(\int_{\mathbb{Z}^k} \beta_y f(x-y)\, dC(y)\right)\\ &= \int_{\mathbb{Z}^k}\int_{\mathbb{Z}^k} \alpha_z\beta_y f(x-y-z)\, dC(y)\, dC(z)\\ &= \int_{\mathbb{Z}^k} \lambda_y f(x-y)\, dC(y),\end{aligned}$$

where we define

$$\lambda_y = \int_{\mathbb{Z}^k}\int_{\mathbb{Z}^k} \alpha_p\beta_q g_y(p, q)\, dC(p)\, dC(q),$$

where $g_y(p, q)$ equals 1 if $p+q = y$ and equals 0 otherwise. Finally, note that for any two elements f and g from $\mathbf{L}_1(\mathbb{R}^k)$ it follows via Fubini's theorem that $(T_\alpha(f)) * (T_\beta(g)) = (T_\alpha \circ T_\beta)(f * g)$.

Recall that the function $|\cos(x_1)\cos(x_2)\cdots\cos(x_k)|$ is expressible as a multiple Fourier series given by $\int_{\mathbb{Z}^k} c_y \exp(\imath\langle x, y\rangle)\, dC(y)$, where it follows easily that $c_y = a_{y_1}a_{y_2}\cdots a_{y_k}$, where $a_n = 0$ if n is odd and $a_n = (2/\pi)\left[((-1)^{n/2})/(1-n^2)\right]$ if n is even. Further, if we define

$$f_1(x) = \frac{1}{2}(|\cos(x_1)\cos(x_2)\cdots\cos(x_k)| + [\cos(x_1)\cos(x_2)\cdots\cos(x_k)])$$

and

$$f_2(x) = \frac{1}{2}(|\cos(x_1)\cos(x_2)\cdots\cos(x_k)| - [\cos(x_1)\cos(x_2)\cdots\cos(x_k)]),$$

then $f_1(x)f_2(x) = 0$,

$$f_1(x) = \int_{\mathbb{Z}^k} \alpha_y \exp(\imath\langle x, y\rangle)\, dC(y),$$

and

$$f_2(x) = \int_{\mathbb{Z}^k} \beta_y \exp(\imath\langle x, y\rangle)\, dC(y)$$

where

$$\alpha_y = \begin{cases} \frac{c_y}{2}, & \text{if } y \notin \{-1, 1\}^k \\ 4^{-k}, & \text{if } y \in \{-1, 1\}^k \end{cases}$$

and

$$\beta_y = \begin{cases} \frac{c_y}{2}, & \text{if } y \notin \{-1, 1\}^k \\ -4^{-k}, & \text{if } y \in \{-1, 1\}^k. \end{cases}$$

But,

$$\begin{aligned} f_1(x)f_2(x) &= 0 \\ &= \int_{\mathbb{Z}^k} \alpha_y \exp(\imath\langle x, y\rangle)\, dC(y) \int_{\mathbb{Z}^k} \beta_y \exp(\imath\langle x, y\rangle)\, dC(y) \\ &= \int_{\mathbb{Z}^k}\int_{\mathbb{Z}^k} \alpha_y\beta_z \exp[\imath(\langle x, y\rangle + \langle x, z\rangle)]\, dC(y)\, dC(z) \\ &= \int_{\mathbb{Z}^k} \lambda_y \exp(\imath\langle x, y\rangle)\, dC(y) \end{aligned}$$

where, as before,

$$\lambda_y = \int_{\mathbb{Z}^k}\int_{\mathbb{Z}^k} \alpha_p\beta_q g_y(p, q)\, dC(p)\, dC(q)$$

and $g_y(p, q)$ equals 1 if $p+q=y$ and equals 0 otherwise. Note that, via Fubini's theorem, $\lambda_y = 0$ for every y. Thus, it follows that $(T_\alpha(f)) * (T_\beta(g)) = (T_\alpha \circ T_\beta)(f * g) = 0$ for any integrable f and g. Finally, to obtain two real-valued, bounded, nowhere zero, Lebesgue integrable functions $T_\alpha(f)$ and $T_\beta(g)$ defined on $\mathbb{R}^k$ whose convolution is identically zero, simply let $f(x) = g(x) = I_S(x)$, where $S = (-1, 1]^k$. □

Example 5.44. *A linear system mapping* $\mathbf{L}_2(\mathbb{R}, \mathcal{M}(\mathbb{R}), \lambda)$ *to* $\mathbf{L}_2(\mathbb{R}, \mathcal{M}(\mathbb{R}), \lambda)$ *described via convolution with a fixed point in* $\mathbf{L}_2(\mathbb{R}, \mathcal{M}(\mathbb{R}), \lambda)$ *that takes on each real value in any nonempty open subset of* $\mathbb{R}$ *may map each element of* $\mathbf{L}_2(\mathbb{R}, \mathcal{M}(\mathbb{R}), \lambda)$ *into the identically zero function.*

Proof: Let h denote the function f given in Example 2.29 and note that h is an element of $\mathbf{L}_2(\mathbb{R}, \mathcal{M}(\mathbb{R}), \lambda)$. For the linear system mapping $\mathbf{L}_2(\mathbb{R}, \mathcal{M}(\mathbb{R}), \lambda)$ into $\mathbf{L}_2(\mathbb{R}, \mathcal{M}(\mathbb{R}), \lambda)$ given by convolution with this function h, the advertised result follows. □

References

Biler, P., and A. Witkowski, *Problems in Mathematical Analysis,* New York: Marcel Dekker, 1990.

Hall, E. B., and G. L. Wise, "An algebraic aspect of linear system theory," *IEEE Transactions on Circuits and Systems,* Vol. 37, pp. 651–653, May 1990.

Hewitt, E., and K. Stromberg, *Real and Abstract Analysis,* Springer-Verlag: New York, 1965.

Langford, E. S., and H. Kestelman, "Problem E2381," *American Mathematical Monthly,* November 1973, pp. 1062–1063.

McShane, E. J., "Linear functionals on certain Banach spaces," *Proceedings of the American Mathematical Society,* Vol. 1, pp. 402–408, 1950.

Natanson, I. P., *Theory of Functions of a Real Variable,* Ungar: New York, 1955.

Oxtoby, J. C., "The category and Borel class of certain subsets of $\mathcal{L}_p^*$," *Bulletin of the American Mathematical Society,* Vol. 43, April 1937, pp. 245–248.

Rudin, W., *Real and Complex Analysis,* Third edition, New York: McGraw Hill, 1988.

Wilansky, A., "Two examples in real variables," *American Mathematical Monthly,* Vol. 60, May 1953, p. 317.

6. Product Spaces

In this chapter we present counterexamples concerning products of topological spaces and products of measure spaces with particular emphasis placed upon examining unexpected properties of $\mathbb{R}^k$.

Example 6.1. *Given any bounded subsets X and Y of $\mathbb{R}^3$ having nonempty interiors, there exists a positive integer n, a partition $\{X_i: 1 \leq i \leq n\}$ of X, and a partition $\{Y_i: 1 \leq i \leq n\}$ of Y such that X_i is congruent to Y_i for each i.*

Proof: The proof of this result due to Banach and Tarski is from Stromberg [1979]. For a point x in $\mathbb{R}^3$, we will let $\|x\|$ denotes the standard Euclidean norm of x. Recall that a square matrix with real entries is said to be orthogonal if it is invertible and if its inverse equals its transpose. A rotation is a mapping from $\mathbb{R}^3$ into $\mathbb{R}^3$ of the form $\rho(x) = Rx$, where R is an orthogonal matrix with unit determinant. (We sometimes will also refer to such a matrix R as a rotation.) A rigid motion is a mapping from $\mathbb{R}^3$ into $\mathbb{R}^3$ of the form $r(x) = \rho(x) + \alpha$, where ρ is a rotation and α is a point in $\mathbb{R}^3$. Two subsets A and B of $\mathbb{R}^3$ are said to be congruent if $B = r(A)$ for some rigid motion r.

Two subsets A and B of $\mathbb{R}^3$ are said to be piecewise congruent if for some positive integer n there exists a partition $\{A_j: 1 \leq j \leq n\}$ of A and a set $\{\rho_j: 1 \leq j \leq n\}$ of rigid motions such that $\{\rho_j(A_j): 1 \leq j \leq n\}$ partitions B. For subsets X, Y, and Z of $\mathbb{R}^3$ recall that if X is piecewise congruent to Y and if Y is piecewise congruent to Z, then X is piecewise congruent to Z, recall that if X is a subset of Y, then X is piecewise congruent to a subset of Y, and recall that if X is piecewise congruent to a subset of Y and if Y is piecewise congruent to a subset of X, then X is piecewise congruent to Y. By a translate of a subset A of $\mathbb{R}^3$ we will mean a set of the form $\{a + b: a \in A\}$ for $b \in \mathbb{R}^3$, and we will denote such a set by $A + b$. For any subset D of $\mathbb{R}^3$ and any positive number δ, we will let δD denote the set $\{\delta x: x \in D\}$.

For a fixed real number θ, let

$$\psi = \begin{pmatrix} -1/2 & -\sqrt{3}/2 & 0 \\ \sqrt{3}/2 & -1/2 & 0 \\ 0 & 0 & 1 \end{pmatrix}$$

and let

$$\phi = \begin{pmatrix} -\cos(\theta) & 0 & \sin(\theta) \\ 0 & -1 & 0 \\ \sin(\theta) & 0 & \cos(\theta) \end{pmatrix}$$

Note that ψ and ϕ are rotations and that $\psi^3 = \phi^2 = I$, where I denotes the 3×3 identity matrix. Let $\mathcal{G}$ denote the group given by the set of all matrices expressible as a finite product of factors each of which is either ψ or ϕ. Note that if $g \in \mathcal{G}$ and if $g \neq I$, then g may be expressed as a product of the form $g = \sigma_1\sigma_2 \cdots \sigma_n$, where n is a positive integer, where $\sigma_i \in \{\phi, \psi, \psi^2\}$ for each i, and where one but not both of σ_i and σ_{i+1} is equal to ϕ for any positive integer $i < n$. Such an expression for g is said to be a reduced word in the letters ϕ, ψ, and ψ^2. If the reduced word expression for g is the unique such expression, then n is denoted by $l(g)$ and is called the length of g and σ_1 is called the first letter of g. We will let $l(I) = 0$.

We will now show that if $\cos(\theta)$ is a transcendental number, then each element in $\mathcal{G}$ different from I may be expressed in only one way as a reduced word in the letters ϕ, ψ, and ψ^2. Note first that the desired result will follow if we show only that no reduced word is equal to I. Consider first a reduced word of the form $\alpha = \sigma_m\sigma_{m-1}\cdots\sigma_2\sigma_1$, where m is a positive integer and where σ_i is either $\psi\phi$ or $\psi^2\phi$ for each i. By induction on m it follows that $\alpha((0,\,0,\,1)) = (\sin(\theta)P_{m-1}(\cos(\theta)),\ \sqrt{3}\sin(\theta)Q_{m-1}(\cos(\theta)),\ R_m(\cos(\theta)))$, where $P_0(x) = -1/2$, $Q_0(x) = \pm 1/2$, $R_1(x) = x$,

$$P_m(x) = \frac{1}{2}xP_{m-1}(x) \pm \frac{3}{2}Q_{m-1}(x) - \frac{R_m(x)}{2},$$
$$Q_m(x) = \mp\frac{1}{2}xP_{m-1}(x) + \frac{1}{2}Q_{m-1}(x) \pm \frac{R_m(x)}{2},$$
$$R_{m+1}(x) = (1-x^2)P_{m-1}(x) + xR_m(x).$$

Note that if $\alpha((0,\,0,\,1)) = (0,\,0,\,1)$, then $R_m(\cos(\theta)) - 1 = 0$, which implies that $\cos(\theta)$ is a root of a polynomial with rational coefficients. Since this is not possible, we conclude that $\alpha \neq I$. Next, consider a reduced word of the form $\beta = \sigma_m\sigma_{m-1}\cdots\sigma_2\sigma_1$, where m is a positive integer and where σ_i is either $\phi\psi$ or $\phi\psi^2$ for each i. Note that β cannot equal I, since such would imply that $\alpha = \phi\beta\phi = \phi I\phi = \phi^2 = I$, in contradiction with our earlier result. Consider next a reduced word of the form $\delta = \psi^{p_1}\phi\psi^{p_2}\phi\cdots\psi^{p_{m-1}}\phi\psi^{m}$, where $m > 1$ is an integer and where $p_i \in \{1,\,2\}$ for each i. Assume that $\delta = I$ and that m is the smallest integer not less than 2 for which that is true. If $p_1 = p_m$, then $\psi^{p_1+p_m}$ is either ψ or ψ^2 and hence $I = \psi^{-p_1}\delta\psi^{p_1} = \phi\psi^{p_2}\cdots\phi\psi^{p_1+p_m}$, which implies that $\beta = I$ for some choice of β as above, in contradiction with our earlier result. If $p_1 + p_m = 3$ and $m > 3$, then $I = \phi\psi^{p_m}\delta\psi^{p_1}\phi = \psi^{p_2}\phi\cdots\psi^{p_m}\phi\psi^{p_{m-1}}$, which has the same form as δ yet with a smaller value of m, contrary to our assumption. Thus, $p_1 + p_m = 3$ with either $m = 2$ or 3. However, if $m = 2$, then $I = \psi^{p_2}\delta\psi^{p_1} = \phi$, and if $m = 3$, then $I = \phi\psi^{p_3}\delta\psi^{p_1}\phi = \psi^{p_2}$, neither of which is possible. Thus, we conclude that $\delta \neq I$. Finally, for reduced words of the form $\gamma = \beta\phi$, it follows that γ cannot equal I, since such would imply that $\delta = \phi\gamma\phi = I$. Thus, the desired result about $\mathcal{G}$ follows when $\cos(\theta)$ is transcendental. Let $\theta = 1$ and note that $\cos(\theta)$ is transcendental.

We will now show that $\mathcal{G}$ may be partitioned into three nonempty sets $\mathcal{G}_1$, $\mathcal{G}_2$, and $\mathcal{G}_3$ such that for each $g \in \mathcal{G}$ it follows that $g \in \mathcal{G}_1$ if and only if $\phi g \in \mathcal{G}_2 \cup \mathcal{G}_3$, that $g \in \mathcal{G}_1$ if and only if $\psi g \in \mathcal{G}_2$, and that $g \in \mathcal{G}_1$ if and only if $\psi^2 g \in \mathcal{G}_3$. To begin, place I in $\mathcal{G}_1$, ϕ in $\mathcal{G}_2$, ψ in $\mathcal{G}_2$, and $\psi^2 \in \mathcal{G}_3$. Assume, for some positive integer n, that all elements σ in $\mathcal{G}$ for which $l(\sigma) \leq n$ have been placed into either $\mathcal{G}_1$, $\mathcal{G}_2$, or $\mathcal{G}_3$. Let σ be an element in $\mathcal{G}$ such that $l(\sigma) = n$. If the first letter of σ is ψ or ψ^2 and if $\sigma \in \mathcal{G}_1$, then place $\phi\sigma$ in $\mathcal{G}_2$. If the first letter of σ is ψ or ψ^2 and if $\sigma \in \mathcal{G}_2 \cup \mathcal{G}_3$, then place $\phi\sigma$ in $\mathcal{G}_1$. If the first letter of σ is ϕ and if $\sigma \in \mathcal{G}_1$, then place $\psi\sigma$ in $\mathcal{G}_2$ and $\psi^2\sigma$ in $\mathcal{G}_3$. If the first letter of σ is ϕ and if $\sigma \in \mathcal{G}_2$, then place $\psi\sigma$ in $\mathcal{G}_3$ and $\psi^2\sigma$ in $\mathcal{G}_1$. If the first letter of σ is ϕ and if $\sigma \in \mathcal{G}_3$, then place $\psi\sigma$ in $\mathcal{G}_1$ and $\psi^2\sigma$ in $\mathcal{G}_2$. Thus, by induction, the desired partition is obtained. Note that the desired properties for this partition hold for any g from $\mathcal{G}$ such that $l(g) \leq 1$. Let g be an element from $\mathcal{G}$ with $l(g) = n$ for $n > 1$ and assume that the desired properties hold for elements in $\mathcal{G}$ with lengths less than n. If the first letter of g is ϕ, then it follows immediately via the construction of our partition that $g \in \mathcal{G}_1$ if and only if $\psi g \in \mathcal{G}_2$ and that $g \in \mathcal{G}_1$ if and only if $\psi^2 g \in \mathcal{G}_3$. Finally, if $l(\phi g) = n - 1$, then by our induction hypothesis we see that $g \notin \mathcal{G}_1$ if and only if $\phi(\phi g) = g \in \mathcal{G}_2 \cup \mathcal{G}_3$, which holds if and only if $\phi g \in \mathcal{G}_1$, which holds if and only if $\phi g \notin \mathcal{G}_2 \cup \mathcal{G}_3$. Thus, the desired properties hold for such a g from $\mathcal{G}$.

If the first letter of g is ψ, then it follows immediately via the construction of our partition that $g \in \mathcal{G}_1$ if and only if $\phi g \in \mathcal{G}_2 \cup \mathcal{G}_3$. Further, since $\psi g = \psi^2\sigma$, where $l(\sigma) = n - 1$ and where the first letter of σ is ϕ, it follows that $\psi g = \psi^2\sigma \in \mathcal{G}_2$ if and only if $\sigma \in \mathcal{G}_3$, which holds if and only if $g = \psi\sigma \in \mathcal{G}_1$, which holds if and only if $\psi^2 g = \sigma \in \mathcal{G}_3$. Thus, for such a g we see that $g \in \mathcal{G}_1$ if and only if $\psi g \in \mathcal{G}_2$ and that $g \in \mathcal{G}_1$ if and only if $\psi^2 g \in \mathcal{G}_3$. If the first letter of g is ψ^2, then as before it follows immediately that $g \in \mathcal{G}_1$ if and only if $\phi g \in \mathcal{G}_2 \cup \mathcal{G}_3$. Since $\psi g = \sigma$, where $l(\sigma) = n - 1$ and where the first letter of σ is ϕ, it follows that $\psi g = \sigma \in \mathcal{G}_2$ if and only if $g = \psi^2\sigma \in \mathcal{G}_1$ which holds if and only if $\sigma \in \mathcal{G}_2$, which holds if and only if $\psi^2 g = \psi\sigma \in \mathcal{G}_3$. Thus, for such a g we see that $g \in \mathcal{G}_1$ if and only if $\psi g \in \mathcal{G}_2$ and that $g \in \mathcal{G}_1$ if and only if $\psi^2 g \in \mathcal{G}_3$. Thus, the partition of $\mathcal{G}$ into $\mathcal{G}_1$, $\mathcal{G}_2$, and $\mathcal{G}_3$ as given satisfies all of the desired properties.

We next show that there exists a partition $\{P, S_1, S_2, S_3\}$ of the unit sphere S in $\mathbb{R}^3$ such that P is countable, such that $\phi(S_1) = S_2 \cup S_3$, such that $\psi(S_1) = S_2$, and such that $\psi^2(S_1) = S_3$. Let P denote the set of all points $p \in S$ such that $g(p) = p$ for some $g \in \mathcal{G}$ different from I. Note that P is countable since $\mathcal{G}$ is countable and since each $g \in \mathcal{G}$ with $g \neq I$ is such that $g(x) = x$ only if x is one of the two poles of g's axis of rotation. For each $x \in S \setminus P$, let $G(x) = \{g(x)\colon\ g \in \mathcal{G}\}$. Note that $G(x) \subset S \setminus P$, since if $g(x) \in P$ for some $g \in \mathcal{G}$, then $\sigma(g(x)) = g(x)$ for some $\sigma \neq I$, which implies that $g^{-1}(\sigma(g(x))) = x$, where $g^{-1}\sigma g \neq I$, which in turn implies that $x \in P$. Note also that $x \in G(x)$ since $I \in \mathcal{G}$ and that $x \in G(y)$ if and only if $y \in G(x)$ for all x and y from $S \setminus P$. Finally, note that if x, $y \in S \setminus P$, then $G(x)$ and $G(y)$ are either equal or disjoint. That is, if $t = g(x) = \sigma(y) \in G(x) \cap G(y)$ and if $z = \tau(x) \in G(x)$, then $z = \tau(x) = \tau(g^{-1}(t)) = \tau(g^{-1}(\sigma(y))) \in G(y)$. Thus, we see that $\mathcal{F} = \{G(x)\colon\ x \in S \setminus P\}$ is a partition of $S \setminus P$.

Let C denote the set obtained by selecting exactly one point from each element in the partition $\mathcal{F}$. Note that C is a subset of $S \setminus P$, that $G(c_1)$ and $G(c_2)$ are disjoint if c_1 and c_2 are distinct elements from C, and that if $x \in S \setminus P$, then $x \in G(c)$ for some $c \in C$. Next, for $j = 1$, 2, and 3, let $S_j = G_j(C) = \{g(c)\colon\ g \in \mathcal{G}_j,\ c \in C\}$, where $\mathcal{G}_j$ is as given. Note that S_j

is a subset of $S \setminus P$ for each j and that $S \setminus P = S_1 \cup S_2 \cup S_3$. Further, note that if $x \in S_j \cap S_i$ for $i \neq j$, then $x = g(c_1) = \sigma(c_2)$ for some $c_1 \in C$, $c_2 \in C$, $g \in G_j$, and $\sigma \in G_i$, which implies that $c_1 = c_2 = c$ and hence that $\sigma^{-1}(g(c)) = c$ for $c \notin P$. From this we see that $\sigma^{-1}g = I$ or that $g = \sigma$, which is not possible, since G_j and G_i are disjoint. We thus conclude that S_j and S_i are disjoint when $j \neq i$, which from above implies that $\{P, S_1, S_2, S_3\}$ partitions S. Finally, via the properties of $\mathcal{G}_1$, $\mathcal{G}_2$, and $\mathcal{G}_3$ we see that $\phi(S_1) = \{\phi(g(c))\colon g \in \mathcal{G}_1, c \in C\} = \{\tau(c)\colon \tau \in \mathcal{G}_2 \cup \mathcal{G}_3, c \in C\} = S_2 \cup S_3$, that $\psi(S_1) = \{\psi(g(c))\colon g \in \mathcal{G}_1, c \in C\} = \{\tau(c)\colon \tau \in \mathcal{G}_2, c \in C\} = S_2$, and that $\psi^2(S_1) = \{\psi^2(g(c))\colon g \in \mathcal{G}_1, c \in C\} = \{\tau(c)\colon \tau \in \mathcal{G}_3, c \in C\} = S_3$, as was desired.

We will next show that if P is a countable subset of the unit sphere S, then there exists a countable set Q and a rotation ω such that $P \subset Q \subset S$ and such that $\omega(Q) = Q \setminus P$. Let $v = (v_1, v_2, v_3)$ be any point in S such that $v_3 = 0$ and such that neither v nor $-v$ is in P. Let $u = (1, 0, 0)$ and note that

$$\sigma = \begin{pmatrix} v_1 & v_2 & 0 \\ -v_2 & v_1 & 0 \\ 0 & 0 & 1 \end{pmatrix}$$

is a rotation such that $\sigma(v) = u$. Note also that neither u nor $-u$ is in the set $\sigma(P)$. For a real number t, consider the rotation

$$\tau_t = \begin{pmatrix} 1 & 0 & 0 \\ 0 & \cos(t) & -\sin(t) \\ 0 & \sin(t) & \cos(t) \end{pmatrix}$$

and note that $\tau_t(u) = u$ for any $t \in \mathbb{R}$. Let $x = (x_1, x_2, x_3)$ and $y = (y_1, y_2, y_3)$ be points in $\sigma(P)$ and let n be a positive integer. Notice that if $x_1 = y_1$, then there exist exactly n values of t in $[0, 2\pi)$ such that $\tau_t^n(x) = y$ and that if $x_1 \neq y_1$, then there exist no such values of t. Thus, since there are only countably many choices for x, y, and n, it follows that there are only countably many values of t for which $\sigma(P) \cap \bigcup_{n=1}^{\infty} \tau_t^n(\sigma(P))$ is not empty. Let t_0 be any value of t for which the previous intersection is empty and let τ denote τ_{t_0}. Further, let ω denote $\sigma^{-1}\tau\sigma$ and let $Q = P \cup \bigcup_{n=1}^{\infty} \omega^n(P)$. Since $\tau^n\sigma = \sigma\omega^n$ for each $n \in \mathbb{N}$, it follows that

$$\sigma(P \cap \omega(Q)) = \sigma\left(P \cap \bigcup_{n=1}^{\infty} \omega^n(P)\right) = \emptyset$$

and hence that $P \cap \omega(Q)$ is empty. The desired result now follows since $Q = P \cup \omega(Q)$.

Consider the set Q and the rotation ω that were just defined. We will now show that there exists a partition $\{T_j\colon 1 \le j \le 10\}$ of the unit sphere S and a set $\{\rho_j\colon 1 \le j \le 10\}$ of rotations such that $\{\rho_j(T_j)\colon 1 \le j \le 6\}$ and $\{\rho_j(T_j)\colon 7 \le j \le 10\}$ each partition S. To begin, let $U_1 = \phi(S_2)$, let $U_2 = \psi(\phi(S_2))$, let $U_3 = \psi^2(\phi(S_2))$, let $V_1 = \phi(S_3)$, let $V_2 = \psi(\phi(S_3))$, and let $V_3 = \psi^2(\phi(S_3))$. Recalling the previously proved properties regarding

S_1, S_2, and S_3, we see that $\{U_j, V_j\}$ partitions S_j for $j = 1$, 2, and 3. Thus, it follows that $\{U_1, U_2, U_3, V_1, V_2, V_3, P\}$ partitions S. Next, let $T_7 = U_1$, let $T_8 = U_2$, let $T_9 = U_3$, and let $T_{10} = P$. In addition, let $\rho_7 = \psi^2\phi$, let $\rho_8 = \phi\psi^2$, let $\rho_9 = \psi\phi\psi$, and let $\rho_{10} = I$. Note that $\{\rho_j(T_j)\colon 7 \le j \le 10\}$ partitions S. Next, let $T_1 = \rho_8(S_1 \cap Q)$, let $T_2 = \rho_9(S_2 \cap Q)$, let $T_3 = \rho_7(S_3 \cap Q)$, let $T_4 = \rho_8(S_1 \setminus Q)$, let $T_5 = \rho_9(S_2 \setminus Q)$, and let $T_6 = \rho_7(S_3 \setminus Q)$. Again, using the properties of the S_j's, we see that $\{T_1, T_4\}$ partitions V_1, that $\{T_2, T_5\}$ partitions V_2, and that $\{T_3, T_6\}$ partitions V_3. Thus, it follows that $\{T_j\colon 1 \le j \le 10\}$ partitions S.

Now, let $\rho_4 = \rho_8^{-1}$, let $\rho_5 = \rho_9^{-1}$, let $\rho_6 = \rho_7^{-1}$, and let $\rho_j = \omega^{-1}\rho_{j+3}$ for $j = 1$, 2, and 3. Note that $\rho_{j+3}(T_{j+3}) = S_j \setminus Q$ for $j = 1$, 2, and 3, and that $\bigcup_{n=1}^{3}(S_j \setminus Q) = S \setminus Q$. Further, since $\rho_j(T_j) = \omega^{-1}(\rho_{j+3}(T_j)) = \omega^{-1}(S_j \cap Q)$ for $j = 1$, 2, and 3 and since $\omega^{-1}(Q \setminus P) = Q$, we see that $\{\rho_j(T_j)\colon 1 \le j \le 3\}$ partitions Q. Note that T_7, T_8, and T_9 are rotations of S_1 and that T_1, T_2, T_3, and T_{10} are countable. Finally, let B denote the closed unit ball in $\mathbb{R}^3$ and note that this result remains true if S is replaced by $S' = B \setminus \{0\}$ and T_j is replaced by $T_j' = \{tx\colon x \in T_j \text{ and } 0 < t \le 1\}$ for each positive integer $j \le 10$.

We next show that there exists a partition $\{B_j\colon 1 \le j \le 40\}$ of the closed unit ball B in $\mathbb{R}^3$ and a set $\{r_j\colon 1 \le j \le 40\}$ of rigid motions such that $\{r_j(B_j)\colon 1 \le j \le 24\}$ and $\{r_j(B_j)\colon 25 \le j \le 40\}$ each partition B. Recall from above that if P is a countable subset of the unit sphere S, then there exists a countable set Q and a rotation ω such that $P \subset Q \subset S$ and such that $\omega(Q) = Q \setminus P$. Thus if P is the subset of S given by the singleton set $\{u\}$, where $u = (1, 0, 0)$, then there exists a countable subset Q of S and a rotation ρ_0 such that $u \in Q$ and such that $\rho_0(Q) = Q \setminus \{u\}$. Let $N_1 = \{(q - u)/2\colon q \in Q\}$ and consider the rigid motion r_0 given by $r_0(x) = \rho_0(x + (u/2)) - (u/2)$. Note first that $0 \in N_1$ and that $r_0(N_1) = N_1 \setminus \{0\}$. Let $N_2 = B \setminus N_1$, let $s_1 = r_0$, let $s_2 = I$, let $M_1 = s_1(N_1)$, and let $M_2 = s_2(N_2)$. It follows immediately that $\{N_1, N_2\}$ partitions B and that $\{M_1, M_2\}$ partitions S'. It also follows for each positive integer $j \le 10$ that $\{T_j' \cap \rho_j^{-1}(M_1), T_j' \cap \rho_j^{-1}(M_2)\}$ partitions T_j', that $\{M_1 \cap T_j' \cap \rho_j^{-1}(M_1), M_2 \cap T_j' \cap \rho_j^{-1}(M_1)\}$ partitions $T_j' \cap \rho_j^{-1}(M_1)$, and that $\{M_1 \cap T_j' \cap \rho_j^{-1}(M_2), M_2 \cap T_j' \cap \rho_j^{-1}(M_2)\}$ partitions $T_j' \cap \rho_j^{-1}(M_2)$. Thus, $\{M_k \cap T_j' \cap \rho_j^{-1}(M_i)\colon 1 \le k \le 2, 1 \le i \le 2, 1 \le j \le 10\}$ partitions S' into forty subsets. Further, if $B_{kij} = s_k^{-1}(M_k \cap T_j' \cap \rho_j^{-1}(M_i))$, then $\{B_{kij}\colon 1 \le k \le 2, 1 \le i \le 2, 1 \le j \le 10\}$ partitions B into forty subsets and, for each positive integer $j \le 10$, $\{\rho_j(s_k(B_{kij}))\colon 1 \le k \le 2, 1 \le i \le 2\}$ partitions $\rho_j(T_j')$, where we note that $\rho_j(s_k(B_{kij})) = M_i \cap \rho_j(M_k \cap T_j')$.

Next, recalling the properties of the partition for S' given above, we see that $\{\rho_j(s_k(B_{kij}))\colon 1 \le k \le 2, 1 \le i \le 2, 1 \le j \le 6\}$ and $\{\rho_j(s_k(B_{kij}))\colon 1 \le k \le 2, 1 \le i \le 2, 7 \le j \le 10\}$ each provide a partition of S'. Further, we see that $\{\rho_j(s_k(B_{k1j}))\colon 1 \le k \le 2, 1 \le j \le 6\}$ partitions M_1, that $\{\rho_j(s_k(B_{k2j}))\colon 1 \le k \le 2, 1 \le j \le 6\}$ partitions M_2, that $\{\rho_j(s_k(B_{k1j}))\colon 1 \le k \le 2, 7 \le j \le 10\}$ partitions M_1, and that $\{\rho_j(s_k(B_{k2j}))\colon 1 \le k \le 2, 7 \le j \le 10\}$ partitions M_2. Thus, it follows that $\{s_1^{-1}(\rho_j(s_k(B_{k1j})))\colon$

$1 \le k \le 2, 1 \le j \le 6\}$ partitions N_1, that $\{s_2^{-1}(\rho_j(s_k(B_{k2j}))): 1 \le k \le 2, 1 \le j \le 6\}$ partitions N_2, that $\{s_1^{-1}(\rho_j(s_k(B_{k1j}))): 1 \le k \le 2, 7 \le j \le 10\}$ partitions N_1, and that $\{s_2^{-1}(\rho_j(s_k(B_{k2j}))): 1 \le k \le 2, 7 \le j \le 10\}$ partitions N_2. That is, if $r_{kij} = s_i^{-1}\rho_j s_k$, then $\{r_{kij}(B_{kij}): 1 \le k \le 2, 1 \le i \le 2, 1 \le j \le 6\}$ and $\{r_{kij}(B_{kij}): 1 \le k \le 2, 1 \le i \le 2, 7 \le j \le 10\}$ each provide a partition of B. The desired result follows after we appropriately relabel these forty subsets and forty rigid motions. That is, we obtain a partition $\{B_j: 1 \le j \le 40\}$ of B and a set $\{r_j: 1 \le j \le 40\}$ of rigid motions such that $\{r_j(B_j): 1 \le j \le 24\}$ and $\{r_j(B_j): 25 \le j \le 40\}$ each partition B.

Consider next a closed ball A in $\mathbb{R}^3$. We will now show that if, for some positive integer n, $A_1, A_2, \dots, A_n$ are translates of A, then A is piecewise congruent to their union. Assume without loss of generality that A is a closed ball of radius $\varepsilon > 0$ that is centered at the origin. Fix $a \in \mathbb{R}^3$ such that $\|a\| > 2\varepsilon$ and let $A' = \{y \in \mathbb{R}^3: \|y - a\| \le \varepsilon\}$. Consider again the partition $\{B_j: 1 \le j \le 40\}$ of B and the set $\{r_j: 1 \le j \le 40\}$ of rigid motions that were obtained above. Note that $\{\varepsilon B_k: 1 \le k \le 40\}$ partitions A and consider rigid motions s_k for $1 \le k \le 40$ given by

$$s_k(x) = \begin{cases} \varepsilon r_k\left(\dfrac{x}{\varepsilon}\right), & \text{if } 1 \le k \le 24 \\ \varepsilon r_k\left(\dfrac{x}{\varepsilon}\right) + a, & \text{if } 25 \le k \le 40. \end{cases}$$

Since $\{s_k(\varepsilon B_k): 1 \le k \le 24\}$ partitions A, since $\{s_k(\varepsilon B_k): 25 \le k \le 40\}$ partitions A', and since A and A' are disjoint, it follows that $\{s_k(\varepsilon B_k): 1 \le k \le 40\}$ partitions $A \cup A'$. That is, A is piecewise congruent to $A \cup A'$. Proceeding via induction, note first that the desired result follows immediately if $n = 1$. Let $n > 1$ and assume that A is piecewise congruent to the union of any $n - 1$ of its translates. Let $A_1, \dots, A_n$ be any n translates of A. By assumption it follows that A is piecewise congruent to $\bigcup_{i=1}^{n-1} A_i$. Further, it is clear that $A_n \setminus \bigcup_{i=1}^{n-1} A_i$ is congruent by translation to a subset of A' and hence that $\bigcup_{i=1}^{n} A_i$ is congruent to a subset of $A \cup A'$, which in turn is piecewise congruent to A. Thus, since A is piecewise congruent to a subset of $\bigcup_{i=1}^{n} A_i$, it follows that A and $\bigcup_{i=1}^{n} A_i$ are piecewise congruent.

Finally, we are ready to show that if X and Y are bounded subsets of $\mathbb{R}^3$ with nonempty interiors, then X is piecewise congruent to Y. Let a be an interior point of X and let b be an interior point of Y. Let $A_\varepsilon = \{x \in \mathbb{R}^3: \|x\| \le \varepsilon\}$ and choose $\varepsilon > 0$ so that $A_\varepsilon + a \subset X$ and $A_\varepsilon + b \subset Y$. Note that since X is bounded, there exists a finite collection $\{A_i: 1 \le i \le n\}$ of translates of A such that X is a subset of $\bigcup_{i=1}^{n} A_i$. Note that A is piecewise congruent to a subset of X, which in turn is a subset of $\bigcup_{i=1}^{n} A_i$, which in turn is itself piecewise congruent to A. Thus, it follows that X is piecewise congruent to A and similarly that Y is piecewise congruent to A. Thus, we conclude that X and Y are piecewise congruent. □

Example 6.2. *For any integer $k > 1$, there exist non-Borel measurable convex subsets of $\mathbb{R}^k$.*

Proof: Consider an open ball centered at the origin with unit radius and take the union of this ball with any non-Borel measurable subset of its boundary. The resulting set is convex and not Borel measurable. □

Example 6.3. *For any integer $k > 1$, there exists a bounded real-valued function defined on a convex subset of $\mathbb{R}^k$ that is convex yet is not Borel measurable.*

Proof: Fix an integer $k > 1$ and let U denote the interior of the closed unit ball B in $\mathbb{R}^k$. Further, let T be an arbitrary subset of the boundary V of the set B. Define f: $B \to \mathbb{R}$ such that $f = 0$ on U, $f = 1$ on T, and $f = 2$ on $V \setminus T$. It is clear that f is a convex function on B. Recall, however, that T was an arbitrary subset of V. Hence, T may be chosen to be a subset of V that is not Borel measurable. Such a choice for T results in a convex function f: $B \to \mathbb{R}$ that is not Borel measurable since $f^{-1}(\{1\}) = T$ is not a Borel subset of $\mathbb{R}^k$. □

Example 6.4. *A real-valued function defined on $\mathbb{R}^2$ that is nondecreasing in each variable need not be Borel measurable.*

Proof: We thank Loren Pitt for suggesting this example to us. Consider the line L through the origin of the plane having a slope of -1. Let S be a non-Borel measurable subset of L. Now define g: $\mathbb{R}^2 \to \mathbb{R}$ as follows. If the point (x, y) is to the "lower left" of L, define $g(x, y)$ to be 0. If the point (x, y) is to the "upper right" of L, define $g(x, y)$ to be 3. Finally, for points (x, y) in L, let $g(x, y) = 1$ if $(x, y) \in S$ and let $g(x, y) = 2$ if $(x, y) \in L \setminus S$. Then for x fixed, $g(x, y)$ is nondecreasing in y, for y fixed, $g(x, y)$ is nondecreasing in x, and yet g is not Borel measurable, since the inverse image of the singleton set $\{1\}$ is S, which is not Borel measurable. □

Example 6.5. *There exists a real-valued function defined on the plane that is discontinuous and yet such that both partial derivatives exist and are continuous.*

Proof: It is well known that if f: $\mathbb{R} \to \mathbb{R}$ is differentiable, then f is continuous. What if g: $\mathbb{R}^2 \to \mathbb{R}$ is such that both partial derivatives exist and are continuous? Must g be continuous? Consider the function g defined via

$$g(x, y) = \begin{cases} \dfrac{xy}{x^2 + y^2}, & \text{if } (x, y) \neq (0, 0) \\ 0, & \text{if } (x, y) = (0, 0). \end{cases}$$

Then it trivially follows that each partial derivative of g exists and is continuous. However, note that g is not continuous. Note also that this function g is continuous in each variable but is not continuous. □

Example 6.6. *If f and g are real-valued functions of a real variable, if h: $\mathbb{R}^2 \to \mathbb{R}$ via $h(x, y) = f(x)g(y)$, if $\lim_{x \to 0} h(x, y)$ exists and is uniform in y, if the limit as $y \to 0$ of this limit exists, if $\lim_{y \to 0} h(x, y)$ exists and is* *uniform in x, and if the limit as $x \to 0$ of this limit exists, then it need not be true that*

$$\lim_{(x,\, y)\to(0,\,0)} h(x, y)$$

exists.

Proof: Let $f = g = \mathrm{I}_{\mathbb{R}\setminus\{0\}}$. Then

$$\lim_{x\to 0} h(x,\, y) = \begin{cases} 1, & \text{if } y \neq 0 \\ 0, & \text{if } y = 0, \end{cases}$$

$$\lim_{y\to 0} h(x, y) = \begin{cases} 1, & \text{if } x \neq 0 \\ 0, & \text{if } x = 0, \end{cases}$$

and note that both of these limits are uniform. Further, note that

$$\lim_{y\to 0} \left(\lim_{x\to 0} h(x,\, y) \right)$$

exists and

$$\lim_{x\to 0} \left(\lim_{y\to 0} h(x,\, y) \right)$$

exists. Nevertheless,

$$\lim_{(x,\, y)\to(0,\,0)} h(x,\, y)$$

does not exist. □

Example 6.7. *If A and B are nonempty sets, $\mathcal{A}$ and $\mathcal{B}$ are σ-algebras on A and B, respectively, and f: $A \to B$ is an $(\mathcal{A} \times \mathcal{B})$-measurable subset of $A \times B$, then f need not be a measurable function.*

Proof: Let $A = B = \mathbb{R}$, let $\mathcal{A}$ denote the Borel measurable subsets of $\mathbb{R}$, and let $\mathcal{B}$ denote the Lebesgue measurable subsets of $\mathbb{R}$. Let f: $A \to B$ via $f(x) = x$. Then f is a Borel measurable subset of $\mathbb{R}^2$ and hence is an $(\mathcal{A} \times \mathcal{B})$-measurable subset of $\mathbb{R}^2$. However, f is obviously not a measurable function mapping A into B. This example results in the humorous conclusion that a nonmeasurable function can be measurable. □

Example 6.8. *If A and B are nonempty sets, $\mathcal{A}$ and $\mathcal{B}$ are σ-algebras on A and B, respectively, and f: $A \to B$ is a measurable function, then f need not be an $(\mathcal{A} \times \mathcal{B})$-measurable subset of $A \times B$.*

Proof: Let $A = B = [0, 2]$, and let $\mathcal{A} = \mathcal{B}$ be given by the σ-algebra on $[0, 2]$ generated by the family of subsets $\{[0, 1], (1, 2]\}$. Now, let f: $A \to B$ via $f(x) = x/2$. Then f is a measurable function mapping A into B. However, f is not an $(\mathcal{A} \times \mathcal{B})$-measurable subset of $A \times B$. This example results in the humorous conclusion that a measurable function need not be measurable. □

Example 6.9. *For any positive integer k and any integer $N > 1$, there exist N subsets $S_1, S_2, \ldots, S_N$ of $\mathbb{R}^k$ that partition $\mathbb{R}^k$ and that are such that, for each positive integer $j \leq N$, S_j is a saturated non-Lebesgue measurable set.*

Proof: For $k = 1$, the result follows from Example 1.46. Assume that $k > 1$ and let Λ denote Lebesgue measure on the Borel subsets of $\mathbb{R}^k$. Let $T_1, \ldots, T_N$ be a partition of the real line as provided by Example 1.46. For positive integers $j \leq N$, let $S_j = T_j \times \mathbb{R} \times \cdots \times \mathbb{R} \subset \mathbb{R}^k$. Fix a positive integer $j \leq N$ and assume that there exists an $\mathcal{F}_\sigma$ subset B of $\mathbb{R}^k$ such that $B \subset S_j$ and $\Lambda(B) > 0$. Define a subset $\hat{B}$ of $\mathbb{R}$ via $\hat{B} = \{b_1 \in \mathbb{R}$: $(b_1, b_2, \ldots, b_k) \in B$ for some $(b_2, \ldots, b_k) \in \mathbb{R}^{k-1}\}$. Note that $\hat{B} \in \mathcal{B}(\mathbb{R})$. Further, notice that $\lambda(\hat{B}) > 0$ since $B \subset \hat{B} \times \mathbb{R} \times \cdots \times \mathbb{R} \subset \mathbb{R}^k$ and $\Lambda(B) > 0$. But $\lambda(\hat{B}) = 0$, since $\hat{B} \subset T_j$ and $\lambda_*(T_j) = 0$. This contradiction implies that $\Lambda(B) = 0$ and hence that $\Lambda_*(S_j) = 0$. It follows similarly that $\Lambda_*(S_j^c) = 0$ also. □

Example 6.10. *A subset of $\mathbb{R}^2$ such that each vertical line intersects it in at most one point need not be a Lebesgue null set.*

Proof: This example is from van Douwen [1989]. To begin we note the following simple consequence of Fubini's theorem and Example 1.6: For any compact subset K of $\mathbb{R}^2$ of positive Lebesgue measure, the set $\{x \in \mathbb{R}$: $K_x \neq \emptyset\}$ has cardinality c. Let $\mathcal{K}$ denote the set of all compact subsets of $\mathbb{R}^2$ of positive Lebesgue measure. Note that $\mathcal{K}$ has cardinality c. This follows since there are clearly at least c such subsets and since $\mathbb{R}^2$ is second countable, there are at most c open sets, and consequently at most c closed sets. Consequently, there exists a well ordering $\prec$ on $\mathcal{K}$ such that each element of $\mathcal{K}$ has fewer than c predecessors.

We construct the desired set A by choosing a point $(x_K, y_K) \in K$ for $K \in \mathcal{K}$ as follows: Let $C \in \mathcal{K}$, and assume (x_K, y_K) is chosen for $K \prec C$. From above we note that we can choose $x_C \in \mathbb{R} \setminus \{x_K \in \mathbb{R}$: $K \prec C\}$ such that the vertical line $\{x_C\} \times \mathbb{R}$ intersects C, and next choose $y_C \in \mathbb{R}$ such that $(x_C, y_C) \in C$. Obviously no vertical line contains more than one point of the resulting set $A = \{(x_K, y_K) \in \mathbb{R}^2$: $K \in \mathcal{K}\}$.

Finally, note that if A were measurable, then it would be a null set by Fubini's theorem, and hence there would exist a compact subset of $\mathbb{R}^2$ with positive Lebesgue measure that is disjoint from A. Since A has a nonempty intersection with every compact subset of $\mathbb{R}^2$ having positive Lebesgue measure, it follows that A is not Lebesgue measurable. □

Example 6.11. *Assuming the continuum hypothesis, there exists a countable cover of $\mathbb{R}^2$ into sets of the form $\{(x, f(x))$: $x \in \mathbb{R}$ and f: $\mathbb{R} \to \mathbb{R}\}$ or $\{(f(x), x)$: $x \in \mathbb{R}$ and f: $\mathbb{R} \to \mathbb{R}\}$.*

Proof: This example is from Goffman [1953]. Let $\{x_\alpha$: $\alpha < c\}$ denote a well ordering of the reals. Let S denote the subset of $\mathbb{R}^2$ consisting of all points $(x_\alpha, x_\beta) \in \mathbb{R}^2$ such that $\alpha \leq \beta$. Then, if $\alpha < c$, under CH there are only countably many points in S of the form (y, x_α) for some real number

y and only countably many points in S^c of the form (x_α, z) for some real number z. Thus, for each $y \in \mathbb{R}$, the values of x for which (x, y) is in S may be written as a sequence $\{a_n\}_{n \in \mathbb{N}}$, where we may complete the sequence arbitrarily if there are only finitely many such values of x. Similarly, for each $x \in \mathbb{R}$, the values of y for which (x, y) is in S^c may be written as a sequence $\{b_n\}_{n \in \mathbb{N}}$. For each positive integer n, define functions f_n and g_n mapping $\mathbb{R}$ to $\mathbb{R}$ by setting $f_n(x) = a_n$ and $g_n(y) = b_n$ for each $x \in \mathbb{R}$ and $y \in \mathbb{R}$, as given above. It follows immediately that the sets $\{(x, f_n(x)): x \in \mathbb{R}\}$ and $\{(g_n(y), y): y \in \mathbb{R}\}$ for $n \in \mathbb{N}$ cover the plane. □

Example 6.12. *Assuming the continuum hypothesis, there exists a subset S of $[0, 1]^2$ such that S_x is countable for all $x \in [0, 1]$, such that S^y is the complement of a countable set for all $y \in [0, 1]$, and such that S is non-Lebesgue measurable.*

Proof: Consider the set Ω of all ordinals less than the first uncountable ordinal. Under CH this set is equipotent to $[0, 1]$. Let B denote a bijection from $[0, 1]$ onto Ω. Now, define a subset S of $[0, 1]^2$ as follows. A point (x, y) is in S if and only if $B(y) \leq B(x)$. Now, note that for $x \in [0, 1]$, S_x is countable since the set of all points $y \in [0, 1]$ such that $(x, y) \in S$ are points y such that $B(y) \leq B(x)$ and there are only countably many such points y. Further, note that for any $y \in [0, 1]$, S^y is co-countable since the set of all points $x \in [0, 1]$ such that $(x, y) \in S$ are points x such that $B(y) \leq B(x)$ and this corresponds to a co-countable set.

Now, let f: $[0, 1]^2 \to [0, 1]$ via $f = I_S$. Notice that for each $z \in [0, 1]$, $f(\cdot, z)$ and $f(z, \cdot)$ are Lebesgue measurable since each is almost everywhere constant. However, note that

$$\int_0^1 f(x, y)dx = 1$$

and that

$$\int_0^1 f(x, y)dy = 0.$$

Thus, since the conclusion of Fubini's theorem does not hold, we see that S is not Lebesgue measurable. □

Example 6.13. *A real-valued, bounded, Borel measurable function f defined on $\mathbb{R}^2$ such that for each real number x, $f(\cdot, x)$ and $f(x, \cdot)$ are integrable and each integrates over $\mathbb{R}$ to zero may be such that f is not integrable over $\mathbb{R}^2$ with respect to Lebesgue measure on $\mathcal{B}(\mathbb{R}^2)$.*

Proof: Let $S = \{(x, y) \in \mathbb{R}^2: x - 1 \leq y \leq x + 1\}$. Let f: $\mathbb{R}^2 \to \mathbb{R}$ via $f(x, y) = 0$ if $(x, y) \notin S$, and, for $(x, y) \in S$, via $f(x, y) = 1$ if $x \leq y$ and $f(x, y) = -1$ if $x > y$. Then for each real number x,

$$\int_{\mathbb{R}} f(x, u)\, du = 0 = \int_{\mathbb{R}} f(u, x)\, du,$$

and f is obviously not integrable over $\mathbb{R}^2$ with respect to Lebesgue measure on $\mathcal{B}(\mathbb{R}^2)$. □

Example 6.14. *The product of two complete measure spaces need not be complete.*

Proof: Consider the measure space $(\mathbb{R}, \mathcal{M}(\mathbb{R}), \lambda)$, where λ is Lebesgue measure on $\mathcal{M}(\mathbb{R})$. Recall that $(\mathbb{R}, \mathcal{M}(\mathbb{R}), \lambda)$ is a complete measure space. It follows, however, that the product measure space given by $\mathbb{R}^2$, $\mathcal{M}(\mathbb{R}) \times \mathcal{M}(\mathbb{R})$, and the product measure $\lambda \times \lambda$ is not complete. In particular, let $s \in \mathbb{R}$ and let A be a subset of $\mathbb{R}$ that is not Lebesgue measurable. Then $\{s\} \in A$ is a subset of the $(\lambda \times \lambda)$-null set $\{s\} \times \mathbb{R}$, but $\{s\} \times A$ is not an element of $\mathcal{M}(\mathbb{R}) \times \mathcal{M}(\mathbb{R})$, since fixing s in $\mathbb{R}$ results in a section of $\{s\} \times A$ that is not $\mathcal{M}(\mathbb{R})$-measurable. Hence, the product measure space $(\mathbb{R}^2, \mathcal{M}(\mathbb{R}) \times \mathcal{M}(\mathbb{R}), \lambda \times \lambda)$ is not complete even though $(\mathbb{R}, \mathcal{M}(\mathbb{R}), \lambda)$ is complete. □

Example 6.15. *If T is a locally compact completely normal Hausdorff space, $\mathcal{B}(T) \times \mathcal{B}(T)$ need not equal $\mathcal{B}(T \times T)$.*

Proof: Let $T = \aleph_1$, consider the product space $T \times T$, and note that the subset $S = \{(\alpha, \beta) \in T \times T\colon \alpha \leq \beta\}$ is closed and is thus an element of $\mathcal{B}(T \times T)$. Let ν denote the Dieudonné measure restricted to $\mathcal{B}(T)$ constructed in Example 4.33. Let $\nu \times \nu$ denote the product measure on $\mathcal{B}(T) \times \mathcal{B}(T)$. Assume that S is an element of $\mathcal{B}(T) \times \mathcal{B}(T)$. Then I_S is a $(\mathcal{B}(T) \times \mathcal{B}(T))$-measurable function, I_S is bounded, and $\nu \times \nu$ is a finite measure. Thus I_S is a $(\nu \times \nu)$-integrable function. Then, using Fubini's theorem, we would find that

$$\int_T \int_T \mathrm{I}_S((\alpha, \beta))\, d\nu(\alpha)\, d\nu(\beta) = \int_T \int_T \mathrm{I}_S((\alpha, \beta))\, d\nu(\beta)\, d\nu(\alpha).$$

However,

$$\int_T \int_T \mathrm{I}_S((\alpha, \beta))\, d\nu(\alpha)\, d\nu(\beta) = 1$$

and yet

$$\int_T \int_T \mathrm{I}_S((\alpha, \beta))\, d\nu(\beta)\, d\nu(\alpha) = 0.$$

Thus we see that $\mathcal{B}(T) \times \mathcal{B}(T) \neq \mathcal{B}(T \times T)$. □

Example 6.16. *The product of two Borel measures on locally compact completely normal Hausdorff spaces need not be a Borel measure on the product of the Hausdorff spaces.*

Proof: From Example 6.15, we note that for a locally compact Hausdorff space T, $\mathcal{B}(T) \times \mathcal{B}(T)$ can be a proper subset of $\mathcal{B}(T \times T)$. Further, using the Borel measure ν on $\mathcal{B}(T)$ in Example 6.15, we recall that the domain of $\nu \times \nu$ is $\mathcal{B}(T) \times \mathcal{B}(T)$. However, since $\mathcal{B}(T) \times \mathcal{B}(T)$ is a proper subset of $\mathcal{B}(T \times T)$, we see that the domain of $\nu \times \nu$ is too small to be a measure on $\mathcal{B}(T \times T)$. □

Example 6.17. *If H is a locally compact Hausdorff space, if $S \subset H \times H$, and if for each compact $K \subset H$, the union over x in K of S_x is a Borel subset of H and the union over y in K of S^y is a Borel subset of H, then S need not be an element of $\mathcal{B}(H) \times \mathcal{B}(H)$.*

Proof: Let $H = \aleph_1$ and recall from Example 4.33 that H is a locally compact Hausdorff space. Notice that if a subset A of H is such that for any ordinal $\alpha < \aleph_1$ there exists an ordinal β in A such that $\alpha < \beta < \aleph_1$, then A is not compact. That is, the open cover consisting of the set of all ordinals less than $\aleph_1$ has no finite subcover. Hence, if K is a compact subset of H, then there exists an ordinal $\alpha < \aleph_1$ such that $K \subset \alpha$. Thus compact subsets of H are countable.

Let S be the subset of $H \times H$ given in Example 6.15. Recall that $S \notin \mathcal{B}(H) \times \mathcal{B}(H)$. Nevertheless, note that for any $x \in H$, $S_x = \{(x, y) \in H \times H\colon y \leq x\}$, which is an intersection of three measurable rectangles. Further, for any compact set K in H, $\bigcup_{x \in K} S_x$ is a countable union of sets in $\mathcal{B}(H) \times \mathcal{B}(H)$ and is thus in $\mathcal{B}(H) \times \mathcal{B}(H)$. By analogous reasoning, we see that for any $y \in H$, $\bigcup_{y \in K} S^y$ is in $\mathcal{B}(H) \times \mathcal{B}(H)$. □

Example 6.18. *A measurable set in a product of two σ-finite measure spaces can have positive measure and yet not include any measurable rectangle of positive measure.*

Proof: This example is from Gaudry [1974]. Let λ denote Lebesgue measure on $\mathcal{M}(\mathbb{R})$. Let K be a Cantor set included in $[0, 1]$ such that $\lambda(K) > 0$. Let $S = \{(x, k - x) \in \mathbb{R}^2\colon x \in [0, 1], k \in K\}$. Note that S, being a continuous image of $[0, 1] \times K$, is compact. Thus, S is an $(\mathcal{M}(\mathbb{R}) \times \mathcal{M}(\mathbb{R}))$-measurable subset of $\mathbb{R}^2$. It follows from Fubini's theorem that $(\lambda \times \lambda)(S) = \lambda(K) > 0$. If $A \times B$ is a measurable rectangle with $(\lambda \times \lambda)(A \times B) = \lambda(A)\lambda(B) > 0$ and if $A \times B \subset S$, then it follows from the definition of S that $A \oplus B \subset K$; that is, if the point $(a, b) \in S$, then $a + b \in K$. We now show that this is impossible.

Consider the convolution of I_A and I_B, mapping $\mathbb{R}$ into $\mathbb{R}$ via $(\mathrm{I}_A * \mathrm{I}_B)(x) = \int_{\mathbb{R}} \mathrm{I}_A(x - y)\mathrm{I}_B(y)\, dy$. This function $\mathrm{I}_A * \mathrm{I}_B$ is continuous and vanishes off $A \oplus B$. Further, via Fubini's theorem, we see that $\int_{\mathbb{R}}(\mathrm{I}_A * \mathrm{I}_B)\, d\lambda = \lambda(A)\lambda(B) > 0$. Now, since $\mathrm{I}_A * \mathrm{I}_B$ is continuous, it follows that $A \oplus B$ must include a nonempty open interval. However, since $A \oplus B$ is a subset of a Cantor set, this is impossible. □

Example 6.19. *For two locally compact metric spaces there may exist a closed set in the product space that is not in any product σ-algebra on the product space.*

Proof: In this example, we follow a method suggested in Dudley [1989]. Consider the locally compact metric space $\mathbb{P}(\mathbb{R})$ with the discrete metric. Let $D = \{(x, x) \in \mathbb{P}(\mathbb{R}) \times \mathbb{P}(\mathbb{R})\colon x \in \mathbb{P}(\mathbb{R})\}$ denote the diagonal in $\mathbb{P}(\mathbb{R}) \times \mathbb{P}(\mathbb{R})$. Obviously, D is a closed set in $\mathbb{P}(\mathbb{R}) \times \mathbb{P}(\mathbb{R})$. Let $\mathcal{A}_1$ and $\mathcal{A}_2$ be any two σ-algebras on $\mathbb{P}(\mathbb{R})$, and consider the product σ-algebra $\mathcal{A}_1 \times \mathcal{A}_2$. Recall that an element of a σ-algebra generated by a family of sets must be contained in the σ-algebra generated by a countable subfamily of these sets. Further, the product σ-algebra $\mathcal{A}_1 \times \mathcal{A}_2$ is generated by the family of measurable rectangles $A \times B$, where A and B are sets in $\mathcal{A}_1$ and $\mathcal{A}_2$, respectively. Hence, for any set C in $\mathcal{A}_1 \times \mathcal{A}_2$, there exist sequences $\{A_n\}_{n\in\mathbb{N}}$ taking values in $\mathcal{A}_1$ and $\{B_n\}_{n\in\mathbb{N}}$ taking values in $\mathcal{A}_2$ such that C is in the σ-algebra generated by $\{A_n \times B_n\colon n \in \mathbb{N}\}$. Now, define an equivalence relation on such a set C by saying that x and u are related if and only if for each $n \in \mathbb{N}$, x and u are either both in A_n or are both not in A_n. This equivalence relation then partitions C into equivalence classes, and there are no more than $2^{\aleph_0}$ such equivalence classes. Now, for any points x, y, and u in $\mathbb{P}(\mathbb{R}) \times \mathbb{P}(\mathbb{R})$, if x and u are in the same equivalence class for a subset C of $\mathbb{P}(\mathbb{R}) \times \mathbb{P}(\mathbb{R})$, then $(x, y) \in C$ if and only if $(u, y) \in C$. Assume that D is an element of $\mathcal{A}_1 \times \mathcal{A}_2$. Now, letting $C = D$ and $y = x$, we see that if x and u are in the same equivalence class for D, then $(x, x) \in D$ if and only if $(u, x) \in D$. Hence, each equivalence class is a singleton set. But from this it follows that there are $2^{\mathfrak{c}}$ equivalence classes for D. Since $2^{\aleph_0} < 2^{\mathfrak{c}}$, we see that D is not an element of $\mathcal{A}_1 \times \mathcal{A}_2$. □

Example 6.20. *A continuous real-valued function defined on the product of two locally compact metric spaces need not be measurable with respect to the product of the Borel sets.*

Proof: Let a locally compact metric space be given by $\mathbb{P}(\mathbb{R})$ with the discrete metric. Then any real-valued function defined on $\mathbb{P}(\mathbb{R}) \times \mathbb{P}(\mathbb{R})$ is continuous. Recall from Example 6.19 that the diagonal D in $\mathbb{P}(\mathbb{R}) \times \mathbb{P}(\mathbb{R})$ is not measurable with respect to the product of the Borel sets, $\mathbb{P}(\mathbb{P}(\mathbb{R})) \times \mathbb{P}(\mathbb{P}(\mathbb{R}))$. Now, consider the continuous function $f\colon \mathbb{P}(\mathbb{R}) \times \mathbb{P}(\mathbb{R}) \to \mathbb{R}$ via $f = I_D$. Then f is continuous, but it is not $(\mathbb{P}(\mathbb{P}(\mathbb{R})) \times \mathbb{P}(\mathbb{P}(\mathbb{R})))$-measurable. □

Example 6.21. *There exists a sequence $\{f_n\}_{n\in\mathbb{N}}$ of continuous functions mapping $[0, 1]$ into $[0, 1]^{[0,1]}$ with the product topology that converges pointwise to a function $f\colon [0, 1] \to [0, 1]^{[0,1]}$ such that, for the Borel sets of $[0, 1]^{[0,1]}$, f is not Lebesgue measurable.*

Proof: This example, suggested by Dudley [1971], points out a problem in measure theory on product spaces. For x and y in $[0, 1]$ and $n \in \mathbb{N}$, let $f_n(x)(y) = \max\{0, 1 - n|x - y|\}$. To show that f_n is continuous, note that for any open $U \subset [0, 1]$ and $y \in [0, 1]$, $\{x \in [0, 1]\colon f_n(x)(y) \in U\}$ is open in $[0, 1]$ since for n and y fixed, $\max\{0, 1 - n|x - y|\}$ is a continuous function of x from $[0, 1]$ into $[0, 1]$. For $z \in [0, 1]$ and open $V \subset [0, 1]$, let $S(z, V) = \{g \in [0, 1]^{[0,1]}\colon g(z) \in V\}$. Then the set $\{S(z, V)\colon z \in [0, 1]$ and V is an open subset of $[0, 1]\}$ is a subbase for the product topology on $[0, 1]^{[0,1]}$. Thus, for each $n \in \mathbb{N}$, the inverse of any element in a subbase for the product topology is an open subset of $[0, 1]$. Hence f_n is continuous for each $n \in \mathbb{N}$. Now let $f\colon [0, 1] \to [0, 1]^{[0,1]}$ via

$$f(x)(y) = \begin{cases} 1, & \text{if } x = y \\ 0, & \text{if } x \neq y, \end{cases}$$

and note that $\lim_{n\to\infty} f_n = f$ pointwise. Now, for the coup de grâce, let S be any subset of $[0, 1]$, and let $V = \{g \in [0, 1]^{[0,1]}\colon g(z) > 1/2$ for some $z \in S\}$. Note that V is an open subset of $[0, 1]^{[0, 1]}$ since it is a union of open sets. Thus $f^{-1}(V) = S$, and S can be chosen to be non-Lebesgue measurable. Thus f, a pointwise limit of continuous functions, is not Lebesgue measurable. □

Example 6.22. *There exists a measurable space $(\Omega, \mathcal{F})$ with a finite measure μ and a real-valued bounded function f defined on the product space that is measurable in each coordinate and yet*

$$\int_\Omega \int_\Omega f(x, y)\, d\mu(x)\, d\mu(y) \neq \int_\Omega \int_\Omega f(x, y)\, d\mu(y)\, d\mu(x).$$

Proof: Let Ω denote the set of all ordinals less than the first uncountable ordinal, and let $\mathcal{F}$ be the σ-algebra of countable and co-countable subsets. Let the measure μ be 1 on co-countable sets and 0 on countable sets. Let f be defined via $f(\alpha, \beta) = \mathrm{I}_{\{\alpha \le \beta\}}$. Then note that

$$\int_\Omega \int_\Omega f(x, y)\, d\mu(x)\, d\mu(y) = 1,$$

while

$$\int_\Omega \int_\Omega f(x, y)\, d\mu(y)\, d\mu(x) = 0.$$

□

Example 6.23. *There exist two measures μ and ν on $\mathcal{B}([0, 1])$ such that if D is the diagonal in $[0, 1]^2$, then*

$$\int_{[0, 1]} \left[\int_{[0, 1]} \mathrm{I}_D\, d\mu \right] d\nu = 0$$

and

$$\int_{[0, 1]} \left[\int_{[0, 1]} \mathrm{I}_D\, d\nu \right] d\mu = 1.$$

Proof: Let μ be Lebesgue measure on $\mathcal{B}([0, 1])$ and let ν be counting measure on $\mathcal{B}([0, 1])$. Then the advertised result is immediate. □

Example 6.24. *If for each $x \in \mathbb{R}$, I_x denotes a copy of $[0, 1]$ and U_x denotes a copy of $(0, 1]$ included therein, then $\prod_{x \in \mathbb{R}} U_x$ is not a Borel set in the compact Hausdorff space $\prod_{x \in \mathbb{R}} \mathrm{I}_x$.*

Proof: This example is from Sidney and Djokovic [1977]. For each subset S of $\mathbb{R}$, let $I(S) = \prod_{x \in S} \mathrm{I}_x$ and let $A(S)$ be the set of all subsets of $I(\mathbb{R})$ of the form $P \times I(\mathbb{R} \setminus S)$, where $P \subset I(S)$ and we identify canonically $I(S) \times I(\mathbb{R} \setminus S)$ with $I(\mathbb{R})$. Let $\mathcal{A}$ be the union of $A(S)$ over all countable subsets S of $\mathbb{R}$. Observe that $\{I(\mathbb{R}), \emptyset\} \subset A(S)$ for any countable subset S of $\mathbb{R}$. Further, observe that $\mathcal{A}$ is closed under countable unions. Thus $\mathcal{A}$ is a σ-algebra on $I(\mathbb{R})$. Next, note that $\mathcal{A}$ includes the product topology and thus is obviously a σ-superalgebra of $\mathcal{B}(I(\mathbb{R}))$. Finally, note that $\prod_{x \in \mathbb{R}} U_x$ is not an element of $\mathcal{A}$. Thus, $\prod_{x \in \mathbb{R}} U_x$ is not an element of $\mathcal{B}(I(\mathbb{R}))$. □

References

Dudley, R. M., "On measurability over product spaces," *Bulletin of the American Mathematical Society,* Vol. 77, pp. 271–274, March 1971.

Dudley, R. M., *Real Analysis and Probability,* Wadsworth & Brooks/Cole: Pacific Grove, California, 1989.

Gaudry, G. I., "Sets of positive product measure in which every rectangle is null," *American Mathematical Monthly,* Vol. 81, pp. 889–890, October 1974.

Goffman, C., *Real Functions,* Prindle, Weber, & Schmidt, Boston, 1953.

Sidney, S. J., and D. Z. Djokovic, "Problem 6023," *American Mathematical Monthly,* Vol. 84, p. 67, January 1977.

Stromberg, K., "The Banach–Tarski paradox," *American Mathematical Monthly,* Vol. 86, March 1979, pp. 151–161.

van Douwen, E. K., "Fubini's theorem for null sets," *American Mathematical Monthly,* Vol. 96, October 1989, pp. 718–721.

7. Basic Probability

A probability space is a measure space $(\Omega, \mathcal{F}, P)$ such that $P(\Omega) = 1$. A complete probability space is a probability space such that each subset of a P-null set is measurable. We never assume that a probability space is complete unless we so state. In the context of probability, measurable sets are called events. A random variable is a measurable real-valued function defined on a probability space. A random process is an indexed family of random variables, each of which is defined on the same probability space. In this chapter, for a random process $\{X(t, \omega): t \in \mathbb{R}\}$ defined on a probability space $(\Omega, \mathcal{F}, P)$, we will let $X(t)$ denote $X(t, \cdot)$.

Example 7.1. *A nonnegative Riemann integrable function or a nonnegative Lebesgue integrable function that integrates to one need not be a probability density function.*

Proof: Let $(\Omega, \mathcal{F}, P)$ be a probability space and let X be a random variable defined on $(\Omega, \mathcal{F}, P)$. For each Borel subset B of $\mathbb{R}$, let $\mu(B) = P(X \in B)$. Note that μ is a measure mapping the real Borel sets into $[0, 1]$. This measure μ is known as the distribution of the random variable X. Depending upon the probability space $(\Omega, \mathcal{F}, P)$ and the random variable X, it might be possible to define $P(X \in A)$ for sets A that are not real Borel sets. However, for any probability space and any random variable defined thereon, the probability that the random variable belongs to a Borel set is always well defined. Let m denote Lebesgue measure on the Lebesgue measurable subsets of $\mathbb{R}$, and let λ denote Lebesgue measure restricted to the real Borel sets. Then, if $\mu \ll \lambda$, a Radon–Nikodym derivative $d\mu/d\lambda$ exists, up to almost everywhere equivalence with respect to λ, and any nonnegative version of this Radon–Nikodym derivative is known as a probability density function of the random variable X. Assume that $\mu \ll \lambda$, and let f be a probability density function obtained by taking such a version of $d\mu/d\lambda$. Then, f is Borel measurable. Further, $f(X)$ is a random variable. In particular, the concept of probability density functions arises from the concept of distributions of random variables and from the Radon–Nikodym theorem.

Alternatively, let g: $\mathbb{R} \to [0, \infty)$ be Borel measurable and be such that $\int_{\mathbb{R}} g\, d\lambda = 1$. For a Borel set B, let $Q(B) = \int_B g\, d\lambda$. Then for the probability space consisting of the reals, the Borel subsets of the reals, and the measure Q, a random variable given by the identity map has this function g as a probability density function. Hence, any nonnegative real-valued Borel measurable function defined on the real line whose Lebesgue integral over the reals is one is a probability density function.

Now, if a function that maps a bounded interval of $\mathbb{R}$ into $[0, \infty)$ is Riemann integrable and has a Riemann integral of one, must such a function be a probability density function? Notwithstanding many books and papers to the contrary, the answer is no. Recall that a bounded real-valued function defined on a bounded interval of real numbers is Riemann integrable if and only if its set of discontinuity points has Lebesgue measure zero. Let C denote the Cantor ternary set. Recall that C has Lebesgue measure zero and cardinality c. Also, recall that the family of real Borel sets has cardinality c. Since Lebesgue measure on the Lebesgue measurable subsets of $\mathbb{R}$ is complete, any subset of C is Lebesgue measurable, and there are 2^c such subsets. Let Z be a subset of $C \cap [1/2, 1]$ that is not Borel measurable. Hence, since C is closed, the function h: $\mathbb{R} \to [0, \infty)$ via

$$h(x) = \begin{cases} 2, & \text{if } x \in [1/2, 1] \setminus Z \\ 0, & \text{otherwise} \end{cases}$$

is such that its set of discontinuity points is a subset of C, which has Lebesgue measure zero. Further, since $h^{-1}(\{2\}) = [1/2, 1] \setminus Z$, h is not Borel measurable. However, h is Riemann integrable, and its Riemann integral is one. Now, for each real Borel set B, let $\nu(B) = \int_B 2 \mathrm{I}_{[1/2, 1]}\, dm$. Consider the probability space consisting of the reals, the real Borel sets, and the measure ν. Further, let Y denote the identity map on $\mathbb{R}$. Then Y is a random variable uniformly distributed over $[1/2, 1]$, yet h is not a probability density function for Y. Note, for instance, that $-\int_{-\infty}^{\infty} h(x) \ln[h(x)]\, dx$ (an integral that frequently arises in the context of differential entropy) exists as a Riemann integral, but $\mathrm{E}[-\ln[h(Y)]]$ does not exist. Furthermore, if f: $\mathbb{R} \to [0, \infty)$ is not Borel measurable but is Lebesgue integrable with a Lebesgue integral of unity, and if X is a random variable defined on some probability space such that f is viewed as a "probability density function" of X, then there could exist sets $A \subset \mathbb{R}$ such that $\int_A f\, dm$ exists, but such that $P(X \in A)$ does not exist. □

Example 7.2. *The set of all probability density functions defined up to almost everywhere equivalence is a nowhere dense subset of* $\mathbf{L}_1(\mathbb{R}, \mathcal{B}(\mathbb{R}), \lambda|_{\mathcal{B}(\mathbb{R})})$.

Proof: The result of this example follows immediately from the development of Example 5.15. □

Example 7.3. *Two absolutely continuous probability distribution functions F and G may have disjoint supports and yet be such that* $\sup_{x \in \mathbb{R}} |F(x) - G(x)|$ *is less than any preassigned positive real number.*

Proof: This example was suggested in Devroye [1982]. Let $n > 3$ be an integer and let F be the probability distribution function corresponding to the uniform distribution on the set $\bigcup_{m=1}^{n} [2m - (3/2), 2m - 1]$, and let G be the probability distribution function corresponding to the uniform distribution on the set $\bigcup_{m=1}^{n} [2m - (1/2), 2m]$. Note that the support of F is disjoint from the support of G. Further, note that $\sup_{x \in \mathbb{R}} |F(x) - G(x)| = 2/n$, which, by choice of n, may be made arbitrarily small. □

Example 7.4. *There exists a probability distribution μ on $\mathcal{B}(\mathbb{R})$ such that $\mu(\mathcal{B}(\mathbb{R}))$ is the Cantor ternary set.*

Proof: Let $\{r_n: n \in \mathbb{N}\}$ be an enumeration of the rationals. Considering the probability distribution μ on $\mathcal{B}(\mathbb{R})$ given by

$$\mu(B) = \sum_{\{n:\ r_n \in B \cap \mathbb{Q}\}} \frac{2}{3^{-n}},$$

the result follows. □

Example 7.5. *For a probability distribution μ on $\mathcal{B}(\mathbb{R})$, the family of μ-continuity sets need not be a σ-algebra.*

Proof: Let μ denote standard Gaussian measure on $\mathcal{B}(\mathbb{R})$, and note that each rational is a μ-continuity set but that $\mathbb{Q}$ is not a μ-continuity set. □

Example 7.6. *A probability distribution on the Borel subsets of a separable metric space need not be tight.*

Proof: The example we present is now clearly widespread and belongs to the category of folklore; however, this general technique was suggested by Lucien Le Cam. Let S be a saturated non-Lebesgue measurable subset of $\mathbb{R}$, and consider S as a metric subspace of $\mathbb{R}$. Then the metric space S is separable. Let G denote standard Gaussian measure on $\mathcal{B}(\mathbb{R})$. As in Example 4.30, define a probability measure μ on $\mathcal{B}(S)$ via $\mu(B) = G(A)$, where $A \in \mathcal{B}(\mathbb{R})$ is such that $B = S \cap A$. Note that a subset of S is compact if and only if it is finite. Further, any finite subset of S has μ-measure zero. Thus, the probability distribution μ is not tight. □

Example 7.7. *Given any positive integer k, any partition $\{S_n\colon 1 \le n \le N\}$ of $\mathbb{R}^k$ into saturated non-Lebesgue measurable subsets, any probability measure P on $(\mathbb{R}^k, \mathcal{B}(\mathbb{R}^k))$ that is equivalent to Lebesgue measure, there exists a probability space $(\mathbb{R}^k, \mathcal{F}, \mu)$ such that $\mathcal{F}$ includes $\mathcal{B}(\mathbb{R}^k)$ and $\sigma(\{S_n\colon 1 \le n \le N\})$, such that μ agrees with P on $\mathcal{B}(\mathbb{R}^k)$, and such that $\mathcal{B}(\mathbb{R}^k)$ is independent of $\sigma(\{S_n\colon 1 \le n \le N\})$.*

Proof: This example is from Hall and Wise [1993]. To begin, recall from Example 6.9 that such a partition $\mathbb{R}^k$ into saturated non-Lebesgue measurable subsets exists for any integer $N > 1$. Next, consider the subset $\mathcal{F}$ of $\mathbb{P}(\mathbb{R}^k)$ given by $\mathcal{F} = \{(S_1 \cap A_1) \cup \cdots \cup (S_N \cap A_N)\colon A_i \in \mathcal{B}(\mathbb{R}^k)$ for $1 \le i \le N\}$. We will show that $\mathcal{F}$ is a σ-algebra on $\mathbb{R}^k$. Choosing $A_1 = \cdots = A_N = \emptyset$ implies that $\emptyset \in \mathcal{F}$. Let A be an element of $\mathcal{F}$. Then $A = (S_1 \cap A_1) \cup \cdots \cup (S_N \cap A_N)$ for some choice of the A_i's from $\mathcal{B}(\mathbb{R}^k)$. Further, $A^c = (S_1 \cap A_1)^c \cap \cdots \cap (S_N \cap A_N)^c$. Since

$$S_i^c = \bigcup_{\substack{j=1 \\ i \neq j}}^{N} S_j,$$

it follows that

$$A^c = \bigcap_{i=1}^{N} \bigcup_{\substack{j=1 \\ i \neq j}}^{N} S_j \cup A_i^c.$$

Hence, A^c is a finite union of sets, each of which is of one of the following three forms:

(1) $S_{n_1} \cap \cdots \cap S_{n_k} \cap B$ where $1 \le n_1 < \cdots < n_k \le N$, $k > 1$, and $B \in \mathcal{B}(\mathbb{R}^k)$;

(2) $S_j \cap B$ for $1 \le j \le N$ and $B \in \mathcal{B}(\mathbb{R}^k)$;

(3) $B \in \mathcal{B}(\mathbb{R}^k)$.

Every set of the form given by (1) is empty since the S_i's are disjoint. Further, any set $B \in \mathcal{B}(\mathbb{R}^k)$ may be expressed as $B = (S_1 \cap B) \cup \cdots \cup (S_N \cap B)$. Hence, A^c is an element of $\mathcal{F}$. Finally, if $B_1, B_2, \ldots$ are in $\mathcal{F}$, then for some choice of the $A_{i,j}$'s from $\mathcal{B}(\mathbb{R}^k)$,

$$\bigcup_{i=1}^{\infty} B_i = \bigcup_{i=1}^{\infty} \bigcap_{j=1}^{N} (S_j \cap A_{i,j}) = \bigcap_{j=1}^{N} S_j \cap \left(\bigcup_{i=1}^{\infty} A_{i,j} \right) \in \mathcal{F}.$$

Thus, $\mathcal{F}$ is a σ-algebra on $\mathbb{R}^k$. Further, note that $\mathcal{F}$ includes $\mathcal{B}(\mathbb{R}^k)$ since $(S_1 \cap A) \cup \cdots \cup (S_N \cap A) = A$ for any $A \in \mathcal{B}(\mathbb{R}^k)$. Further, since $(S_1 \cap \emptyset) \cup \cdots \cup (S_j \cap \mathbb{R}^k) \cup \cdots \cup (S_N \cap \emptyset) = S_j$, it follows that $S_j \in \mathcal{F}$ for any positive integer $j \leq N$.

Let $A_1, \ldots, A_N$ and $B_1, \ldots, B_N$ be sets from $\mathcal{B}(\mathbb{R}^k)$ such that $(S_1 \cap A_1) \cup \cdots \cup (S_N \cap A_N) = (S_1 \cap B_1) \cup \cdots \cup (S_N \cap B_N)$. We will show that $P(A_i \triangle B_i) = 0$ for any positive integer $i \leq N$. Fix a positive integer $i \leq N$. By assumption, $(S_1 \cap A_1) \cup \cdots \cup (S_N \cap A_N) = (S_1 \cap B_1) \cup \cdots \cup (S_N \cap B_N)$. Intersecting each side with S_i implies that $(S_i \cap A_i) = (S_i \cap B_i)$, which implies that $(S_i \cap A_i) \cap (S_i \cap B_i)^c = (S_i \cap A_i) \cap (S_i^c \cup B_i^c) = (S_i \cap A_i \cap S_i^c) \cup (S_i \cap A_i \cap B_i^c) = (S_i \cap A_i \cap B_i^c) = \emptyset$ and similarly that $(S_i \cap B_i \cap A_i^c) = \emptyset$. Thus, we see that $(S_i \cap A_i \cap B_i^c) \cup (S_i \cap B_i \cap A_i^c) = S_i \cap (A_i \triangle B_i) = \emptyset$. Since $(A_i \triangle B_i) \in \mathcal{B}(\mathbb{R}^k)$, it follows that $P(A_i \triangle B_i) = P^*(S_i \cap (A_i \triangle B_i)) = P^*(\emptyset) = 0$.

Next, define a measure μ on the measurable space $(\mathbb{R}^k, \mathcal{F})$ via

$$\mu((S_1 \cap A_1) \cup \cdots \cup (S_N \cap A_N)) = \frac{1}{N}(P(A_1) + \cdots + P(A_N))$$

for $(S_1 \cap A_1) \cup \cdots \cup (S_N \cap A_N) \in \mathcal{F}$. That μ is well defined follows from the preceding paragraph and that μ is in fact a probability measure that agrees with P on $\mathcal{B}(\mathbb{R}^k)$ is then straightforward. Further, notice that $\mu(S_i) = 1/N$ for each positive integer $i \leq N$ and that, for any set $B \in \mathcal{B}(\mathbb{R}^k)$ and any positive integer $i \leq N$, $\mu(S_i \cap B) = P(B)/N = \mu(S_i)\mu(B)$. Thus, S_i is independent of $\mathcal{B}(\mathbb{R}^k)$ for each positive integer $i \leq N$. Finally, notice that $\mathcal{B}(\mathbb{R}^k)$ is in fact independent of $\sigma(S_1, \ldots, S_N)$ since $\{\emptyset, S_1, \ldots, S_N\}$ is a π-system. □

Example 7.8. *There exist two uncorrelated Gaussian random variables such that they are not independent and their sum is not Gaussian.*

Proof: Let X be a zero mean, unit variance Gaussian random variable and define

$$Y = \begin{cases} X, & \text{if } |X| \leq \alpha \\ -X, & \text{if } |X| > \alpha, \end{cases}$$

where α is some positive real number. Notice that Y is also a zero mean, unit variance Gaussian random variable since, for $B \in \mathcal{B}(\mathbb{R})$, $P(Y \in B) = P(Y \in B \mid |X| \leq \alpha)\ P(|X| \leq \alpha) + P(Y \in B \mid |X| > \alpha)\ P(|X| > \alpha) = P(X \in B \mid |X| \leq \alpha)\ P(|X| \leq \alpha) + P(-X \in B \mid |X| > \alpha)\ P(|X| > \alpha) =$

$P(X \in B \mid |X| \leq \alpha)\ P(|X| \leq \alpha) + P(X \in B \mid |X| > \alpha)\ P(|X| > \alpha) = P(X \in B)$. Let f be a continuous zero mean, unit variance Gaussian probability density function and let F be the associated probability distribution function. Then the correlation coefficient ρ of X and Y is given by

$$\rho = \mathrm{E}[XY] = 2\int_0^\alpha x^2 f(x)\,dx - 2\int_\alpha^\infty x^2 f(x)\,dx,$$

which implies that

$$\rho = 4\int_0^\alpha x^2 f(x)\,dx - 1$$

since $\mathrm{E}[X^2] = 1$. Thus $\rho = 0$ when α is such that

$$g(\alpha) = \int_0^\alpha x^2 f(x)\,dx = \frac{1}{4}.$$

Since $g(0) = 0$, $g(y) \to 1/2$ as $y \to \infty$, and g is continuous, it is clear that such an α exists. Thus, for such a value of α, X and Y are uncorrelated Gaussian random variables even though they are not independent since $P(Y \in A,\ X \in B) = 0$ if A and B are disjoint Borel measurable subsets of $[0, \infty)$. Further, note that

$$X + Y = \begin{cases} 2X, & \text{if } |X| \leq \alpha \\ 0, & \text{if } |X| > \alpha, \end{cases}$$

which implies that $X + Y$ is not Gaussian since it has positive variance and its distribution function is discontinuous at the origin. □

Example 7.9. *A Lebesgue measurable function of a random variable need not be a random variable.*

Proof: Let A be a subset of $[0, 1]$ that is Lebesgue measurable but not Borel measurable and define $f(x) = \mathrm{I}_A(x)$. Define a random variable X on the probability space given by $[0, 1]$, the real Borel subsets of $[0, 1]$, and Lebesgue measure on the Borel subsets of $[0, 1]$ via $X(\omega) = \omega$. It then follows that $f(X(\omega)) = \mathrm{I}_A(X(\omega)) = \mathrm{I}_A(\omega)$, which implies that $f(X)$ is not a random variable since $\{\omega\colon f(X(\omega)) = 1\} = A$ is not a Borel subset of $[0, 1]$. □

Example 7.10. *A random variable can have a continuous probability density function and yet not have an integrable characteristic function.*

Proof: This example is from Billingsley [1986]. Recall that if C is a characteristic function corresponding to a probability density function f and if $\{p_k\}_{k \in \mathbb{N}}$ is a sequence of nonnegative real numbers summing to unity, then $\phi(t) = \sum_{k=1}^{\infty} p_k C(t/k)$ defines a characteristic function ϕ corresponding to the probability density function g defined by $g(x) = \sum_{k=1}^{\infty} p_k k f(kx)$. Now consider the characteristic function C given by $C(t) = (1 - |t|)\mathrm{I}_{(-1,\,1)}(t)$ and the corresponding probability density function f given by

$$f(x) = \frac{1 - \cos(x)}{\pi x^2}.$$

For each positive integer k, let $s_k = 1/[k(k+1)]$ and let $p_k = k(s_k - s_{k+1})$. Note that g is a uniformly converging sum of continuous probability density functions and is thus a continuous probability density function. Note that ϕ is a piecewise affine function. In particular, on $[0, \infty)$, ϕ is decreasing and, for positive integers n,

$$\phi(n) = 1 - \sum_{k=1}^{n} s_k = \frac{1}{n+1}.$$

Thus we see that the characteristic function ϕ is not integrable. □

Example 7.11. *There do not exist two strict sense, independent, identically distributed positive random variables X and Y defined on a common probability space such that* $\mathrm{E}[X/Y] \leq 1$.

Proof: Note that $\mathrm{E}[X/Y] = \mathrm{E}[X]\mathrm{E}[1/X]$, and, by Jensen's inequality, this is greater than one since the random variables are strict sense random variables. □

Example 7.12. *There exists a nonempty convex subset C of $\mathbb{R}^2$ and a convex real-valued function f defined on C and a random variable X taking values in C such that $f(X)$ is not a real-valued random variable.*

Proof: Let C be the convex subset of $\mathbb{R}^2$ given in Example 6.2. Consider the probability space given by the boundary of C, the Borel subsets, and normalized arc length measure. Let the random variable X be the identity map on the boundary of C. Let f be the real-valued convex function given in Example 6.3. Then, the subset of the boundary of C where $f(X) = 2$ is not a Borel subset of the boundary of C and hence $f(X)$ is not a real-valued random variable. Note, in contrast to the above, that for a nonempty convex subset of the reals, a convex real-valued function g defined thereon, and a random variable Y taking values in this convex set, $g(Y)$ is always a random variable. □

Example 7.13. *There exist two independent strict sense random variables X and Y and a function $f\colon \mathbb{R} \to \mathbb{R}$ such that $f(X) = Y$.*

Proof: Let S_1, S_2 be a partition of $\mathbb{R}$ as provided by Example 1.46, and for these two subsets of $\mathbb{R}$ let $(\Omega, \mathcal{F}, \mu)$ be a probability space as provided by Example 7.7, where $k = 1$ and P is standard Gaussian measure on $(\mathbb{R}, \mathcal{B}(\mathbb{R}))$. Let $S = S_1$ and note that $S^c = S_2$. Define random variables X and Y on this space via $X(\omega) = \omega$ and $Y(\omega) = \mathrm{I}_S(\omega)$ for $\omega \in \Omega$. Since $\sigma(X) = \mathcal{B}(\mathbb{R})$ and $\sigma(Y) = \sigma(S)$, it follows from Example 7.7 that X and Y are independent random variables. However, note that $Y = \mathrm{I}_S(X)$. □

Example 7.14. *There exist two uncorrelated positive variance Gaussian random variables X and Y and a function $g\colon \mathbb{R} \to \mathbb{R}$ such that $X = g(Y)$.*

Proof: Let S_1, S_2 be a partition of $\mathbb{R}$ as provided by Example 1.46, and for these two subsets of $\mathbb{R}$ let $(\Omega, \mathcal{F}, \mu)$ be a probability space as provided by Example 7.7, where $k = 1$ and P is standard Gaussian measure on

$(\mathbb{R}, \mathcal{B}(\mathbb{R}))$. Let $S = S_1$ and note that $S^c = S_2$. Define random variables X and Y on this space via $X(\omega) = \omega$ and $Y(\omega) = \omega[\mathrm{I}_S(\omega) - \mathrm{I}_{S^c}(\omega)]$. Note that X is a standard Gaussian random variable. Further, for $B \in \mathcal{B}(\mathbb{R})$, note that $\mu(Y \in B) = \mu([S \cap B] \cup [S^c \cap -B]) = [P(B) + P(-B)]/2 = P(B)$ since standard Gaussian measure is symmetric. Thus, Y is also a standard Gaussian random variable. Finally, note that, even though $Y = X(\mathrm{I}_S(X) - \mathrm{I}_{S^c}(X))$, it follows that X and Y are uncorrelated since $\mathrm{E}[XY] = \mathrm{E}[X^2[\mathrm{I}_S(X) - \mathrm{I}_{S^c}(X)]] = \mathrm{E}[X^2](\mathrm{E}[\mathrm{I}_S(X)] - \mathrm{E}[\mathrm{I}_{S^c}(X)]) = \mathrm{E}[X^2][\mu(S) - \mu(S^c)] = 0$. □

Example 7.15. *Given any integer $n > 2$, there exist n standard Gaussian random variables $\{X_1, \ldots, X_n\}$ that are not mutually independent and not mutually Gaussian yet such that the random variables in any proper subset of $\{X_1, \ldots, X_n\}$ containing at least two elements are mutually independent and mutually Gaussian.*

Proof: This example is from Pierce and Dykstra [1969]. Let $X_1, \ldots, X_n$ be random variables possessing a joint probability density function given by

$$f(x_1, \ldots, x_n) = \frac{1}{(2\pi)^{n/2}} e^{-\frac{1}{2}\sum_{i=1}^n x_i^2} \left[1 + \prod_{i=1}^n \left(x_i e^{-\frac{1}{2}x_i^2}\right)\right].$$

Note that for any positive integer $j \leq n$ it follows that

$$\begin{aligned}
&\int_{\mathbb{R}} f(x_1, \ldots, x_n)\, dx_j \\
&= \int_{\mathbb{R}} \frac{1}{(2\pi)^{n/2}} e^{-\frac{1}{2}\sum_{i=1}^n x_i^2}\, dx_j \\
&\quad + \int_{\mathbb{R}} \frac{1}{(2\pi)^{n/2}} e^{-\frac{1}{2}\sum_{i=1}^n x_i^2} \left[\prod_{i=1}^n \left(x_i e^{-\frac{1}{2}x_i^2}\right)\right] dx_j \\
&= \frac{1}{(2\pi)^{(n-1)/2}} e^{-\frac{1}{2}\sum_{\substack{i=1 \\ i\neq j}}^n x_i^2} \\
&\quad + \frac{1}{(2\pi)^{n/2}} e^{-\frac{1}{2}\sum_{\substack{i=1 \\ i\neq j}}^n x_i^2} \left[\prod_{\substack{i=1 \\ i\neq j}}^n x_i e^{-\frac{1}{2}x_i^2}\right] \int_{\mathbb{R}} x_j e^{-x_j^2}\, dx_j \\
&= \frac{1}{(2\pi)^{(n-1)/2}} e^{-\frac{1}{2}\sum_{\substack{i=1 \\ i\neq j}}^n x_i^2}.
\end{aligned}$$

Thus, since any subset of $\{X_1, \ldots, X_n\}$ containing $n-1$ random variables is composed of mutually independent standard Gaussian random variables, it follows that any proper subset of $\{X_1, \ldots, X_n\}$ containing at least two random variables is also composed of mutually independent standard Gaussian random variables. However, it is clear that the random variables in $\{X_1, \ldots, X_n\}$ are neither mutually independent nor mutually Gaussian. □

Example 7.16. *If X and Y are two random variables defined on a probability space $(\Omega, \mathcal{F}, P)$, if $C: \mathbb{R}^2 \to \mathbb{R}$ is continuous at all but one point, and if $C(X, y)$ is independent of Y and has a fixed distribution D for each real number y, then it need not follow that $C(X, Y)$ is independent of Y or that $C(X, Y)$ has the distribution D.*

Proof: This example is from Perlman and Wichura [1975]. Let the probability space be given by [0, 1], the Lebesgue measurable subsets, and Lebesgue measure. Let $X(\omega) = Y(\omega) = \omega$. Let C be defined via $C(x, y) = y\mathrm{I}_{\{y\}}(x)$. Then $C(X, y) = 0$ a.s. for any real number y, and is thus independent of Y. However, $C(X, Y) = Y$ pointwise on Ω. Thus $C(X, Y)$ is uniformly distributed on [0, 1] and is not independent of Y. □

Example 7.17. *There exist simple random variables X and Y defined on a common probability space that are not independent and yet the moment generating function of $X + Y$ is the product of the moment generating function of X and the moment generating function of Y.*

Proof: This example was suggested by Flury [1986]. Let X and Y be random variables defined on some probability space such that $P(X = 1$ and $Y = 1) = 1/9$, $P(X = 1$ and $Y = 2) = 1/18$, $P(X = 1$ and $Y = 3) = 1/6$, $P(X = 2$ and $Y = 1) = 1/6$, $P(X = 2$ and $Y = 2) = 1/9$, $P(X = 2$ and $Y = 3) = 1/18$, $P(X = 3$ and $Y = 1) = 1/18$, $P(X = 3$ and $Y = 2) = 1/6$, $P(X = 3$ and $Y = 3) = 1/9$. Then $P(X = 1) = 1/3$, $P(X = 2) = 1/3$, $P(X = 3) = 1/3$, $P(Y = 1) = 1/3$, $P(Y = 2) = 1/3$, and $P(Y = 3) = 1/3$. Thus the common moment generating function M of X and Y is given by $M(t) = (e^t + e^{2t} + e^{3t})/3$. Also, the moment generating function of $X+Y$ is given by $(e^{2t} + 2e^{3t} + 3e^{4t} + 2e^{5t} + e^{6t})/9$, which happens to equal $[M(t)]^2$. □

Example 7.18. *There exists a standard Gaussian random variable X defined on a probability space $(\Omega, \mathcal{F}, P)$ such that $X(\Omega)$ is a saturated non-Lebesgue measurable set.*

Proof: Let S be a saturated non-Lebesgue measurable subset of $\mathbb{R}$. With respect to the relative topology, the Borel subsets of S are those subsets B of S that can be written as $B = A \cap S$, where $A \in \mathcal{B}(\mathbb{R})$. Let G denote standard Gaussian measure on $\mathcal{B}(\mathbb{R})$, and for a Borel subset B of S, define $P(B) = G(A)$, where A is a real Borel subset of $\mathbb{R}$ such that $B = A \cap S$. Recall (see Example 4.30) that the outer G measure of $S \cap A$ equals $G(A)$ for any $A \in \mathcal{B}(\mathbb{R})$. Thus, $P(B) = G^*(B)$ for $B \in \mathcal{B}(S)$. Recall from Example 4.30 that P is a probability measure on $(S, \mathcal{B}(S))$. Now define a random variable X on the probability space $(S, \mathcal{B}(S), P)$ via $X(\omega) = \omega$. Observe that for a real Borel set A, $X^{-1}(A) = S \cap A \in \mathcal{B}(S)$. Further, for a real Borel set A, $P(X^{-1}(A)) = G(A)$. Thus, X is a standard Gaussian random variable, and $X(S) = S$. □

Example 7.19. *For any integer $N > 1$, there exists a probability space and N standard Gaussian random variables $X_1, X_2, \ldots, X_N$ such that $P(X_i \in A_i$ for each $i \in I) = \prod_{i \in I} P(X_i \in A_i)$ holds whenever the sets in*

$\{A_i\colon i \in I\}$ are Borel sets but does not hold for all sets $\{A_i\colon i \in I\}$ for which the indicated probabilities are well defined where I is any subset of $\{1, 2, \dots, N\}$ having cardinality at least two.

Proof: In 1948 it was shown by Jessen and Doob that there exist an underlying probability space and two random variables X and Y defined on this probability space such that $P(X \in A \text{ and } Y \in B) = P(X \in A)$ $P(Y \in B)$ holds for all Borel sets A and B but not for all sets for which the indicated probabilities are defined. For any integer $N > 2$, we could obviously define random variables $X_1 = X$, $X_2 = Y$, and $X_3, \dots, X_N$ all identically zero, and obtain an analogous result. If we wanted to obtain an analogous result using nonconstant random variables, we could take copies of the spaces used in the Jessen result or the Doob result and construct a product space and obtain an analogous result for any even positive integer, and by producting with yet another probability space, we could extend it to any integer greater than one. However, if $N > 2$, due to the properties of product measures, these efforts suffer from the deficiency that there exists a pair of these random variables that do not exhibit this property. In this example we extend this earlier result to the case of N standard Gaussian random variables, where N is any integer greater than one, and where any two of these random variables also have this property.

Let N be an integer greater than one, let $S_1, S_2, \dots, S_N$ be a partition of $\mathbb{R}$ as provided by Example 1.46, and, for this partition, let $\mathcal{F}$ be the σ-algebra on $\mathbb{R}$ provided by Example 7.7. Recall that any set A in $\mathcal{F}$ may be represented as $A = (S_1 \cap B_1) \cup \cdots \cup (S_N \cap B_N)$, where the sets $B_1, B_2, \dots, B_N$ are real Borel sets. Let G be standard Gaussian measure on $(\mathbb{R}, \mathcal{B}(\mathbb{R}))$. For each positive integer i not greater than N, let μ_i be the measure on $(\mathbb{R}, \mathcal{F})$ given by $\mu_i((S_1 \cap B_1) \cup \cdots \cup (S_N \cap B_N)) = G(B_i)$. (That these measures are well defined follows as it did in Example 7.7.) Now, consider the following product measures on $\mathcal{G} = \mathcal{F} \times \mathcal{F} \times \mathcal{F} \times \cdots \times \mathcal{F}$, the product σ-algebra of N copies of $\mathcal{F}$:

$$\lambda_1 = \mu_1 \times \mu_2 \times \mu_3 \times \mu_4 \times \cdots \times \mu_{N-1} \times \mu_N,$$
$$\lambda_2 = \mu_2 \times \mu_1 \times \mu_3 \times \mu_4 \times \cdots \times \mu_{N-1} \times \mu_N,$$
$$\lambda_3 = \mu_2 \times \mu_3 \times \mu_1 \times \mu_4 \times \cdots \times \mu_{N-1} \times \mu_N, \dots$$

and,

$$\lambda_N = \mu_2 \times \mu_3 \times \mu_4 \times \mu_5 \times \cdots \times \mu_N \times \mu_1.$$

Let P be a probability measure defined on $(\mathbb{R}^N, \mathcal{G})$ via $P = (\lambda_1 + \lambda_2 + \cdots + \lambda_N)/N$. Note that P is not a product measure since

$$P(A_1 \times A_2 \times \cdots \times A_N) = \begin{cases} \dfrac{1}{N}, & \text{if } A_1 = S_1 \text{ and } A_j = \mathbb{R} \text{ for } j > 1 \\ \dfrac{j-1}{N}, & \text{if } A_j = S_j \text{ for } j > 1 \\ & \text{and } A_i = \mathbb{R} \text{ for } i \neq j \end{cases}$$

and

$$P(S_1 \times S_2 \times \cdots \times S_N) = \frac{1}{N} \neq \frac{(N-1)!}{N^N}.$$

Now, consider the probability space $(\mathbb{R}^N, \mathcal{G}, P)$, and N random variables $X_1, X_2, \ldots, X_N$ defined on this probability space via $X_i(\omega_1, \omega_2, \ldots, \omega_N) = \omega_i$, for positive integers i not greater than N. Note that for any N real Borel sets $B_1, B_2, \ldots, B_N$, we have $P(X_1 \in B_1, X_2 \in B_2, \ldots, X_N \in B_N) = P(B_1 \times B_2 \times \cdots \times B_N) = G(B_1)G(B_2)\cdots G(B_N) = P(X_1 \in B_1)P(X_2 \in B_2)\cdots P(X_N \in B_N)$. Thus the random variables $X_1, X_2, \ldots, X_N$ are mutually independent and each has a standard Gaussian distribution. However, note that $P(X_1 \in S_1, X_2 \in S_2, \ldots, X_N \in S_N) = P(S_1 \times S_2 \times \cdots \times S_N) = 1/N \neq (N-1)!/N^N = P(X_1 \in S_1) P(X_2 \in S_2)\cdots P(X_N \in S_N)$.

Now, we show that this property holds for any pair of these random variables. For the two random variables X_1 and X_j, with $j > 1$, note that $P(X_1 \in S_1 \text{ and } X_j \in S_j) = P(A_1 \times A_2 \times \cdots \times A_N) = 1/N$, where $A_1 = S_1$, $A_j = S_j$, and $A_k = \mathbb{R}$ for positive integers k not greater than N and not equal to 1 or j, yet $P(X_1 \in S_1)P(X_j \in S_j) = (j-1)/N^2$. For two random variables X_j and X_k with $1 < j < k \leq N$, note that $P(X_j \in S_j \text{ and } X_k \in S_k) = P(A_1 \times A_2 \times \cdots \times A_N) = (j-1)/N$, where $A_j = S_j$, $A_k = S_k$, and $A_m = \mathbb{R}$ for positive integers m not greater than N and not equal to j or k, yet $P(X_j \in S_j)P(X_k \in S_k) = [(j-1)(k-1)]/N^2$. Finally, note that the random variables $X_1, X_2, \ldots, X_N$ are mutually Gaussian, each having zero mean and unit variance. □

Example 7.20. *For any integer $N > 1$ and any diffuse distribution on $\mathbb{R}$, there exists a probability space and N random variables $X_1, X_2, \ldots, X_N$, each with the given distribution such that $P(X_i \in A_i$ for each $i \in I) = \prod_{i\in I} P(X_i \in A_i)$ holds whenever the sets in $\{A_i\colon i \in I\}$ are Borel sets but does not hold for all sets $\{A_i\colon i \in I\}$ for which the indicated probabilities are well defined where I is any subset of $\{1, 2, \ldots, N\}$ having cardinality at least two.*

Proof: Recall that if $X_1, X_2, \ldots, X_N$ are mutually independent random variables defined on a common probability space, and if $g_1, g_2, \ldots, g_N$ are Borel measurable functions mapping $\mathbb{R}$ into $\mathbb{R}$, then it follows that $g_1(X_1), g_2(X_2), \ldots, g_N(X_N)$ are mutually independent random variables. However, if the g_i's are chosen as any functions that map $\mathbb{R}$ into $\mathbb{R}$ such that the $g_i(X_i)$'s are random variables, the resulting set of random variables need not be mutually independent. For instance, in Example 7.19, simply let g_i denote the indicator function of S_i for each positive integer i not greater than N.

Recall that if F is a continuous probability distribution function and if $F^{-1}\colon(0, 1) \to \mathbb{R}$ via $F^{-1}(x) = \inf\{y \in \mathbb{R}\colon F(y) = x\}$, we have that $F(F^{-1}(x)) = x$ for all $x \in (0, 1)$. Now, let F be a continuous probability distribution function, let H be the probability distribution function corresponding to standard Gaussian measure on $(\mathbb{R}, \mathcal{B}(\mathbb{R}))$, and note that for any nonempty subset S of $\mathbb{R}$, $F^{-1}(H(x)) \in F^{-1}(H(S))$ if and only if $x \in S$, where F^{-1} is as given above. From this observation and Example 7.19 it immediately follows that for any integer $N > 1$, if $D_1, D_2, \ldots, D_N$ are diffuse probability distributions on $\mathcal{B}(\mathbb{R})$, then there exists a probability space $(\mathbb{R}^N, \mathcal{G}, P)$ and N random variables $Y_1, Y_2, \ldots, Y_N$ such that for each positive integer i not greater than N, Y_i has distribution D_i, and such that $P(Y_1 \in A_1, Y_2 \in A_2, \ldots, Y_N \in A_N) = P(Y_1 \in A_1)P(Y_2 \in A_2)\cdots P(Y_N \in A_N)$ holds for any real Borel sets $A_1, A_2, \ldots, A_N$, but not for all sets for which the indicated probabilities are well defined. Furthermore, this result holds for all subsets of $\{Y_1, Y_2, \ldots, Y_N\}$ having cardinality at least two. □

Example 7.21. *There exists a sequence $\{X_n\}_{n\in\mathbb{N}}$ of mutually independent, identically distributed, nonnegative random variables defined on a common probability space such that $\sup_{n\in\mathbb{N}} \mathrm{E}[X_n^p] < \infty$ for all positive real numbers p and yet $P(\sup_{j\in\mathbb{N}} X_{n_j} < \infty) = 0$ whenever $\{n_j\}_{j\in\mathbb{N}}$ is a subsequence of the positive integers.*

Proof: This example is due to Emmanuele et al. [1985]. Consider a sequence $\{X_n\}_{n\in\mathbb{N}}$ of mutually independent, identically distributed, nonnegative random variables defined on a common probability space for which $P(X_n \le x) = 1 - e^{-x}$ if $x \ge 0$. Then $\mathrm{E}[X_n^p] = \int_0^\infty x^p e^{-x}\,dx = \Gamma(p+1) < \infty$ for all positive real numbers p. Note that

$$P\left(\max_{1\le j\le k} X_{n_j} \le x\right) = (1 - e^{-x})^k$$

for nonnegative real values of x and for positive integers k. Taking the limit as $k \to \infty$, we see that

$$P\left(\sup_{j\in\mathbb{N}} X_{n_j} \le x\right) = 0$$

for all nonnegative real numbers x. Finally, letting $x \to \infty$, we see that

$$P\left(\sup_{j\in\mathbb{N}} X_{n_j} < \infty\right) = 0. \qquad \square$$

Example 7.22. *There exists a non-Markovian random process that satisfies the Chapman–Kolmogorov equation.*

Proof: This example is from Feller [1971]. Consider a set Ω_1 consisting of all permutations of (1, 2, 3) combined with (1, 1, 1), (2, 2, 2), and (3, 3, 3) and define a probability measure P on the power set of Ω_1 such that $P(\{\omega\}) = 1/9$ for each $\omega \in \Omega_1$. If $\omega = (a, b, c)$, then define $Y_1(\omega) = a$, $Y_2(\omega) = b$, and $Y_3(\omega) = c$. Notice that Y_1, Y_2, and Y_3 are pairwise independent but not mutually independent. Now consider the set Ω_2 consisting of all infinite sequences of the form $(a_1, b_1, c_1, a_2, b_2, c_2, \ldots)$, where for each positive integer n, (a_n, b_n, c_n) is an element of Ω_1. Sampling from Ω_1 with replacement countably infinitely many times thus results in a sequence $(a_1, b_1, c_1, a_2, b_2, c_2, \ldots)$ that is an element from Ω_2. For each positive integer n, define a triple $(X_{3n-2}, X_{3n-1}, X_{3n})$ of random variables via $X_{3n-2} = a_n$, $X_{3n-1} = b_n$, and $X_{3n} = c_n$, where each random variable X_i is defined on the probability space given by Ω_2, the power set of Ω_2, and the obvious probability measure induced via sampling with replacement from Ω_1. Notice that the elements in the sequence of triples $\{(X_{3n-2}, X_{3n-1}, X_{3n})\}$ are mutually independent. In addition, notice that, as above, the random variables X_{3n-2}, X_{3n-1}, and X_{3n} are pairwise independent yet not mutually independent for each positive integer n. Hence, $\{X_1, X_2, X_3, \ldots\}$ is a sequence of random variables that are pairwise, yet not mutually, independent and identically distributed, with X_1 taking the values 1, 2, or 3 each with probability 1/3.

Notice that the random variables $\{X_1, X_2, X_3, \ldots\}$ define a random process with three "states" such that $p_{jk} \equiv P(X_{n+1} = k \,|\, X_n = j) = 1/3$ for each j and k in $\{1, 2, 3\}$. Further, notice that $p_{jk}^{(n)} \equiv P(X_{m+n} = k \,|\, X_m = j) = 1/3$ for each j and k in $\{1, 2, 3\}$ via the pairwise independence of the family $\{X_1, X_2, X_3, \ldots\}$. Thus, the Chapman–Kolmogorov equation $p_{jk}^{(m+n)} = \sum_{v=1}^{3} p_{jv}^{(m)} p_{vk}^{(n)}$ is satisfied. However, this random process is not Markov since $P(X_3 = i \,|\, X_2) = 1/3$ (via pairwise independence) and $P(X_3 = i \,|\, X_1, X_2) = \mathrm{I}_{\{X_3=i\}}$ since X_1 and X_2 precisely determine X_3. □

Example 7.23. *If $\{X(t)\colon t \in T\}$ is a stochastic process defined on some underlying probability space $(\Omega, \mathcal{F}, P)$, where T is any uncountable index set, then for any random variable Y defined on this probability space such that Y is $\sigma(\{X(t)\colon t \in T\})$-measurable, there exists a countable subset K of T such that Y is $\sigma(\{X(t)\colon t \in K\})$-measurable.*

Proof: Recall that $\sigma(\{X(t)\colon t \in T\})$ contains all sets of the form $\{\omega \in \Omega\colon X(t) \in B\}$, where $t \in T$ and $B \in \mathcal{B}(\mathbb{R})$. Thus, by Example 4.2, for each rational number r, there exists a countable subset $\mathcal{A}_r$ of $\mathcal{B}(\mathbb{R})$ and a countable subset K_r of T such that $\{\omega \in \Omega\colon Y(\omega) \le r\} \in \sigma(\{\{\omega \in \Omega\colon X(t) \in B\}\colon B \in \mathcal{A}_r \text{ and } t \in K_r\})$. Now, let $\mathcal{A} = \bigcup_{r\in\mathbb{Q}} \mathcal{A}_r$ and let $K = \bigcup_{r\in\mathbb{Q}} K_r$. Then for any rational number r, $\{\omega \in \Omega\colon Y(\omega) \le r\} \in \sigma(\{\{\omega \in \Omega\colon X(t) \in B\}\colon B \in \mathcal{A} \text{ and } t \in K\})$. Hence, for any rational number r, $\{\omega \in \Omega\colon Y(\omega) \le r\} \in \sigma(\{X(t)\colon t \in K\})$. Thus, Y is $\sigma(\{X(t)\colon t \in K\})$-measurable, where K is countable. □

Example 7.24. *If $\{X(t)\colon t \in \mathbb{R}\}$ is a stochastic process defined on a complete probability space $(\Omega, \mathcal{F}, P)$ and if almost all sample paths of $\{X(t)\colon t \in \mathbb{R}\}$ are continuous, then the stochastic process need not be $(\mathcal{B}(\mathbb{R}) \times \mathcal{F})$-measurable.*

Proof: Let $\Omega = \mathbb{R}$, let $\mathcal{F} = \mathcal{M}(\mathbb{R})$, and let P be defined via $P(A) = \lambda(A \cap [0, 1])$. Note that the resulting probability space is complete. Let S be a subset of the Cantor ternary set that is not Borel measurable. [Such sets exist in profusion since the family of all subsets of the Cantor ternary set has cardinality 2^c and $\mathrm{card}(\mathcal{B}(\mathbb{R})) = c$.] Define $\{X(t)\colon t \in \mathbb{R}\}$ via $X(t) = I_S(t) I_{\mathbb{R}\setminus[0,1]}(\omega)$. Then almost all sample paths are identically zero. However, $\{(t, \omega) \in \mathbb{R} \times \Omega\colon X(t, \omega) \in \{1\}\} = S \times (\mathbb{R} \setminus [0, 1])$, which is not an element of $(\mathcal{B}(\mathbb{R}) \times \mathcal{F})$. □

Example 7.25. *If $\{X(t)\colon t \in \mathbb{R}\}$ is a stochastic process defined on a complete probability space $(\Omega, \mathcal{F}, P)$ and if each sample path of $\{X(t)\colon t \in \mathbb{R}\}$ is a.e. $[\lambda]$ continuous, then the stochastic process need not be $(\mathcal{B}(\mathbb{R})\times\mathcal{F})$-measurable.*

Proof: The stochastic process constructed in the proof to Example 7.24 is such that each sample path is a.e. $[\lambda]$ continuous and yet it is not $(\mathcal{B}(\mathbb{R})\times\mathcal{F})$-measurable. □

Example 7.26. *There exists a random process $\{X(t)\colon t \in \mathbb{R}\}$ whose sample functions are each positive and Lebesgue integrable yet whose mean is not integrable.*

Proof: A version of Fubini's theorem (see, for instance, Doob [1953, p. 62]) states that if $\{X(t)\colon t \in \mathbb{R}\}$ is a measurable random process and if $\mathrm{E}[|X(t)|]$ is Lebesgue integrable, then almost all sample functions of the random process are Lebesgue integrable. In this example it is shown that these conditions are by no means necessary conditions.

Consider the probability space given by $\mathbb{R}$, $\mathcal{B}(\mathbb{R})$, and standard Gaussian measure on $(\mathbb{R}, \mathcal{B}(\mathbb{R}))$. Further, for $t \in \mathbb{R}$, define a random process $\{X(t)\colon t \in \mathbb{R}\}$ via $X(t, \omega) = \exp\left[-\omega t - (t^2/2)\right]$. A simple change of variable shows that

$$\begin{aligned}\mathrm{E}[|X(t)|] &= \int_{\mathbb{R}} \frac{1}{\sqrt{2\pi}} \exp\left(-\omega t - \frac{t^2}{2}\right) \exp\left(-\frac{\omega^2}{2}\right) d\omega \\ &= \int_{\mathbb{R}} \frac{1}{\sqrt{2\pi}} \exp\left(-\frac{\tau^2}{2}\right) d\tau = 1.\end{aligned}$$

Hence, for any real number t, $\mathrm{E}[|X(t)|]$ is equal to one, a positive nonzero constant, and thus $\mathrm{E}[|X(t)|]$ is not Lebesgue integrable. Notice, however, that not only are the sample functions of $X(t)$ each Lebesgue integrable, but they are also infinitely differentiable, everywhere positive, and such that they vanish at $-\infty$ and ∞. □

Example 7.27. *A modification $\{Y(t)\colon t \in [0, \infty)\}$ of a random process $\{X(t)\colon t \in [0, \infty)\}$ that is adapted to a filtration $\{\mathcal{F}_t\colon t \in [0, \infty)\}$ need not itself be adapted to this filtration even if $\mathcal{F}_0$ is complete with respect to the underlying probability measure.*

Proof: Let $\Omega = \{1, 2, 3\}$, $\mathcal{F} = \mathbb{P}(\Omega)$, and a probability measure P be defined on $(\Omega, \mathcal{F})$ via $P(\{1\}) = P(\{2\}) = 1/2$. Further, define $\{X(t)\colon t \in [0, \infty)\}$ via $X(t, \omega) = 1$ if $\omega = 1$ and $X(t, \omega) = 2$ otherwise. Then $\{X(t)\colon t \in [0, \infty)\}$ is adapted to the filtration defined via $\mathcal{F}_t = \sigma(\{1\})$ for each $t \in [0, \infty)$. Notice that $\mathcal{F}_0$ is complete. Now, let $\{Y(t)\colon t \in [0, \infty)\}$ be defined via $Y(t, \omega) = \omega$. Notice that $\{Y(t)\colon t \in [0, \infty)\}$ is a modification of $\{X(t)\colon t \in [0, \infty)\}$ since for every $t \in [0, \infty)$, $P(X(t) = Y(t)) = 1$. However, for any $t \in [0, \infty)$, $Y(t)$ is not $\mathcal{F}_t$-measurable, since $Y(t)$ takes on two different values on the atom $\{2, 3\}$. Hence, $\{Y(t)\colon t \in [0, \infty)\}$ is not adapted to the filtration $\{\mathcal{F}_t\colon t \in [0, \infty)\}$. □

Example 7.28. *The σ-algebra generated by a random process $\{X(t): t \in \mathbb{R}\}$ defined on a probability space such that $X(t) = 0$ a.s. need not be countably generated.*

Proof: Consider the probability space consisting of the real line, the Borel subsets, and standard Gaussian measure. Define the random process $\{X(t): t \in \mathbb{R}\}$ via $X(t, \omega) = I_{\{\omega\}}(t)$. Note that the σ-algebra generated by this random process is the σ-algebra of countable and co-countable sets. Recall from Example 4.1 that this σ-algebra is not countably generated. □

Example 7.29. *The sample-path continuity of a random process does not imply the continuity of its generated filtration.*

Proof: This example is from Morrison and Wise [1987]. Let $\Omega = [0, 1]$, let $\mathcal{F}$ denote the Borel subsets of $[0, 1]$, and let P denote Lebesgue measure on the Borel subsets of $[0, 1]$. Let f be an infinitely differentiable real-valued function defined on $[0, \infty)$ such that $f(x) = 0$ for $x \leq 1/2$ and such that f is strictly increasing on $[1/2, \infty)$. Define a random process $\{X(t): t \in [0, \infty)\}$ on $(\Omega, \mathcal{F}, P)$ via $X(t, \omega) = \omega f(t)$ for $\omega \in \Omega$ and $t \geq 0$. Note that

$$\sigma(X(s) : s \leq t) = \begin{cases} \{\emptyset, \Omega\}, & \text{if } t \leq 1/2 \\ \mathcal{B}([0, 1]), & \text{if } t > 1/2. \end{cases}$$

Thus, despite the discontinuity in the filtration generated by the random process $X(t)$, the sample paths of the random process are infinitely differentiable. □

Example 7.30. *There exists a partition $\{S_t: t \in \mathbb{R}\}$ of the real line, a probability space $(\Omega, \mathcal{F}, P)$, and a stationary Gaussian random process $\{X(t): t \in \mathbb{R}\}$ defined thereon composed of mutually independent standard Gaussian random variables such that for any real number t, $X(t, \Omega) = S_t$.*

Proof: Let $\{S_t: t \in \mathbb{R}\}$ be a partition of the real line as given in Example 1.46, and for each $t \in \mathbb{R}$ let $(\Omega_t, \mathcal{F}_t, P_t)$ be a probability space as provided by Example 7.18 for the saturated non-Lebesgue measurable set S_t. Let $(\Omega, \mathcal{F}, P)$ be the product space of $(\Omega_t, \mathcal{F}_t, P_t)$ over all $t \in \mathbb{R}$, and let $\omega = \{\omega_t\}_{t\in\mathbb{R}}$ denote a generic element of Ω. Note that $\omega_t \in S_t$ for each $t \in \mathbb{R}$. Define a stochastic process $\{X(t): t \in \mathbb{R}\}$ via $X(t, \omega) = \omega_t$. The desired result now follows from Example 7.18 and the basic properties of product measures. □

Example 7.31. *There exists a random process $\{X(t): t \in \mathbb{R}\}$ such that every sample path is non-Lebesgue measurable on any Lebesgue measurable set having positive Lebesgue measure, every sample path is nonzero in each uncountable closed subset of the reals, and $P(X(t) = 0) = 1$ for all real numbers t.*

Proof: Recall from Example 1.46 that there exists a partition $\{S_y: y \in \mathbb{R}\}$ of $\mathbb{R}$ into saturated non-Lebesgue measurable subsets. Consider the probability space consisting of $\mathbb{R}$, $\mathcal{B}(\mathbb{R})$, and standard Gaussian measure on $\mathcal{B}(\mathbb{R})$. Define the random process $\{X(t): t \in \mathbb{R}\}$ via $X(t, \omega) = I_{S_\omega}(t)$, where we note that $X(t, \omega) = 1$ if and only if ω is such that $t \in S_\omega$. Thus, for each real number t, there is a unique point ω_t such that $X(t, \omega_t) = 1$. Hence, $\{X(t)\colon t \in [0, \infty)\}$ is a well-defined random process. Further, since $P(\omega = \omega_t) = 0$, we see that $P(X(t) = 0) = 1$ for all real numbers t. Finally, we note that each sample path of this random process is the indicator function of a saturated non-Lebesgue measurable subset of $\mathbb{R}$. □

Example 7.32. *For any random process indexed by the reals, there exists a probability space and a random process indexed by the reals and defined on that probability space such that the two random processes have the same family of finite-dimensional distributions yet such that almost all of the sample paths of the second random process are not locally integrable.*

Proof: Consider a probability space $(\Omega, \mathcal{F}, P)$ and any random process $\{X(t)\colon t \in \mathbb{R}\}$ defined on $(\Omega, \mathcal{F}, P)$. Let $(\tilde{\Omega}, \tilde{\mathcal{F}}, \tilde{P})$ denote the probability space in Example 7.31, and let $\{\tilde{X}(t)\colon t \in \mathbb{R}\}$ denote the random process constructed therein. Now, consider the probability space $(\Omega\times\tilde{\Omega}, \mathcal{F}\times\tilde{\mathcal{F}}, P\times\tilde{P})$. Let (ω_1, ω_2) denote a generic point in $\Omega \times \tilde{\Omega}$, and consider the random process $\{Y(t)\colon t \in \mathbb{R}\}$ defined on $(\Omega\times\tilde{\Omega}, \mathcal{F}\times\tilde{\mathcal{F}}, P\times\tilde{P})$ via $Y(t, (\omega_1, \omega_2)) = X(t, \omega_1) + \tilde{X}(t, \omega_2)$. Then the finite-dimensional distributions of $\{X(t)\colon t \in \mathbb{R}\}$ defined on $(\Omega, \mathcal{F}, P)$ and of $\{Y(t)\colon t \in \mathbb{R}\}$ defined on $(\Omega \times \tilde{\Omega}, \mathcal{F} \times \tilde{\mathcal{F}}, P \times \tilde{P})$ are the same, and yet the sample paths of $\{Y(t)\colon t \in \mathbb{R}\}$ are such that almost all sample paths are not locally integrable. □

Example 7.33. *There exists a zero mean wide sense stationary random process that does not possess a spectral representation.*

Proof: Consider a random process $\{X(t)\colon t \in \mathbb{R}\}$ defined on a probability space $(\Omega, \mathcal{F}, P)$. Recall that $\{X(t)\colon t \in \mathbb{R}\}$ is said to possess a spectral representation if

$$X(t) = \int_{\mathbb{R}} \exp(2\pi \imath t\lambda)\, dY(\lambda),$$

where $\{Y(t)\colon t \in \mathbb{R}\}$ is a random process with orthogonal increments and where the previous stochastic integral is defined in the usual manner as a mean-square limit. Associated with the random process $\{Y(t)\colon t \in \mathbb{R}\}$ is a nondecreasing function $F\colon \mathbb{R} \to \mathbb{R}$ such that $\mathrm{E}[|Y(t)-Y(s)|^2] = F(t)-F(s)$ for all $s < t$, such that F is right continuous, and such that $F(x)$ vanishes as $x \to -\infty$. The function F is called the spectral distribution function of the random process $\{X(t)\colon t \in \mathbb{R}\}$. A wide sense stationary random process $\{X(t)\colon t \in \mathbb{R}\}$ possesses a spectral representation as above if and only if its autocorrelation function $R(\tau)$ is of the form

$$R(\tau) = \int_{\mathbb{R}} \exp(2\pi \imath \tau\lambda)\, dF(\lambda),$$

where F is as above. Notice that applying the dominated convergence theorem to the previous integrand implies that $R(\tau)$ is continuous.

Consider a zero mean wide sense stationary random process $\{Z(t)\colon t \in \mathbb{R}\}$ that possesses a continuous autocorrelation function $R_1(\tau)$. Define a new autocorrelation function $R_2(\tau)$ via $R_2(\tau) = R_1(\tau)$ if $\tau \neq 0$ and $R_2(0) = 1 + R_1(0)$. Notice that there exists a zero mean random process $\{X(t)\colon t \in \mathbb{R}\}$ that possesses the autocorrelation function $R_2(\tau)$ and that is also wide sense stationary. However, since $R_2(\tau)$ is not continuous at the origin, the result of the previous paragraph implies that $\{X(t)\colon t \in \mathbb{R}\}$ possesses no spectral representation. Further, notice that this random process $\{X(t)\colon t \in \mathbb{R}\}$ could be chosen to be a zero mean, strictly stationary, Gaussian random process. □

Example 7.34. *There exists a wide sense stationary random process having periodic sample paths yet whose autocorrelation function is not periodic.*

Proof: Let $X(t) = A\cos(\lambda t + \theta)$, where $A \in \mathbb{R} \setminus \{0\}$ and λ and θ are independent random variables such that λ has a symmetric distribution and θ has a uniform distribution on the interval $(0, 2\pi)$. Then

$$\begin{aligned}
\mathrm{E}[X(t)X(t+\tau)] =& A^2\,\mathrm{E}[\cos(\lambda t + \theta)\cos[\lambda(t+\tau) + \theta]] \\
=& \frac{A^2}{2}\,\mathrm{E}[\cos(\lambda\tau) + \cos[\lambda(2t+\tau) + 2\theta]] \\
=& \frac{A^2}{2}(\mathrm{E}[\cos(\lambda\tau)] + \mathrm{E}[\cos[\lambda(2t+\tau)]]\mathrm{E}[\cos(2\theta)] \\
&- \mathrm{E}[\sin[\lambda(2t+\tau)]]\mathrm{E}[\sin(2\theta)]) \\
=& \frac{A^2}{2}\mathrm{E}[\cos(\lambda\tau)],
\end{aligned}$$

since $\mathrm{E}[\cos(2\theta)] = \mathrm{E}[\sin(2\theta)] = 0$. Now, let $C_\lambda(t)$ denote the characteristic function of λ. Since λ is symmetric, it follows that $\mathrm{E}[X(t)X(t+\tau)] = (A^2/2)C_\lambda(\tau)$. If λ is a zero mean Gaussian random variable with a positive variance σ^2, then $C_\lambda(t) = \exp(-\sigma^2 t^2/2)$. Hence, $\mathrm{E}[X(t)X(t+\tau)]$ is not periodic even though $X(t)$ has periodic sample paths. □

Example 7.35. *For a zero mean, wide sense stationary random process, the existence of an ordinary derivative pointwise on the underlying probability space does not imply the existence of a mean-square derivative.*

Proof: Let $X(t) = A\cos(\lambda t + \theta)$, where $A \in \mathbb{R} \setminus \{0\}$ and λ and θ are independent random variables such that λ has a symmetric distribution and θ has a uniform distribution on the interval $(0, 2\pi)$. Note that $\mathrm{E}[X(t)X(t+\tau)] = (A^2/2)\mathrm{E}[\cos(\lambda\tau)]$. Consequently, since

$$R(\tau) = \int_{\mathbb{R}} \exp(2\pi\imath\tau s)\, dF(s),$$

λ may be chosen in such a way that $X(t)$ possesses any preassigned spectral distribution function $F(s)$. Choosing λ such that $\int_{\mathbb{R}} s^2\, dF(s) = \infty$ results in a random process $X(t)$ that is not mean-square differentiable. Notice, however, that each sample path of $X(t)$ is infinitely differentiable. □

Example 7.36. *Almost sure continuity does not imply almost sure sample continuity of a random process.*

Proof: Consider a random process $\{X(t)\colon t \in \mathbb{R}\}$ defined on a probability space $(\Omega, \mathcal{F}, P)$. Recall that $\{X(t)\colon t \in \mathbb{R}\}$ is said to be almost surely continuous at $t \in \mathbb{R}$ if

$$P\left(\limsup_{h\to 0} |X(t+h) - X(t)| = 0\right) = 1.$$

Further, $\{X(t)\colon t \in \mathbb{R}\}$ is said to be almost surely continuous if it is almost surely continuous at s for all s in $\mathbb{R}$. Finally, recall that $\{X(t)\colon t \in \mathbb{R}\}$ is said to be almost surely sample continuous if

$$\bigcup_{t\in\mathbb{R}} \left\{\omega \in \Omega\colon \limsup_{h\to 0} |X(t+h, \omega) - X(t, \omega)| \neq 0\right\}$$

is a subset of a null set. The following example shows that almost sure continuity does not imply almost sure sample continuity of a random process.

Let Z be a random variable defined on $(\Omega, \mathcal{F}, P)$ and possessing a probability density function and let $X(t) = \mathrm{I}_{[0,\infty)}(t - Z)$ for $t \in \mathbb{R}$. Then, for any fixed real number t,

$$P\left(\limsup_{h\to 0} |X(t+h) - X(t)| = 0\right) = P(Z \neq t) = 1.$$

Hence, $X(t)$ is almost surely continuous. Notice, however, that

$$\bigcup_{t\in\mathbb{R}} \left\{\omega \in \Omega\colon \limsup_{h\to 0} |X(t+h, \omega) - X(t, \omega)| \neq 0\right\} = \Omega$$

since for each $\omega' \in \Omega$ there exists a $t' \in \mathbb{R}$, namely $t' = Z(\omega')$, such that $X(\cdot)(\omega')$ is discontinuous at t'. Thus $\{X(t)\colon t \in \mathbb{R}\}$ is not almost surely sample continuous even though it is almost surely continuous. □

Example 7.37. *If $(\Omega, \mathcal{F}, P)$ is a probability space such that $\mathcal{F}$ is separable with respect to the pseudometric d defined via $d(A, B) = P(A \triangle B)$, then no random process of strict sense mutually independent random variables defined on this probability space indexed by an uncountable parameter set exists.*

Proof: This example was inspired by Varadarayan [1962]. For $(\Omega, \mathcal{F}, P)$ as above, let $\{X(t)\colon t \in I\}$ be a random process of strict sense random variables defined on $(\Omega, \mathcal{F}, P)$, where the index set I is uncountable. Assume that this random process is composed of mutually independent random variables. For any subset J of I, let $\mathcal{F}_J = \sigma(X(t)\colon t \in J)$, and let P_J denote the restriction of P to $\mathcal{F}_J$. Assume that the probability space $(\Omega, \mathcal{F}_I, P_I)$ is such that $\mathcal{F}_I$ is separable with respect to the pseudometric d defined via $d(A, B) = P_I(A \triangle B)$. Consequently there exists a sequence $\{A_n\}_{n\in\mathbb{N}}$

of events in $\mathcal{F}_I$ such that for any set A in $\mathcal{F}_I$, there exists a subsequence $\{A_{n_k}\}_{k\in\mathbb{N}}$ of $\{A_n\}_{n\in\mathbb{N}}$ such that $P_I\left(A_{n_k} \triangle A\right) \to 0$ as $k \to \infty$. Recall from Example 4.2 that for any set $B \in \mathcal{F}_I$ there exists a countable subset C of I such that $B \in \mathcal{F}_C$. For each positive integer i, let I_i be a countable subset of I such that $A_i \in \mathcal{F}_{I_i}$ and let $I_0 = \bigcup_{i\in\mathbb{N}} I_i$.

Now, let $\{B_m\}_{m\in\mathbb{N}}$ be a sequence of elements from $\mathcal{F}$ that converges with respect to the pseudometric defined above. Let $\{m_i\}_{i\in\mathbb{N}}$ be an increasing sequence of positive integers such that $P(B_k \triangle B_j) < 2^{-i}$ for all integers j and k not less than m_i. Let $\{F_i\}_{i\in\mathbb{N}}$ be the subsequence of the B_m's given by $F_i = B_{m_i}$ for each positive integer i and note that $P(F_i \triangle F_j) < 2^{-i}$ for all $j > i$. Let $H_i = \bigcap_{j=i}^{\infty} F_j$ for each positive integer i and let $B = \bigcup_{i=1}^{\infty} H_i$. Note that $B \triangle F_i$ is a subset of $\bigcup_{n=i}^{\infty}(F_n \triangle F_{n+1})$ for each positive integer i. Thus,

$$P(B \triangle F_i) \leq \sum_{n=i}^{\infty} P(F_n \triangle F_{n+1}) < \sum_{n=i}^{\infty} \frac{1}{2^n} = \frac{1}{2^{i-1}}$$

for each positive integer i. If $m \geq m_i$, then $P(B \triangle B_m) = P((B \triangle F_i) \triangle (B_{m_i} \triangle B_m)) \leq P(B \triangle F_i) + P(B_{m_i} \triangle B_m) < 2^{i-1} + 2^i$. Thus B_m converges to B as $m \to \infty$. Constructing a limit in this way we see that, for any set $A \in \mathcal{F}_I$ there exists a set $A_0 \in \mathcal{F}_{I_0}$ such that $P_I(A \triangle A_0) = 0$.

Fix $s \in I \setminus I_0$ and let $A \in \mathcal{F}_{\{s\}}$ be such that $0 < P_I(A) < 1$. [Such a set exists since the $X(t)$'s are strict sense random variables.] Let $A_0 \in \mathcal{F}_{I_0}$ be such that $P_I(A \triangle A_0) = 0$. Since $X(s)$ and $\{X(t)\colon t \in I_0\}$ are independent, we see that $P_I(A \cap A_0) = P_I(A)P_I(A_0)$. Further, since $P_I(A \triangle A_0) = 0$, we see that $P_I(A_0) = P_I(A \cap A_0)$ and hence that $P_I(A_0) = 0$. But, similarly, we see that $P_I(A) = P_I(A \cap A_0)$, which results in a contradiction, since $P_I(A) > 0$. Thus, we conclude that $(\Omega, \mathcal{F}_I, P_I)$ is not separable with respect to d and hence that $(\Omega, \mathcal{F}, P)$ is not separable with respect to d. This final contradiction implies that no such random process may be defined on $(\Omega, \mathcal{F}, P)$. □

Example 7.38. *There exists a Gaussian random process indexed by the reals whose autocorrelation function is almost everywhere zero yet whose variance takes on every nonnegative real number in any nonempty open interval of real numbers.*

Proof: Let the random process $\{X(t)\colon t \in \mathbb{R}\}$ be given by $X(t) = Gf(t)$, where G is a Gaussian random variable with a standard Gaussian distribution and f is the function given in Example 2.29, and note that the desired result follows. □

Example 7.39. *There exists a strictly stationary, mean-square continuous, zero mean Gaussian random process $\{X(t)\colon t \in \mathbb{R}\}$ whose spectral measure has compact support and yet is such that no sample path is Lebesgue measurable on any Lebesgue measurable set of positive Lebesgue measure.*

Proof: Consider any strictly stationary, zero mean, mean-square continuous, Gaussian random process $\{Y(t)\colon t \in \mathbb{R}\}$ whose spectral measure has compact support and that has continuous sample paths, defined on a suitable probability space. Further, consider the probability space and the random process, here called $\{Z(t): t \in \mathbb{R}\}$, given in Example 7.31. By considering the product space as in Example 7.32, we can now define a random process having the desired properties. □

Example 7.40. *There exists a Gaussian random process $\{X(t): t \in \mathbb{R}\}$ defined on a probability space $(\Omega, \mathcal{F}, P)$ such that each sample path takes on every nonzero real number c times on any nonempty perfect set.*

Proof: Let the probability space be given by the reals, the Borel subsets, and standard Gaussian measure. Let $G: \mathbb{R} \to \mathbb{R}$ via $G(x) = x + I_{\{0\}}(x)$. Note that G is a random variable that has a standard Gaussian distribution but does not take on the value zero. Further, let f denote the function mapping the reals into the reals given in Example 2.30, and let $g: \mathbb{R} \to \mathbb{R}$ via $g(x) = f(x)+I_{\{0\}}(x)$. Then the random process given by $X(t) = Gg(t)$ has the desired properties. □

Example 7.41. *There exists a probability space $(\Omega, \mathcal{F}, P)$ and a stationary zero mean Gaussian random process $\{X(t): t \in \mathbb{R}\}$ defined thereon such that for any real number t, the range of $X(t)$ is a saturated non-Lebesgue measurable set.*

Proof: Recall from Example 1.46 that there exists a partition $\{S_t: t \in \mathbb{R}\}$ of the real line into saturated non-Lebesgue measurable sets. Further, recall from Example 7.18 that for any saturated non-Lebesgue measurable set S, there exists a probability measure P on $\mathcal{B}(S)$ such that the natural injection from S into $\mathbb{R}$ is a standard Gaussian random variable defined on $(S, \mathcal{B}(S), P)$ such that $X(S) = S$. Now, for each $t \in \mathbb{R}$, let $(S_t, \mathcal{B}(S_t), P_t)$ be such a probability space and let Y_t be such a random variable. Further, consider the probability space $(\Omega, \mathcal{F}, P)$ given by the product of the probability spaces $(S_t, \mathcal{B}(S_t), P_t)$. Define a random process $\{X(t): t \in \mathbb{R}\}$ via $X(t, \omega) = Y_t(\omega_t)$, where ω_t is the tth coordinate in Ω. It then follows that $\{X(t): t \in \mathbb{R}\}$ is a zero mean Gaussian random process such that the range of $X(t)$ is S_t, a saturated non-Lebesgue measurable set. □

Example 7.42. *There exists a probability space $(\Omega, \mathcal{F}, P)$, and a stationary Gaussian random process $\{X(t): t \in \mathbb{R}\}$ defined thereon and composed of mutually independent standard Gaussian random variables such that, given any real number v, there exists one and only one real number t for which $X(t, \omega) = v$ for some $\omega \in \Omega$.*

Proof: Consider the partition of the real line $\{S_t: t \in \mathbb{R}\}$, the probability space $(\Omega, \mathcal{F}, P)$, and the stochastic process $\{X(t): t \in \mathbb{R}\}$ defined thereon that were given in Example 7.30. As before, let $\omega = \{\omega_t\}_{t \in \mathbb{R}}$ denote a generic element of Ω. Since for each real number t, the set $X(t, \Omega)$ is equal to S_t, a set in the partition given above for $\mathbb{R}$, the result follows immediately. □

Example 7.43. *There exists a probability space $(\Omega, \mathcal{F}, P)$, and a stationary Gaussian random process $\{X(t): t \in \mathbb{R}\}$ defined thereon and composed* *of mutually independent standard Gaussian random variables such that, given any fixed $\omega \in \Omega$, there exist no distinct real numbers t_1 and t_2 such that $X(t_1, \omega) = X(t_2, \omega)$.*

Proof: Consider the partition of the real line $\{S_t: t \in \mathbb{R}\}$, the probability space $(\Omega, \mathcal{F}, P)$, and the stochastic process $\{X(t): t \in \mathbb{R}\}$ defined thereon that were given in Example 7.30. As before, let $\omega = \{\omega_t\}_{t \in \mathbb{R}}$ denote a generic element of Ω. Note that $X(t_1, \omega) \in S_{t_1}$ and $X(t_2, \omega) \in S_{t_2}$. Since S_{t_1} and S_{t_2} are disjoint, the result follows. □

Example 7.44. *There exists a probability space $(\Omega, \mathcal{F}, P)$, and a stationary Gaussian random process $\{X(t): t \in \mathbb{R}\}$ defined thereon and composed of mutually independent standard Gaussian random variables such that for any real number t, $\bigcup_{y \in \mathbb{R} \setminus \{t\}} X(y, \Omega)$ is a proper subset of the real line.*

Proof: Consider the partition of the real line $\{S_t: t \in \mathbb{R}\}$, the probability space $(\Omega, \mathcal{F}, P)$, and the stochastic process $\{X(t): t \in \mathbb{R}\}$ defined thereon that were given in Example 7.30. As before, let $\omega = \{\omega_t\}_{t \in \mathbb{R}}$ denote a generic element of Ω. Note that $\bigcup_{y \in \mathbb{R} \setminus \{t\}} X(y, \Omega) = \bigcup_{y \in \mathbb{R} \setminus \{t\}} S_y = \mathbb{R} \setminus S_t$. The result now follows since S_t is nonempty. □

Example 7.45. *There exists a probability space $(\Omega, \mathcal{F}, P)$, and a stationary Gaussian random process $\{X(t): t \in \mathbb{R}\}$ defined thereon and composed of mutually independent standard Gaussian random variables such that for any disjoint subsets A_1 and A_2 of $\mathbb{R}$ and any $\omega \in \Omega$, it follows that $X(A_1, \omega) \cap X(A_2, \omega) = \emptyset$.*

Proof: Consider the partition of the real line $\{S_t: t \in \mathbb{R}\}$, the probability space $(\Omega, \mathcal{F}, P)$, and the stochastic process $\{X(t): t \in \mathbb{R}\}$ defined thereon that were given in Example 7.30. As before, let $\omega = \{\omega_t\}_{t \in \mathbb{R}}$ denote a generic element of Ω. Note that $X(A_1, \omega)$ is a subset of $\bigcup_{t \in A_1} S_t$ and $X(A_2, \omega)$ is a subset of $\bigcup_{t \in A_2} S_t$. The result now follows immediately since A_1 and A_2 are disjoint and since the S_t's partition $\mathbb{R}$. □

Example 7.46. *A Gaussian process indexed by $[0, \infty)$ with continuous sample paths and with stationary, mutually independent increments need not be a Brownian motion.*

Proof: Let G be a standard Gaussian random variable defined on some probability space. Then define the random process $\{X(t): t \in \mathbb{R}\}$ via $X(t) = G$, and note that the result follows. □

Example 7.47. *The Skorohod imbedding theorem may fail for $\mathbb{R}^2$-valued Brownian motion.*

Proof: This example was suggested by Robert Dalang. Recall that Brownian motion $\{B(t): t \in [0, \infty)\}$ is a zero mean Gaussian random process such that all of its sample paths are continuous and such that $\mathrm{E}[B(t)B(s)] = \min\{t, s\}$. The Skorohod imbedding theorem states that for any zero mean second-order random variable X there exists a stopping time T for the Brownian motion $\{B(t): t \in [0, \infty)\}$ such that $B(T)$ has the same distribution as X.

Now consider $\mathbb{R}^2$-valued Brownian motion $\{B(t): t \in [0, \infty)\}$ and an $\mathbb{R}^2$-valued zero mean second-order random variable X. Recall that the components of $\{B(t): t \in [0, \infty)\}$ are real-valued Brownian motions. Let X be such that $X = 0$ with probability 1/2 and such that X has a Euclidean norm greater than one with probability 1/2. Recalling that the Bessel process associated with $\{B(t): t \in [0, \infty)\}$ is almost surely strictly positive for all nonzero t, we see that for any stopping time T, $B(T)$ cannot have the distribution of X. □

Example 7.48. *There exists a Gaussian random process $\{X(t): t \in [0, \infty)\}$ taking values in $\mathbb{R}^2$ such that the inner product of $X(t)$ with any point in $\mathbb{R}^2$ other than the origin is a standard real-valued Brownian motion, yet $\{X(t): t \in [0, \infty)\}$ is not standard Brownian motion.*

Proof: This example is from Hardin [1985]. Let $\{B_1(t): t \in [0, \infty)\}$ and $\{B_2(t): t \in [0, \infty)\}$ be two independent standard Brownian motion processes and let $W_1(t) = B_1(2t/3) - B_2(t/3)$ and $W_2(t) = B_1(t/3) + B_2(2t/3)$ for $t \in [0, \infty)$. Consider any point in $\mathbb{R}^2$ that is not equal to the origin and denote its first and second coordinates as λ_1 and λ_2, respectively. Now, let $Y(t) = \lambda_1 W_1(t) + \lambda_2 W_2(t)$ for $t \in [0, \infty)$ and notice that

$$\begin{aligned} \mathrm{E}[Y(s)Y(t)] =& \lambda_1^2 [\min(2t/3, 2s/3) + \min(t/3, s/3)] \\ &+ \lambda_1\lambda_2 [\min(2t/3, s/3) - \min(t/3, 2s/3)] \\ &+ \lambda_2\lambda_1 [\min(2s/3, t/3) - \min(s/3, 2t/3)] \\ &+ \lambda_2^2 [\min(t/3, s/3) + \min(2t/3, 2s/3)] \\ =& (\lambda_1^2 + \lambda_2^2) \min(t, s). \end{aligned}$$

That is, the random process $\{W(t): t \in [0, \infty)\} = \{(W_1(t), W_2(t)): t \in [0, \infty)\}$ is such that the inner product of $W(t)$ with any point in $\mathbb{R}^2$ behaves as if $\{W(t): t \in [0, \infty)\}$ were a standard Brownian motion process in $\mathbb{R}^2$. But $\{W_1(t): t \in [0, \infty)\}$ and $\{W_2(t): t \in [0, \infty)\}$ are not independent random processes, and hence $\{W(t): t \in [0, \infty)\}$ cannot be a standard Brownian motion process in $\mathbb{R}^2$. □

References

Billingsley, P., *Probability and Measure,* Second edition, John Wiley, New York, 1986.

Devroye, L., "A note on random variate generation," *Journal of Statistical Computation and Simulation,* Vol. 14, February 1982, pp. 149–158.

Doob, J. L., *Stochastic Processes,* John Wiley, New York, 1953.

Doob, J. L., "On a problem of Marczewski," *Colloquium Mathematicum,* Vol. 1, pp. 216–217, 1948.

Emmanuele, G., A. Villani, and L. Takács, "Problem 6452," *American Mathematical Monthly,* Vol. 92, August–September, 1985, p. 515.

Feller, W., *An Introduction to Probability Theory and its Applications,* Vol. 2, Second edition, John Wiley, New York, 1971.

Flury, B. K., "On sums of random variables and independence," *The American Statistician,* Vol. 40, August 1986, pp. 214–215.

Hall, E. B., and G. L. Wise, "A result on multidimensional quantization," *Proceedings of the American Mathematical Society,* Vol. 118, June 1993, pp. 609–613.

Hardin, C. D., Jr., "A spurious Brownian motion," *Proceedings of the American Mathematical Society,* Vol. 93, February 1985, p. 350.

Jessen, B., "On two notions of independent functions," *Colloquium Mathematicum,* Vol. 1, pp. 214–215, 1948.

Morrison, J. M., and G. L. Wise, "Continuity of filtrations of sigma algebras," *Statistics and Probability Letters,* Vol. 6, No. 1, September 1987, pp. 55–60.

Perlman, M. D., and M. J. Wichura, "A note on substitution in conditional distribution," *The Annals of Statistics,* Vol. 3, No. 5, 1975, pp. 1175–1179.

Pierce, D. A., and R. L. Dykstra, "Independence and the normal distribution," *The American Statistician,* Vol. 23, No. 4, October 1969, p. 39.

Varadarayan, V., "On an existence theorem for probability spaces," *Selected Translations in Mathematical Statistics and Probability,* Vol. 2, pp. 237–240, 1962.

8. Conditioning

Let $(\Omega, \mathcal{F}, P)$ be a probability space. Let $\mathcal{A}$ be a σ-subalgebra of $\mathcal{F}$ and let X be a random variable such that either X^+ or X^- is integrable. Define a signed measure μ on $\mathcal{A}$ via $\mu(A) = \int_A X\,dP$. Note that μ is absolutely continuous with respect to $P|_{\mathcal{A}}$. The conditional expectation $\mathrm{E}[X \mid \mathcal{A}]$ is defined to be the Radon–Nikodym derivative $d\mu/dP|_{\mathcal{A}}$. Any $\mathcal{A}$-measurable random variable equal almost surely to $\mathrm{E}[X \mid \mathcal{A}]$ is said to be a version of $\mathrm{E}[X \mid \mathcal{A}]$. The conditional probability $P(F \mid \mathcal{A})$ of an event F conditioned on the σ-subalgebra $\mathcal{A}$ is defined as $P(F \mid \mathcal{A}) = \mathrm{E}[\mathrm{I}_F \mid \mathcal{A}]$. A function Q: $\Omega \times \mathcal{B}(\mathbb{R}) \to [0, 1]$ is called a regular conditional probability for X given $\mathcal{A}$ if $Q(\omega, \cdot)$ is a probability measure for each fixed $\omega \in \Omega$, and if for each fixed $B \in \mathcal{B}([0, 1])$, $Q(\cdot, B)$ is almost surely equal to $P(X \in B \mid \mathcal{A})$. Since conditional expectation is a Radon–Nikodym derivative, the standard properties of conditional expectation follow from basic results in integration theory. In particular, if X is $\mathcal{A}$-measurable, then X is a version of $\mathrm{E}[X \mid \mathcal{A}]$. If Z is a random variable defined on $(\Omega, \mathcal{F}, P)$, $\mathrm{E}[X \mid Z]$ will mean $\mathrm{E}[X \mid \sigma(Z)]$. It is a consequence of the Radon–Nikodym theorem that there exists a Borel measurable function g: $\mathbb{R} \to \overline{\mathbb{R}}$ such that a version of $\mathrm{E}[X \mid Z]$ can be written as $g(Z)$.

It should be obvious that a random variable equal almost surely to a conditional expectation need not be a version of that conditional expectation; indeed, it is if and only if it is $\mathcal{A}$-measurable. Also, $\mathrm{E}[X + Y \mid \mathcal{A}]$ need not be expressible as $\mathrm{E}[X \mid \mathcal{A}] + \mathrm{E}[Y \mid \mathcal{A}]$; indeed, $\mathrm{E}[X + Y \mid \mathcal{A}]$ could almost surely equal zero when the sum $\mathrm{E}[X \mid \mathcal{A}] + \mathrm{E}[Y \mid \mathcal{A}]$ is not defined. Furthermore, when X and Y are second-order random variables, $\mathrm{E}[X \mid Y]$ could equal X pointwise even though the covariance between X and Y is zero.

Let $(\Omega, \mathcal{F})$ be a measurable space and let $\mathcal{P}$ be a nonempty family of probability measures on $(\Omega, \mathcal{F})$. Then $(\Omega, \mathcal{F}, \mathcal{P})$ is called a probability structure; it is also sometimes called a statistical experiment. A σ-subalgebra $\mathcal{S}$ of $\mathcal{F}$ is said to be sufficient if for each $\mathcal{F}$-measurable bounded real-valued function f defined on Ω, there exists an $\mathcal{S}$-measurable bounded real-valued function g defined on Ω such that $\int_A f\,dP = \int_A g\,dP$ for each $A \in \mathcal{S}$ and for all $P \in \mathcal{P}$; that is, g is a.e. $[P]$ equal to the conditional expectation of f conditioned on $\mathcal{S}$ when P is the relevant probability measure, and g is not dependent on P, but the associated set of P measure zero might depend on P. Intuitively, there may be sets in $\mathcal{F}$ that are not in $\mathcal{S}$, but these are irrelevant to drawing inferences about the unknown probability measure $P \in \mathcal{P}$. A real-valued sufficient statistic is a real-valued measurable function X defined on $(\Omega, \mathcal{F})$ such that $\sigma(X) = X^{-1}(\mathcal{B}(\mathbb{R}))$ is sufficient.

Consider a probability space $(\Omega, \mathcal{F}, P)$ and a σ-subalgebra $\mathcal{H}$ of $\mathcal{F}$. Further, let $\mathcal{H}_1$ and $\mathcal{H}_2$ be two families each composed of elements from $\mathcal{F}$. The families $\mathcal{H}_1$ and $\mathcal{H}_2$ are said to be conditionally independent given $\mathcal{H}$ if $P(A_1 \cap A_2 \mid \mathcal{H}) = P(A_1 \mid \mathcal{H})P(A_2 \mid \mathcal{H})$ a.s. for all $A_1 \in \mathcal{H}_1$ and for all $A_2 \in \mathcal{H}_2$. Further, two random variables X and Y defined on $(\Omega, \mathcal{F}, P)$

are said to be conditionally independent given $\mathcal{H}$ if $\sigma(X)$ and $\sigma(Y)$ are conditionally independent given $\mathcal{H}$.

Example 8.1. *Conditional probability and knowledge need not be related.*

Proof: The topic of σ-algebras is basic to the subject of conditioning since conditioning is conventionally taken with respect to a σ-algebra. In many cases the σ-algebra of interest is that generated by some random variables representing data. Hence, in applications, it is common to treat σ-algebras as somehow representing knowledge or information associated with data. The following example from Billingsley [1986] shows that associating σ-algebras with knowledge or information, as commonly understood, may lead to incorrect conclusions.

Consider the probability space $([0, 1], \mathcal{B}([0, 1]), \lambda)$, where λ denotes Lebesgue measure on $\mathcal{B}([0, 1])$ and consider the σ-subalgebra $\mathcal{G}$ given by the family of all subsets of $[0, 1]$ that are either countable or co-countable. Now, for $B \in \mathcal{B}([0, 1])$, consider the conditional probability $P(B \mid \mathcal{G})$. Since $\mathcal{G}$ contains all singletons $\{\omega\}$, and hence might be seen as being completely informative, one might suppose that $P(B \mid \mathcal{G})$ is equal to I_B. In other words, one might rationalize that to know the sets in $\mathcal{G}$ implies that one *knows* ω itself and hence *knows* whether or not ω is contained in B, leading to the conclusion that $P(B \mid \mathcal{G})$ should be one when ω is contained in B and zero otherwise. It follows quickly, however, from the definition of conditional probability that $P(B \mid \mathcal{G}) = P(B)$, except possibly off a countable subset of $[0, 1]$. □

Example 8.2. *Conditional probabilities need not be measures.*

Proof: Conditional probability has traditionally occupied a prominent position in much of the theory and applications of probability. Unfortunately, conditional probabilities do not necessarily possess many of the properties of a probability measure. Indeed, as the following example shows, conditional probabilities need not even be measures.

Consider a subset H of the interval $[0, 1]$ with the properties that the outer Lebesgue measure of H is 1 and the inner Lebesgue measure of H is 0. (See Example 1.46.) Further, let $\Omega = [0, 1]$ and let λ denote Lebesgue measure on $\mathcal{B}([0, 1])$. Define $\mathcal{F} = \{(H \cap B_1) \cup (H^c \cap B_2)\colon B_1, B_2 \in \mathcal{B}([0, 1])\}$ and note that $\mathcal{F}$ is a σ-algebra on Ω and that $\mathcal{B}([0, 1])$ is a σ-subalgebra of $\mathcal{F}$. Now, define a probability measure $P\colon \mathcal{F} \to [0, 1]$ on the measurable space $(\Omega, \mathcal{F})$ via $P((H \cap B_1) \cup (H^c \cap B_2)) = [\lambda(B_1) + \lambda(B_2)]/2$ to obtain a probability space $(\Omega, \mathcal{F}, P)$. (That P is well-defined follows from the properties of H.)

Consider now this probability space $(\Omega, \mathcal{F}, P)$. The following example, adapted from Billingsley [1986], shows that conditional probabilities need not be measures.

Since $P(H) = 1/2$ and $P(B) = \lambda(B)$ for $B \in \mathcal{B}([0, 1])$ imply that $P(H \cap B) = \lambda(B)/2 = P(H)P(B)$, it follows that H is independent of $\mathcal{B}([0, 1])$. Let F be a set in $\mathcal{F}$ with probability zero and assume that $P(\cdot \mid \mathcal{B}([0, 1]))(\omega)$ is a probability measure on $\mathcal{F}$ for each ω outside the null set F. Note that there exists a collection $\{A_n: n \in \mathbb{N}\}$ of subsets of $[0, 1]$ such that $\mathcal{B}([0, 1]) = \sigma(\{A_n: n \in \mathbb{N}\})$ and such that $\{A_n: n \in \mathbb{N}\}$ is closed under finite intersections. Define $K_n = \{\omega \in \Omega: P(A_n \,|\, \mathcal{B}([0, 1]))(\omega) = \mathrm{I}_{A_n}(\omega)\}$ and note that $K_n \in \mathcal{B}([0, 1])$ and $P(K_n) = 1$ for all $n \in \mathbb{N}$ since $P(A_n \,|\, \mathcal{B}([0, 1])) = \mathrm{I}_{A_n}$ a.s. Now let $K = \bigcap_{n \in \mathbb{N}} K_n \cap F^c$ and note that $P(K) = 1$. Further, note that the function that, for a fixed ω in K, maps an element B of $\mathcal{B}([0, 1])$ to $\mathrm{I}_B(\omega)$ is a probability measure on $\mathcal{B}([0, 1])$, which agrees with $P(B \,|\, \mathcal{B}([0, 1]))(\omega)$ whenever $B \in \{A_n: n \in \mathbb{N}\}$. Thus, the Dynkin system theorem implies that for $\omega \in K$, $P(B \,|\, \mathcal{B}([0, 1]))(\omega)$ is uniquely determined to be $\mathrm{I}_B(\omega)$ for any set B in $\mathcal{B}([0, 1])$, and hence, in particular, if $\omega \in K$, then $P(\{\omega\} \,|\, \mathcal{B}([0, 1]))(\omega) = 1$. Recalling the assumption that $P(\cdot \,|\, \mathcal{B}([0, 1]))(\omega)$ is a probability measure on $\mathcal{F}$ for each ω outside the null set F, it follows that if $\omega \in H \cap K$, then $P(H \,|\, \mathcal{B}([0, 1]))(\omega) \geq P(\{\omega\} \,|\, \mathcal{B}([0, 1]))(\omega) = 1$, and if $\omega \in H^c \cap K$, then $P(H \,|\, \mathcal{B}([0, 1]))(\omega) \leq P(\{\omega\}^c \,|\, \mathcal{B}([0, 1]))(\omega) = 0$. Thus, if $\omega \in K$, then $P(H \,|\, \mathcal{B}([0, 1]))(\omega) = \mathrm{I}_H(\omega)$. But H and $\mathcal{B}([0, 1])$ are independent, and hence $P(H \,|\, \mathcal{B}([0, 1])) = P(H) = 1/2$ a.s. This contradiction implies that $P(\cdot \,|\, \mathcal{B}([0, 1]))(\omega)$ is not almost surely a probability measure on $\mathcal{F}$. Hence, a conditional probability is not necessarily a measure. □

Example 8.3. *Regular conditional probabilities need not exist.*

Proof: A regular conditional probability allows one to sidestep many of the undesirable aspects of conditional probability since a regular conditional probability is by definition required to be a measure for each fixed $\omega \in \Omega$. Unfortunately, however, regular conditional probabilities do not always exist. In fact, the situation described in Example 8.2, in addition to showing that a conditional probability need not be a measure, also provides an example in which a regular conditional probability does not exist. □

Example 8.4. *Independence need not imply conditional independence.*

Proof: The concept of conditional independence arises frequently in many aspects of probability theory. For example, the concept plays an important role in the study of Markov processes. Unfortunately, misconceptions often arise regarding the relationship between conditional independence and independence. As the following example, suggested by Chow and Teicher [1988], indicates, independence need not imply conditional independence.

Consider a probability space $(\Omega, \mathcal{F}, P)$ and a σ-subalgebra $\mathcal{H}$ of $\mathcal{F}$. Further, let $\mathcal{H}_1$ and $\mathcal{H}_2$ be two families each composed of elements from $\mathcal{F}$. The families $\mathcal{H}_1$ and $\mathcal{H}_2$ are said to be conditionally independent given $\mathcal{H}$ if $P(A_1 \cap A_2 \,|\, \mathcal{H}) = P(A_1 \,|\, \mathcal{H})P(A_2 \,|\, \mathcal{H})$ a.s. for all $A_1 \in \mathcal{H}_1$ and for all $A_2 \in \mathcal{H}_2$. Further, two random variables X and Y defined on $(\Omega, \mathcal{F}, P)$ are said to be conditionally independent given $\mathcal{H}$ if $\sigma(X)$ and $\sigma(Y)$ are conditionally independent given $\mathcal{H}$.

Let X_1 and X_2 be two independent, identically distributed random variables such that $P(X_1 = 1) = P(X_1 = -1) = 1/2$. Further, let $Z = X_1 + X_2$ and let $A_i = X_i^{-1}(\{1\})$ for $i = 1$ and 2. In this case, $P(A_i \,|\, Z) = 1/2$ on $Z^{-1}(\{0\})$ for $i = 1$ or 2, and $P(A_1 \cap A_2 \,|\, Z) = 0$ on $Z^{-1}(\{0\})$.

In particular, $P(A_1 \cap A_2 \mid Z) \neq P(A_1 \mid Z)P(A_2 \mid Z)$ on an event of positive probability. Thus, the independent random variables X_1 and X_2 are not conditionally independent given $\sigma(Z)$. □

Example 8.5. *Conditional independence need not imply independence.*

Proof: The following example, adapted from Chow and Teicher [1988], shows that conditional independence need not imply independence. Consider three mutually independent random variables Y_1, Y_2, and Y_3 such that each random variable takes on only integer values and such that $P(Y_i = m) < 1$ for all integers m and for $i = 1, 2$, and 3. Further, let $S_2 = Y_1 + Y_2$ and $S_3 = Y_1 + Y_2 + Y_3$ and notice that Y_1 and S_3 are dependent random variables. Let $B_i = S_2^{-1}(\{i\})$ for each integer i. There exists an integer k such that B_k has positive probability. On such a set B_k it follows that

$$\begin{aligned}
P(Y_1 = i,\, S_3 = j \mid S_2 = k) &= P(Y_1 = i,\, S_2 = k,\, S_3 = j)/P(S_2 = k) \\
&= P(Y_1 = i)P(Y_2 = k - i)P(Y_3 = j - k)/P(S_2 = k) \\
&= [P(Y_1 = i)P(Y_2 = k - i)/P(S_2 = k)]P(Y_3 = j - k) \\
&= [P(Y_1 = i,\, S_2 = k)/P(S_2 = k)]P(Y_3 = j - k) \\
&= P(Y_1 = i \mid S_2 = k)P(Y_3 = j - k) \\
&= P(Y_1 = i \mid S_2 = k)[P(Y_3 = j - k)P(S_2 = k)/P(S_2 = k)] \\
&= P(Y_1 = i \mid S_2 = k)[P(S_2 = k,\, S_3 = j)/P(S_2 = k)] \\
&= P(Y_1 = i \mid S_2 = k)P(S_3 = j \mid S_2 = k).
\end{aligned}$$

Thus, even though Y_1 and S_3 are dependent random variables, Y_1 and S_3 are conditionally independent given $\sigma(S_2)$. □

Example 8.6. *Conditional expectation need not be a smoothing operator.*

Proof: A commonly occurring misconception regarding conditional expectation is that it is a "smoothing" operator. Consider, for example, a random process $\{X(t): t \in \mathbb{R}\}$ defined on a probability space $(\Omega, \mathcal{F}, P)$ and a σ-subalgebra $\mathcal{H}$ of $\mathcal{F}$. It has been argued by some that $\mathrm{E}[X(t) \mid \mathcal{H}]$ is "smoother" than $X(t)$ as a function of t. To dispel this notion simply let $\{X(t): t \in \mathbb{R}\}$ be a random process that is discontinuous everywhere, such that for each $t \in \mathbb{R}$, $X(t)$ is an $\mathcal{H}$-measurable random variable. The version of $\mathrm{E}[X(t) \mid \mathcal{H}]$ given by $X(t)$ obviously retains this same property.

Perhaps a little less obvious is the fact that, for a random variable X on $(\Omega, \mathcal{F}, P)$ and a σ-subalgebra $\mathcal{G}$ of $\mathcal{F}$, $\mathrm{E}[X \mid \mathcal{G}]$ need not be as smooth a function of ω as X, where the set Ω is assumed to have some associated topology. Consider, for instance, the probability space given by the interval $[0, 1]$, the σ-algebra $\mathcal{G}$ given by the countable and co-countable subsets of $[0, 1]$, and Lebesgue measure on $\mathcal{G}$. If $X = 1$, then a version of $\mathrm{E}[X \mid \mathcal{G}]$ is given by $1 - I_B$, where B equals the set of rationals in $[0, 1]$. Hence, for this choice of X, even though X is everywhere continuous, there exists a version of $\mathrm{E}[X \mid \mathcal{G}]$ that is everywhere discontinuous. □

Example 8.7. *The existence of* $\mathrm{E}[\mathrm{E}[X|Y]]$ *does not imply that the mean of* X *exists.*

Proof: A commonly encountered property of conditional expectation is the so-called nesting property. Unfortunately, this property is sometimes misapplied. In this example, from Enis [1973], it is shown that $\mathrm{E}[\mathrm{E}[X \mid Y]]$ may exist even when the expectation of X does not exist. In other words, before calculating $\mathrm{E}[\mathrm{E}[X \mid Y]]$ and claiming one has found the mean of X, it is necessary first to ascertain that the mean of X actually exists.

Consider random variables X and Y defined on the same probability space such that Y possesses a probability density function $g(y)$ given by

$$g(y) = \frac{1}{\sqrt{2\pi y}} \exp\left(\frac{-y}{2}\right), \; y > 0,$$

and, for each $y > 0$, X and Y are such that a conditional density function of X given $Y = y$, denoted by $f(x \mid y)$, exists and is given by

$$f(x \mid y) = \sqrt{\frac{y}{2\pi}} \exp\left(\frac{-yx^2}{2}\right), \; y > 0 \text{ and } x \in \mathbb{R}.$$

It follows easily that $\mathrm{E}[X \mid Y] = 0$ a.s. and thus that $\mathrm{E}[\mathrm{E}[X \mid Y]] = 0$. Notice, however, that the mean of X does not exist since X has a Cauchy probability density function $h(x)$ given by

$$h(x) = \int_{\mathbb{R}} f(x \mid y) g(y)\, dy = \frac{1}{\pi(1 + x^2)}$$

for $x \in \mathbb{R}$. □

Example 8.8. *Conditional expectation of an integrable random variable need not be obtainable from a corresponding conditional probability distribution.*

Proof: Let X be an integrable random variable defined on a probability space $(\Omega, \mathcal{F}, P)$ and let $\mathcal{H}$ be a σ-subalgebra of $\mathcal{F}$. A common misconception is that the conditional expectation $\mathrm{E}[X \mid \mathcal{H}]$ can always be expressed as $\int_\Omega X\, dP(\cdot \mid \mathcal{H})$, where $P(\cdot \mid \mathcal{H})$ denotes conditional probability given $\mathcal{H}$. In fact, recalling the result in Example 8.2, $P(\cdot \mid \mathcal{H})$ might not be a measure, and hence the preceding integral might not even be defined. □

Example 8.9. $\mathrm{E}[X \mid \mathcal{F}]$ *and* $\mathrm{E}[Y \mid \mathcal{F}]$ *need not be independent if* X *and* Y *are independent.*

Proof: Let S and T be independent, zero mean, unit variance Gaussian random variables defined on a common probability space and let $X = S + T$ and $Y = S - T$. Further, let $\mathcal{F} = \sigma(S)$. Note that $\mathrm{E}[X \mid \mathcal{F}] = S$ a.s. and $\mathrm{E}[Y \mid \mathcal{F}] = S$ a.s. □

Example 8.10. *There exist random variables X, Y_1, and Y_2 such that* $\mathrm{E}[X \mid Y_1] = \mathrm{E}[X \mid Y_2] = 0$, *and yet* $\mathrm{E}[X \mid Y_1, Y_2] = X$.

Proof: This example is from Hall, Wessel, and Wise [1991]. Let $\Omega = [0, 1]$, $\mathcal{F}$ denote the Borel subsets of Ω, and P denote Lebesgue measure on $\mathcal{F}$. Let $Y_1(\omega) = \mathrm{I}_{[0, 1/2)}(\omega)$, $Y_2(\omega) = \mathrm{I}_{[1/4, 3/4)}(\omega)$, and $X(\omega) = 1$ for $\omega \in [0, 1/4) \cup [1/2, 3/4)$ and $X(\omega) = -1$ for $\omega \in [1/4, 1/2) \cup [3/4, 1]$. Then it straightforwardly follows that $\mathrm{E}[X \mid Y_1] = \mathrm{E}[X \mid Y_2] = 0$ a.s., but $\mathrm{E}[X \mid Y_1, Y_2] = X$ a.s. Recalling that $\mathrm{E}[X \mid Y_1]$ and $\mathrm{E}[X \mid Y_2]$, respectively, are $\sigma(Y_1)$-measurable and $\sigma(Y_2)$-measurable, we see that $\mathrm{E}[X \mid Y_1] = \mathrm{E}[X \mid Y_2] = 0$ pointwise in ω. Similarly, we see that $\mathrm{E}[X \mid Y_1, Y_2] = X$ pointwise in ω. Thus, in this situation, it is fruitless to attempt to approximate X based on any function of $\mathrm{E}[X \mid Y_1]$ and $\mathrm{E}[X \mid Y_2]$. □

Example 8.11. *Conditional expectation need not minimize mean-square error.*

Proof: One of the most common misconceptions in probability theory is that conditional expectation minimizes mean-square error. This mistaken concept arises in many $\mathbf{L}_2$ minimization problems in probability and statistics as well as in estimation and filtering applications in engineering. As the following example from Wise [1985] indicates, even for bounded random variables, conditional expectation may not even come close to minimizing the mean-square error even though there exists a function mapping the reals into the reals by which the random variable of interest may be estimated precisely.

Consider a subset H of the interval $[0, 1]$ with the properties that the outer Lebesgue measure of H is 1 and the inner Lebesgue measure of H is 0. (See Example 1.46.) Further, let $\Omega = [0, 1]$ and let λ denote Lebesgue measure on $\mathcal{B}([0, 1])$. Define $\mathcal{F} = \{(H \cap B_1) \cup (H^c \cap B_2) \colon B_1, B_2 \in \mathcal{B}([0, 1])\}$ and note that $\mathcal{F}$ is a σ-algebra on Ω and that $\mathcal{B}([0, 1])$ is a σ-subalgebra of $\mathcal{F}$. Now, define a probability measure $P\colon \mathcal{F} \to [0, 1]$ on the measurable space $(\Omega, \mathcal{F})$ via $P((H \cap B_1) \cup (H^c \cap B_2)) = [\lambda(B_1) + \lambda(B_2)]/2$ to obtain a probability space $(\Omega, \mathcal{F}, P)$ as in Example 7.7. Further, let A be a fixed nonzero real number and define two random variables X and Y on $(\Omega, \mathcal{F}, P)$ via $X(\omega) = \omega$ and $Y(\omega) = A\mathrm{I}_H(\omega)$. Notice that $\sigma(X) = \mathcal{B}([0, 1])$ and that $\sigma(Y) = \{\Omega, \emptyset, H, H^c\}$. Further, since $P(H) = 1/2$ and $P(B) = \lambda(B)$ for $B \in \mathcal{B}([0, 1])$, it follows that $P(H \cap B) = \lambda(B)/2 = P(H)P(B)$, or that X is independent of Y. Hence $\mathrm{E}[Y \mid X] = \mathrm{E}[Y] = A/2$ a.s., which implies that $\mathrm{E}[(Y - \mathrm{E}[Y \mid X])^2] = \mathrm{E}[[Y - (A/2)]^2] = A^2/4$. But, $Y(\omega) = A\mathrm{I}_H(X(\omega))$ for all $\omega \in \Omega$. Thus, $\mathrm{E}[[Y - A\mathrm{I}_H(X)]^2] = 0$. In other words, for this example there exists a function $f\colon \mathbb{R} \to \mathbb{R}$ such that $Y(\omega) = f(X(\omega))$ for all $\omega \in \Omega$, yet, by choice of A, $\mathrm{E}[(Y - \mathrm{E}[Y \mid X])^2]$ may be arbitrarily large. Note further that in this case $\sigma(Y)$ is finite, $\sigma(X)$ contains all singletons and is countably generated, and all moments of X and Y exist. It follows straightforwardly that this same phenomenon can be exhibited when Y is as above and X has a standard Gaussian distribution. □

Example 8.12. *The existence of a joint probability density function need* *not imply that the regression function satisfies any regularity property beyond Borel measurability.*

Proof: Given two random variables X and Y defined on the same probability space, a common problem concerns the determination of the form of the regression function $\mathrm{E}[X \mid Y = y]$. For example, Vitale and Pipkin [1976] considered this problem when both X and Y are uniformly distributed. In this example, from Hall and Wise [1991], it is shown that the existence of a joint probability density function for X and Y in no way guarantees that the regression function will obey any regularity property other than Borel measurability.

Let g: $\mathbb{R} \to \mathbb{R}$ be Borel measurable and define

$$f(x, y) = \frac{1}{4}\exp[-\exp(|y|)|x - g(y)|].$$

Note that $f(x, y)$ is a joint probability density function since

$$\begin{aligned}\int_{\mathbb{R}}\int_{\mathbb{R}} f(x, y)\,dx\,dy &= \int_{\mathbb{R}}\int_{\mathbb{R}} \frac{1}{4}\exp[-\exp(|y|)|x - g(y)|]\,dx\,dy \\ &= \int_{\mathbb{R}}\int_{\mathbb{R}} \frac{1}{4}\exp[-\exp(|y|)|z|]\,dz\,dy \\ &= \int_{\mathbb{R}} \frac{1}{2}\exp(-|y|)\,dy = 1.\end{aligned}$$

Let X and Y be random variables such that the pair (X, Y) has a joint probability density function given by $f(x, y)$. Notice from the above calculation that a second marginal probability density function of $f(x, y)$ is given by $f_Y(y) = \exp(-|y|)/2$. Recall that a version of $\mathrm{E}[X \mid Y = y]$ is given by $\int_{\mathbb{R}} x\,[f(x, y)/f_Y(y)]\,dx$. This version will be used throughout the remainder of this example. Substituting for $f_Y(y)$ implies that

$$\begin{aligned}\mathrm{E}[X \mid Y = y] &= 2\exp(|y|)\int_{\mathbb{R}} \frac{x}{4}\exp[-\exp(|y|)|x - g(y)|]\,dx \\ &= 2\exp(|y|)\int_{\mathbb{R}} \frac{z + g(y)}{4}\exp[-\exp(|y|)|z|]\,dz \\ &= 2\exp(|y|)\,g(y)\frac{1}{2\exp(|y|)} = g(y).\end{aligned}$$

Hence, the random variables X and Y with the joint probability density function $f(x, y)$ are such that $\mathrm{E}[X \mid Y = y] = g(y)$, where g was an arbitrarily preselected Borel measurable function. □

Example 8.13. *A σ-subalgebra that includes a sufficient σ-subalgebra need not be sufficient.*

Proof: Let $(\Omega, \mathcal{F})$ be a measurable space and let $\mathcal{P}$ be a family of probability measures on $(\Omega, \mathcal{F})$. The triple $(\Omega, \mathcal{F}, \mathcal{P})$ is called a probability structure. If $\mathcal{S}$ is a σ-subalgebra of $\mathcal{F}$, then $\mathcal{S}$ is said to be sufficient if for each $\mathcal{F}$-measurable bounded real-valued function f defined on Ω, there exists an $\mathcal{S}$-measurable bounded real-valued function g defined on Ω such that $\int_A f\,dP = \int_A g\,dP$ for each A in $\mathcal{S}$ and for all P in $\mathcal{P}$. That is, g is almost surely $[P]$ equal to the conditional expectation of f conditioned on $\mathcal{S}$ when P is the relevant probability measure. (Note that although g does not depend on P, the set of P-measure zero might depend on P.) A common misconception is that if $\mathcal{S}$ is a sufficient σ-subalgebra, then any σ-subalgebra of $\mathcal{F}$ that includes $\mathcal{S}$ as a subset is also sufficient. The following example from Burkholder [1961] presents a nonsufficient σ-subalgebra that includes a sufficient σ-subalgebra.

Let $\mathcal{P}$ denote the family of probability measures P on $(\mathbb{R}, \mathcal{B}(\mathbb{R}))$ such that $P(B) = P(-B)$ for any set B in $\mathcal{B}(\mathbb{R})$, where, for any subset B of $\mathbb{R}$, $-B$ is defined to be $\{x \in \mathbb{R}\colon -x \in B\}$. Let $\mathcal{A} = \{B \in \mathcal{B}(\mathbb{R})\colon B = -B\}$ and note that $\mathcal{A}$ is a σ-subalgebra of $\mathcal{B}(\mathbb{R})$. Further, $\mathcal{A}$ is a sufficient σ-subalgebra since, given any bounded Borel measurable function f, $g(x) = [f(x) + f(-x)]/2$ is an $\mathcal{A}$-measurable function for which $\int_A f\,dP = \int_A g\,dP$ for any $A \in \mathcal{A}$ and any $P \in \mathcal{P}$.

Suppose now that Z is a subset of $\mathbb{R}$ that contains 0 and for which $Z = -Z$. Also, define $\mathcal{D} = \{B \cup A\colon B \in \mathcal{B}(\mathbb{R}),\ B \subset Z, \text{ and } A \in \mathcal{A}\}$. A straightforward examination shows that $\mathcal{D}$ is a σ-subalgebra of $\mathcal{B}(\mathbb{R})$ that includes $\mathcal{A}$.

Assume that $\mathcal{D}$ is a sufficient σ-subalgebra and let f be a bounded Borel measurable function. Then there exists a $\mathcal{D}$-measurable function g for which $\int_D f\,dP = \int_D g\,dP$ for any $D \in \mathcal{D}$ and any $P \in \mathcal{P}$. Let $x \in Z$ and note that $\{x\} \in \mathcal{D}$. Choosing $D = \{x\}$ above then implies that $f(x)P(\{x\}) = g(x)P(\{x\})$ for any measure P in $\mathcal{P}$. Now, let $x \in Z^c$ and note that $\{x, -x\} \in \mathcal{D}$, $\{x\} \notin \mathcal{D}$, and $\{-x\} \notin \mathcal{D}$. Letting $D = \{x, -x\}$ above implies that $[f(x) + f(-x)]P(\{x\}) = 2g(x)P(\{x\})$ since $P(\{x\}) = P(\{-x\})$, by definition of $\mathcal{P}$, and $g(x) = g(-x)$ since g is $\mathcal{D}$-measurable. Given any $x \in \mathbb{R}$, there exists a measure P in $\mathcal{P}$ for which $P(\{x\}) > 0$. Thus, it follows that $g(x) = f(x)$ if $x \in Z$ and $g(x) = [f(x) + f(-x)]/2$ if $x \in Z^c$. Let $f(x) = -1$ if $x < 0$ and $f(x) = 1$ if $x \geq 0$. This choice for f implies that g, as defined above, is nonzero on Z and zero on Z^c. Hence, $Z = \{g^{-1}(\{0\})\}^c \in \mathcal{D}$. Now choose a subset Z_0 of $\mathbb{R}$ that contains 0, is such that $Z_0 = -Z_0$, and is not an element of $\mathcal{B}(\mathbb{R})$. (Such sets abound!) Substituting Z_0 for Z thus implies, based on the above discussion, that $Z_0 \in \mathcal{D}$. But Z_0 cannot be in $\mathcal{D}$ since $Z_0 \notin \mathcal{B}(\mathbb{R})$. This contradiction implies that $\mathcal{D}$ is not a sufficient σ-subalgebra even though it includes the sufficient σ-subalgebra $\mathcal{A}$. □

Example 8.14. *There exists a sufficient statistic that is a function of a random variable that is not a sufficient statistic.*

Proof: The following example demonstrates that a nonsufficient statistic may be as "informative" as a sufficient statistic. In particular, a situation from Burkholder [1961] is presented in which a sufficient statistic is a function of a random variable that is not a sufficient statistic.

Let $S(x)$ and $T(x)$ be functions mapping $\mathbb{R}$ into $\mathbb{R}$ defined via $S(x) =$

$|x|$ and

$$T(x) = \begin{cases} x, & \text{if } x \in Z \\ |x|, & \text{if } x \in Z^c, \end{cases}$$

where Z, as in Example 8.13, is a subset of $\mathbb{R}$ that contains 0, is such that $Z = -Z$, and is not an element of $\mathcal{B}(\mathbb{R})$. Also, as in the prior example, let $\mathcal{A} = \{B \in \mathcal{B}(\mathbb{R})\colon B = -B\}$, let $\mathcal{D} = \{B \cup A\colon B \in \mathcal{B}(\mathbb{R}),\ B \subset Z,$ and $A \in \mathcal{A}\}$, and consider the probability structure $(\mathbb{R}, \mathcal{B}(\mathbb{R}), \mathcal{P})$, where $\mathcal{P}$ denotes the family of probability measures P on $(\mathbb{R}, \mathcal{B}(\mathbb{R}))$ such that $P(B) = P(-B)$ for any set B in $\mathcal{B}(\mathbb{R})$. Recall that $\mathcal{A}$ is a sufficient σ-subalgebra that is included in $\mathcal{D}$, a σ-subalgebra that is not sufficient.

Notice that $\sigma(S) = \mathcal{A}$ and that $\sigma(T) = \mathcal{D}$. That is, $\mathcal{A}$ is the smallest σ-subalgebra with respect to which S is measurable and $\mathcal{D}$ is the smallest σ-subalgebra with respect to which T is measurable. Recall that a statistic W is said to be sufficient if and only if $\sigma(W)$ is a sufficient σ-subalgebra. Hence, even though $S(x) = |T(x)|$, S is a sufficient statistic, whereas T is not. □

Example 8.15. *The smallest sufficient σ-algebra including two sufficient σ-algebras need not exist.*

Proof: This example is from Burkholder [1962]. Let Ω be the set of all ordered pairs $x = (x_1, x_2)$ of real numbers satisfying $|x_1| = |x_2| > 0$. Let $\mathcal{F}$ be the σ-algebra generated by the singleton sets in Ω and the set $D = \{x \in \Omega\colon x_1 = x_2\}$. For $x = (x_1, x_2)$ from Ω, let P_x denote the probability measure on $\mathcal{F}$ that assigns probability 1/4 to each of the points x, $(x_1, -x_2)$, $(-x_1, x_2)$, and $(-x_1, -x_2)$. Let $\mathcal{P} = \{P_x\colon x \in \Omega\}$.

If $x \in \Omega$, let $a_{0x} = \{x, (x_1, -x_2), (-x_1, x_2), (-x_1, -x_2)\}$. Let $S = \{x \in \Omega\colon |x_1| + |x_2| < 1\}$. Note that if $x \in S$, then $a_{0x} \subset S$, and if $x \notin S$, then $a_{0x} \subset (\Omega \setminus S)$. For $x \in \Omega$ let

$$a_{1x} = \begin{cases} a_{0x}, & \text{if } x \in S \\ \{x, (x_1, -x_2)\}, & \text{if } x \notin S \end{cases}$$

and let

$$a_{2x} = \begin{cases} a_{0x}, & \text{if } x \in S \\ \{x, (-x_1, x_2)\}, & \text{if } x \notin S. \end{cases}$$

If $i = 1, 2$, let $\mathcal{F}_i$ be the smallest σ-algebra containing each set a_{ix} for $x \in \Omega$. Clearly, $\mathcal{F}_i \subset \mathcal{F}$, $i = 1, 2$.

To show that $\mathcal{F}_1$ and $\mathcal{F}_2$ are sufficient, it is enough, by symmetry, to show that $\mathcal{F}_1$ is sufficient. Suppose that f is a bounded real-valued $\mathcal{F}$-measurable function. Let $g\colon \Omega \to \mathbb{R}$ via

$$g(x) = \begin{cases} \dfrac{f(x) + f((x_1, -x_2)) + f((-x_1, x_2)) + f((-x_1, -x_2))}{4}, & \text{if } x \in S \\[2ex] \dfrac{f(x) + f((x_1, -x_2))}{2}, & \text{if } x \in \Omega \setminus S. \end{cases}$$

Then g is constant on each set a_{1x}. Also, since f is $\mathcal{F}$-measurable, there is a real number c_1 such that the set $\{x \in \Omega\colon f(x) \neq c_1\} \cap D$ is countable and a real number c_2 such that $\{x \in \Omega\colon f(x) \neq c_2\} \cap (\Omega \setminus D)$ is countable, implying that $\{x \in \Omega\colon g(x) \neq (c_1 + c_2)/2\}$ is countable. Thus, g is $\mathcal{F}_1$-measurable. Let $A_1 \in \mathcal{F}_1$ and let $h = \mathrm{I}_{A_1}$. Then h is constant on each set a_{1x} and

$$\begin{aligned}\int_\Omega fh\,dP_x &= \frac{1}{4}[f(x) + f((x_1, -x_2)) + f((-x_1, x_2)) + f((-x_1, -x_2))]h(x) \\ &= g(x)h(x) = \int_\Omega gh\,dP_x\end{aligned}$$

if $x \in S$, and

$$\begin{aligned}\int_\Omega fh\,dP_x =&\frac{1}{4}[f(x) + f((x_1, -x_2))]h(x) \\ &+\frac{1}{4}[f((-x_1, x_2)) + f((-x_1, -x_2))]h((-x_1, x_2)) \\ =&\frac{1}{2}[g(x)h(x) + g((-x_1, x_2))h((-x_1, x_2))] = \int_\Omega gh\,dP_x\end{aligned}$$

if $x \in (\Omega \setminus S)$. Therefore, $\int_{A_1} f\,dP = \int_{A_1} g\,dP$ for all $P \in \mathcal{P}$, implying that $\mathcal{F}_1$ is sufficient.

Now let $\mathcal{G}$ denote the σ-algebra generated by $\mathcal{F}_1$ and $\mathcal{F}_2$. Clearly $\mathcal{G}$ is the smallest σ-algebra containing each set a_{0x}, $x \in S$ and each set $\{x\}$, $x \in (\Omega \setminus S)$. Suppose that $\mathcal{G}$ is sufficient. Then there is a $\mathcal{G}$-measurable real-valued function g such that $P(B \cap D) = \int_B g\,dP$ for $B \in \mathcal{G}$ and $P \in \mathcal{P}$. Therefore, $P_x(a_{0x} \cap D) = \int_{a_{0x}} g\,dP_x = g(x)$ if $x \in S$, $P_x(\{x\}) = \int_{\{x\}} g\,dP_x = g(x)/4$ if $x \in (\Omega \setminus S)$, implying that

$$g(x) = \begin{cases} \dfrac{1}{2}, & \text{if } x \in S \\ 0 \text{ or } 1, & \text{if } x \in \Omega \setminus S. \end{cases}$$

Since g is $\mathcal{G}$-measurable, $g^{-1}(\{1/2\}) = S$ must belong to $\mathcal{G}$. This contradicts the fact that no uncountable set whose complement is uncountable can belong to $\mathcal{G}$. Accordingly, the σ-subalgebra $\mathcal{G}$ is not sufficient.

For each x in S, let $\mathcal{G}_x = \{A \in \mathcal{F}\colon$ either $a_{0x} \subset A$ or $a_{0x} \subset \Omega \setminus A\}$. It follows easily from the properties of $\mathcal{P}$ and from the definition of $\mathcal{G}$ that $\mathcal{G}_x$ is a sufficient σ-algebra satisfying $\mathcal{G} \subset \mathcal{G}_x$ for $x \in S$. Suppose that a smallest sufficient σ-algebra $\mathcal{H}$ including $\mathcal{F}_1$ and $\mathcal{F}_2$ exists. Then $\mathcal{G} \subset \mathcal{H} \subset \mathcal{G}_x$ for $x \in S$, implying that $\mathcal{G} \subset \mathcal{H} \subset \bigcap_{x \in S} \mathcal{G}_x$.

Let $A \in \bigcap_{x \in S} \mathcal{G}_x$. If $x \in A \cap S$, then $x \in S$ implies that $A \in \mathcal{G}_x$, and this together with $x \in A$ implies that $a_{0x} \subset A$. Accordingly, if $A \cap S$ is uncountable, then both $A \cap D$ and $A \cap (\Omega \setminus D)$ are uncountable, implying that $\Omega \setminus A$ is countable. Hence, in this case, $A \in \mathcal{G}$. If $A \cap S$ is countable, then $(\Omega \setminus A) \cap S$ is uncountable, implying that $\Omega \setminus A$, hence A, belongs to $\mathcal{G}$. Thus $\bigcap_{x \in S} \mathcal{G}_x \subset \mathcal{G}$. Consequently, $\mathcal{G} = \mathcal{H}$, contradicting the fact that $\mathcal{G}$ is not sufficient. This implies that no smallest sufficient σ-algebra including $\mathcal{F}_1$ and $\mathcal{F}_2$ exists. □

Example 8.16. *Uniform integrability and almost sure convergence of a sequence of random variables need not imply convergence of the corresponding conditional expectations, and Fatou's lemma need not hold for conditional expectations.*

Proof: Fatou's lemma and uniform integrability are powerful tools in analysis and are often relied upon in the area of probability theory. Recall that if a sequence of random variables is uniformly integrable, then almost sure convergence implies convergence of the corresponding expectations. Convergence of conditional expectations with respect to an arbitrary σ-subalgebra, however, does not follow in general. The following example, adapted from Zheng [1980], describes a situation in which Fatou's lemma does not hold and in which uniform integrability and almost sure convergence do not imply that the corresponding conditional expectations converge.

Let $\Omega = (0, 1) \times (0, 1)$, let $\mathcal{H} = \{B \times (0, 1)\colon B \in \mathcal{B}((0, 1))\}$, and note that $\mathcal{H}$ is a σ-algebra on Ω. Let μ denote Lebesgue measure on $\mathcal{B}((0, 1))$ and let P denote Lebesgue measure on $\mathcal{B}((0, 1) \times (0, 1))$. For each positive integer n, let $B_n = (0, 1/n)$ and let A_n denote the nth term in the sequence $(0, 1/2)$, $(1/2, 1)$, $(0, 1/4)$, $(1/4, 1/2)$, $(1/2, 3/4)$, $(3/4, 1)$, $(0, 1/8)$, $(1/8, 1/4), \ldots$. Note that $(\Omega, \mathcal{B}((0, 1) \times (0, 1)), P)$ is a probability space and that $\mathcal{H}$ is a σ-subalgebra of $\mathcal{B}((0, 1) \times (0, 1))$. Now, for each positive integer n, define a random variable $X_n(x, y) = \mathrm{I}_{A_n \times B_n}(x, y)/\mu(B_n)$. Let $B \in \mathcal{B}((0, 1))$ and note that

$$\int_{B\times(0,1)} X_n \, dP = \int_{B\times(0,1)} \frac{1}{\mu(B_n)} \mathrm{I}_{A_n\times B_n} \, dP$$

$$= \mu(A_n \cap B) = \int_{B\times(0,1)} \mathrm{I}_{A_n\times(0,1)} \, dP,$$

which thus implies that $\mathrm{E}[X_n \mid \mathcal{H}] = \mathrm{I}_{A_n\times(0,1)}$ a.s. Next, note that $X_n \geq 0$ for each positive integer n and that $X_n \to 0$ as $n \to \infty$. Note also that, since $\mathrm{E}[|X_n|] = \mu(A_n) \to 0$, the random variables $\{X_n\colon n \in \mathbb{N}\}$ are uniformly integrable. Further, note that $\limsup_{n\to\infty} \mathrm{I}_{A_n\times(0,1)} = 1$ and that $\liminf_{n\to\infty} \mathrm{I}_{A_n\times(0,1)} = 0$. Thus, even though the random variables $\{X_n\colon n \in \mathbb{N}\}$ are uniformly integrable, $\limsup_{n\to\infty} \mathrm{E}[X_n \mid \mathcal{H}] = 1$ a.s. and $\liminf_{n\to\infty} \mathrm{E}[X_n \mid \mathcal{H}] = 0$ a.s. In particular, since $\limsup_{n\to\infty} \mathrm{E}[X_n \mid \mathcal{H}] = 1$ a.s. and $\mathrm{E}\left[\limsup_{n\to\infty} X_n \mid \mathcal{H}\right] = 0$ a.s., Fatou's lemma does not hold and the conditional expectations do not converge. □

Example 8.17. *There exists a martingale $\{X_n, \mathcal{F}_n\colon n \in \mathbb{N}\}$ defined on a probability space $(\Omega, \mathcal{F}, P)$ and an almost surely finite stopping time T relative to the filtration $\{\mathcal{F}_n\colon n \in \mathbb{N}\}$ such that $T \geq 1$ a.s. and yet such that $\mathrm{E}[X_T \mid \mathcal{F}_1]$ is not almost surely equal to X_1.*

Proof: This example is from Dudley [1989]. Let $\{Y_n\}_{n\in\mathbb{N}}$ be a sequence of mutually independent random variables such that $P(Y_n = 2^n) = P(Y_n = -2^n) = 1/2$ for each positive integer n and let $\{\mathcal{F}_n\colon n \in \mathbb{N}\}$ be defined via

$\mathcal{F}_n = \sigma(Y_1, \ldots, Y_n)$. Let T be the least n for which $X_n > 0$. Then T is a stopping time relative to $\{\mathcal{F}_n: n \in \mathbb{N}\}$ and is finite almost surely. Note that $X_T = 2$ almost surely. So, although 1 and T are stopping times with $1 \leq T$, $P(\mathrm{E}[X_T \mid \mathcal{F}_1] = X_1) = 1/2$. □

References

Billingsley, P., *Probability and Measure,* Second edition, John Wiley, New York, 1986.

Burkholder, D. L., "Sufficiency in the undominated case," *The Annals of Mathematical Statistics,* Vol. 32, No. 4, pp. 1191–1200, Dec., 1961.

Burkholder, D. L., "On the order structure of the set of sufficient subfields," *The Annals of Mathematical Statistics,* Vol. 33, No. 2, pp. 596–599, June, 1962.

Chow, Y. S., and H. Teicher, *Probability Theory: Independence, Interchangeability, Martingales,* Second edition, New York: Springer-Verlag, 1988.

Dudley, R. M., *Real Analysis and Probability,* Wadsworth & Brooks/Cole: Pacific Grove, California, 1989.

Enis, P., "On the relation $\mathrm{E}(X) = \mathrm{E}\{\mathrm{E}(X \mid Y)\}$," *Biometrika,* Vol. 60, pp. 432–433, 1973.

Hall, E. B., A. E. Wessel, and G. L. Wise, "Some aspects of fusion in estimation theory," *IEEE Transactions on Information Theory,* Vol. 37, No. 2, pp. 420–422, March 1991.

Hall, E. B., and G. L. Wise, "On optimal estimation with respect to a large family of cost functions," *IEEE Transactions on Information Theory,* Vol. 37, pp. 691–693, May 1991.

Vitale, A. R., and A. C. Pipkin, "Conditions on the regression function when both variables are uniformly distributed," *The Annals of Probability,* Vol. 4, No. 5, pp. 869–873, Oct., 1976.

Wise, G. L., "A note on a common misconception in estimation," *Systems and Control Letters,* Vol. 5, pp. 355–356, April, 1985.

Zheng, W., "A note on the convergence of sequences of conditional expectations of random variables," *Zeitschrift für Wahrscheinlichkeitstheorie und Verwandte Gebiete,* Vol. 53, pp. 291–292, Sept., 1980.

9. Convergence in Probability Theory

In this chapter we present counterexamples related to several different concepts involving convergence in probability theory. In particular, we consider infinite products of random variables, various forms of convergence of sequences of random variables, convergence of characteristic functions, convergence of density functions and distribution functions, the central limit theorem, the laws of large numbers, and martingales.

Let M be a metric space, and let $\mathcal{P}$ denote the family of all probability measures defined on $\mathcal{B}(M)$. The weak* topology on $\mathcal{P}$ is the topology with a neighborhood base consisting of all sets of the form

$$\left\{m \in \mathcal{P} : \left|\int_M f_i\, dm - \int_M f_i\, d\mu\right| < \varepsilon_i,\ i = 1, 2, \ldots, k\right\}$$

with $\mu \in \mathcal{P}$, $k \in \mathbb{N}$, $\varepsilon_i > 0$, and $f_i \in C_b(M)$. If, for each $n \in \mathbb{N}$, μ_n and μ belong to $\mathcal{P}$, then we say that $\mu_n \to \mu$ in the weak* topology if

$$\int_M f\, d\mu_n \to \int_M f\, d\mu$$

as $n \to \infty$ for all $f \in C_b(M)$. Indeed, this is the definition of the weak* topology in functional analysis; nevertheless, there is a strong tradition in probability and statistics favoring the term weak topology.

Example 9.1. *A sequence $\{Y_1, Y_2, \ldots\}$ of integrable random variables may be mutually independent yet not be such that* $\mathrm{E}[Y_1Y_2\cdots]$ *is equal to* $\mathrm{E}[Y_1]\mathrm{E}[Y_2]\cdots$.

Proof: Recall that if $\{Y_1, \ldots, Y_n\}$ is a family of mutually independent, integrable random variables, then $\mathrm{E}[Y_1\cdots Y_n] = \mathrm{E}[Y_1]\cdots\mathrm{E}[Y_n]$. In this example, suggested by Billingsley [1986], it is shown that this property may not be extended to include an infinite sequence of integrable random variables.

Consider a probability space $(\Omega, \mathcal{F}, P)$ and a sequence $\{Y_1, Y_2, \ldots\}$ of mutually independent integrable random variables defined on this space such that for each positive integer n, Y_n assumes the values $1/2$ and $3/2$ each with probability $1/2$. Define $X_n = Y_1Y_2\cdots Y_n$ and note that $\{X_n\colon n \in \mathbb{N}\}$ is a martingale with respect to $\sigma(Y_1, \ldots, Y_n)$ since X_n is $\sigma(Y_1, \ldots, Y_n)$-measurable, $\mathrm{E}[|X_n|] = \mathrm{E}[Y_1\cdots Y_n] = \mathrm{E}[Y_1]\cdots\mathrm{E}[Y_n] = 1 < \infty$, and

$$\begin{aligned}\mathrm{E}[X_{n+1}|\sigma(Y_1, \ldots, Y_n)] &= \mathrm{E}[Y_1\cdots Y_{n+1}|\sigma(Y_1, \ldots, Y_n)]\\ &= Y_1\cdots Y_n\mathrm{E}[Y_{n+1}] = X_n \text{ a.s.,}\end{aligned}$$

since Y_{n+1} is independent of $\sigma(Y_1, \ldots, Y_n)$ and $\mathrm{E}[Y_{n+1}] = 1$. Further, since $\mathrm{E}[|X_n|] = 1 < \infty$, it follows that X_n converges with probability one to some integrable random variable X. That $X = 0$ a.s. will now be shown.

Define S_n to be the number of times Y_k is equal to $3/2$ for $k \le n$. Then, by definition, it is clear that

$$X_n = \left(\frac{1}{2}\right)^{n-S_n}\left(\frac{3}{2}\right)^{S_n} = \frac{3^{S_n}}{2^n}.$$

Taking the natural log of each side, we see that

$$\ln(X_n) = n\left(\frac{1}{n}[S_n \ln(3)] - \ln(2)\right).$$

The strong law of large numbers implies that S_n/n converges to $\mathrm{E}[S_1] = 1/2$ a.s. Hence, $\ln(X_n)/n \to [\ln(3)/2] - \ln(2)$ a.s. as $n \to \infty$, where $[\ln(3)/2] - \ln(2) < 0$. Let $F \in \mathcal{F}$ be a set of measure one such that $\ln(X_n)/n \to [\ln(3)/2] - \ln(2)$ pointwise on F as $n \to \infty$. Fix any $\omega \in F$ and choose ε so that $0 < \varepsilon < \ln(2) - [\ln(3)/2]$. In this case, it follows that there exists some positive integer N such that for this ω,

$$\left|\frac{\ln[X_n(\omega)]}{n} - \left(\frac{\ln(3)}{2} - \ln(2)\right)\right| < \varepsilon$$

for all $n > N$. But this implies that

$$\frac{\ln[X_n(\omega)]}{n} < K \equiv \frac{\ln(3)}{2} - \ln(2) + \varepsilon < 0.$$

Hence, $X_n(\omega) < \exp(nK) \to 0$ as $n \to \infty$ since $K < 0$. Thus, $X_n(\omega) \to X(\omega) \equiv 0$ a.s., and hence, $\mathrm{E}[Y_1 Y_2 \cdots] = \mathrm{E}[X] = 0$ even though $\mathrm{E}[Y_1]\ \mathrm{E}[Y_2] \cdots = 1$. □

Example 9.2. *For any positive real number B, there exists a sequence $\{X_n\}_{n\in\mathrm{N}}$ of integrable random variables and an integrable random variable X such that X_n converges to X pointwise and yet $\mathrm{E}[X_n] = -B$ and $\mathrm{E}[X] = B$.*

Proof: Consider the probability space given by the interval [0, 1], $\mathcal{B}([0, 1])$, and Lebesgue measure on $\mathcal{B}([0, 1])$. Let $X(\omega) = 2B\mathrm{I}_{[1/2,\, 1]}(\omega)$, let $X_1 = -B$, and, for integers $n > 1$, let $X_n(\omega) = -2nB\mathrm{I}_{(1/2-1/n,\, 1/2)}(\omega) + X(\omega)$. Finally, note that $\mathrm{E}[X_n] = -B$, $\mathrm{E}[X] = B$, and yet $X_n \to X$ pointwise. □

Example 9.3. *If a random variable X and a sequence of random variables $\{X_n\}_{n\in\mathrm{N}}$ are defined on a common probability space, if X_n has a unique median for each positive integer n, and if $X_n \to X$ almost surely as $n \to \infty$, then it need not follow that the sequence of medians of the X_n's converges.*

Proof: Let G be a standard Gaussian random variable defined on some probability space. Define $X = \mathrm{I}_{\{G\leq 0\}}$ and for each positive integer n, let $X_n = \mathrm{I}_{\{G\leq(-1)^n/n\}}$. Then we see that $X_n \to X$ almost surely as $n \to \infty$. Further, note that for n even, X_n has a unique median of one and if n is odd, X_n has a unique median of zero. Hence, the sequence of medians does not converge. □

Example 9.4. *There exists a sequence of probability distribution functions $\{F_n\colon n \in \mathbb{N}\}$ such that the corresponding sequence of characteristic functions $\{C_n\colon n \in \mathbb{N}\}$ converges to a function $C(t)$ for all t, yet $C(t)$ is not a characteristic function.*

Proof: Recall that if $\{F_n\}_{n\in\mathrm{N}}$ is a uniformly tight sequence of distributions whose corresponding characteristic functions C_n converge to a function $C(t)$

for all t, then $F_n \to F$ as $n \to \infty$, where F is a distribution with characteristic function C. The following example points out that a similar result does not hold for a general sequence of distributions whose characteristic functions converge for all t.

Let F_n be a zero mean Gaussian distribution with a variance given by n for each positive integer n. Note that the characteristic function of F_n is given by $C_n = \exp\left(-nt^2/2\right)$. Further, $C_n(t) \to C(t) \equiv \mathrm{I}_{\{0\}}(t)$ for all t as $n \to \infty$. However, $C(t)$ cannot be a characteristic function since it is not continuous. □

Example 9.5. *A density of a centered normalized sum of a sequence of second-order, mutually independent, identically distributed random variables may exist yet not converge to a standard Gaussian probability density function.*

Proof: This example was inspired by Feller [1971]. For a density h we will let h^{*n} denote the $(n-1)$-fold convolution of h with itself; that is, h^{*n} is a density of the sum of n mutually independent, identically distributed random variables each with the density h. For $x > 0$ and $p \geq 1$, let $u_p(x) = (x[\ln(x)]^{2p})^{-1}$. Let

$$g(x) = \begin{cases} 6[\ln(4)]^4 u_2(x)/[3 + 2\ln(4)], & \text{if } x \in (0,\, 1/4) \\ 48(1 - 2x)/[3 + 2\ln(4)], & \text{if } x \in [1/4,\, 1/2] \\ 0, & \text{otherwise,} \end{cases}$$

and notice that g is a probability density function. There exists a positive number h_2 such that for $x \in (0,\, h_2)$, $u_2(x)$ decreases monotonically and $g(x) > u_2(x)$. Then, for $x \in (0,\, h_2)$,

$$\begin{aligned} g^{*2}(x) &= \int_{-\infty}^{\infty} g(x-y)g(y)\,dy \\ &> \int_0^x u_2(x-y)u_2(y)\,dy > x[u_2(x)]^2 = u_4(x). \end{aligned}$$

Similarly, there exists $h_3 > 0$ such that for $x \in (0,\, h_3)$,

$$g^{*3}(x) > \int_0^x u_2(x-y)u_4(y)\,dy > xu_2(x)u_4(x) = u_6(x).$$

Assume there exists $h_n > 0$ such that for $x \in (0,\, h_n)$, $g^{*n}(x) > u_{2n}(x)$. Obviously, there then exists $h_{n+1} > 0$ such that for $x \in (0,\, h_{n+1})$, $g(x) > u_2(x)$, $g^{*n}(x) > u_{2n}(x)$, and both $u_2(x)$ and $u_{2n}(x)$ decrease monotonically. For $x \in (0,\, h_{n+1})$,

$$g^{*(n+1)}(x) > \int_0^x u_2(x-y)u_{2n}(y)\,dy > xu_2(x)u_{2n}(x) = u_{2(n+1)}(x).$$

Thus, by induction, for each n there exists an open interval $(0,\, h_n)$ in which $g^{*n}(x) > u_{2n}(x)$, and hence, $g^{*n}(x) \to \infty$ as $x \downarrow 0$. Let $v(x) =$

$[g(x)+g(-x)]/2$, and let $f(x) = [v(x+1)+v(x)+v(x-1)]/3$. Notice that f is a symmetric probability density function supported on $[-3/2, 3/2]$ and continuous everywhere except at the origin and ± 1. Notice that $f^{*n}(x)$ is unbounded at $\{0, \pm 1, \pm 2, \ldots, \pm n\}$. Let σ^2 denote the variance associated with the density f.

Let $\{X_n\}_{n\in\mathbb{N}}$ denote a sequence of mutually independent random variables each with the density f. For positive integers n, let $S_n = \sum_{i=1}^{n} X_n$ and note that the random variable $S_n/\sigma\sqrt{n}$ has a probability density function given by $\sigma\sqrt{n}f^{*n}(x\sigma\sqrt{n})$. This density is continuous except at the $(2n+1)$ points of the form $k/\sigma\sqrt{n}$ for $k = 0, \pm 1, \ldots, \pm n$. Notice that for any rational r there are an infinite number of pairs of integers (k, n) with $|k| \leq n$, such that $r = k/\sqrt{n}$. Let $\Lambda = \bigcup_{r\in\mathbb{Q}} r/\sigma$. It follows that $S_n/\sigma\sqrt{n}$ converges in distribution to a standard Gaussian random variable, but the densities $\sigma\sqrt{n}f^{*n}(x\sigma\sqrt{n})$ do not converge at any point of Λ, and in no interval of positive length is the sequence of densities bounded. Note that Λ is dense in $\mathbb{R}$. In particular, between any two distinct real numbers there are an infinite number of points where the densities do not converge. □

Example 9.6. *There exists a sequence $\{\mu_n\}_{n\in\mathbb{N}}$ of probability distributions and a probability distribution μ such that each of these probability distributions has a probability density function and such that μ_n converges in the weak* topology to μ and yet the probability density function of μ_n does not converge to that of μ as $n \to \infty$.*

Proof: Let $f = \mathrm{I}_{[0,1]}$ and for each positive integer n, let f_n be given by $f_n(x) = [1+\sin(2\pi nx)]\mathrm{I}_{[0,1]}(x)$. From this, the result is evident. □

Example 9.7. *There exists a sequence $\{\mu_n\}_{n\in\mathbb{N}}$ of probability distributions and a probability distribution μ such that each of these probability distributions has an infinitely differentiable probability density function, such that μ_n converges in the weak* topology to μ and yet for any $\varepsilon \in (0, 1)$ there exists a closed set C such that $\mu(C) = 1 - \varepsilon$ and such that $\mu_n(C)$ does not converge to $\mu(C)$ as $n \to \infty$.*

Proof: Recall that there exists an infinitely differentiable real-valued function B of a real variable that is zero on $\mathbb{R} \setminus [0, 1]$, that is nonnegative, and that has an integral of unity. Let f be the probability density function given by $\mathrm{I}_{[0,1)}$. Let $\{r_n\colon n \in \mathbb{N}\}$ be an enumeration of the rationals in $[0, 1)$. For any $\varepsilon \in (0, 1)$, as in Example 1.4, let U be an open subset of $[0, 1)$ containing each of the rationals in $[0, 1)$ and having Lebesgue measure less than ε. For each positive integer n, let f_n be the probability density function given by placing contracted and scaled translates of B over the rationals $r_1, r_2, \ldots, r_n$ such that the support of each of these translates is included in U and such that the integral of each of these translates is $1/n$. Now, each such density function f_n is infinitely differentiable. The corresponding sequence of probability distributions converges in the weak* topology to the probability distribution of a uniform measure on $[0, 1)$ due to the denseness of the rationals, and yet for the closed set $C = \mathbb{R} \setminus U$, $\mu_n(C)$ does not converge to $\mu(C) = 1 - \varepsilon$. □

Example 9.8. *There exists a sequence of atomic probability distributions that converge in the weak* topology to an absolutely continuous probability distribution.*

Proof: For each positive integer n, let μ_n be the probability distribution that places mass $1/n$ at each of the points $0, 1/n, 2/n, \ldots, (n-1)/n$. Let μ denote the uniform distribution on $[0, 1)$. From here the desired result follows immediately. □

Example 9.9. *If $\{X_n\}_{n\in\mathbf{N}}$ is a sequence of mutually independent identically distributed random variables defined on the same probability space such that the common probability distribution function has at least two points of increase, then there does not exist any random variable X defined on that probability space such that X_n converges to X in probability.*

Proof: This example arose from a comment by Warry Millar. Since the probability distribution function of X_1 has at least two points of increase, there must exist two open intervals I_1 and I_2 separated by a positive distance greater than the length of either interval such that $P(X_1 \in I_1) > 0$ and $P(X_1 \in I_2) > 0$. Hence, for any two distinct positive integers i and j, the probability distribution function of $|X_i - X_j|$ has at least two points of increase. This is easily seen by considering the situation where both X_i and X_j are in the same interval I_1 or I_2 and the situation where one is in I_1 and the other is in I_2. Since the probability distribution function of $|X_i - X_j|$ has at least two points of increase, there exists a positive real number ε such that $P(|X_i - X_j| > \varepsilon) > 0$. Further, this positive number $P(|X_i - X_j| > \varepsilon)$ is not dependent on the two different positive integers i and j. Hence, the sequence $\{X_n\}_{n\in\mathbf{N}}$ is not Cauchy in probability. Thus there can exist no random variable X to which it converges in probability. □

Example 9.10. *If $\{X_n\}_{n\in\mathbf{N}}$ is a sequence of mutually independent random variables defined on a common probability space, then the almost sure unconditional convergence of*

$$\sum_{n=1}^{\infty} X_n$$

to a real-valued random variable need not imply the almost sure convergence of

$$\sum_{n=1}^{\infty} |X_n|$$

to a real-valued random variable.

Proof: Let $\{B_n\}_{n\in\mathbf{N}}$ be a sequence of mutually independent random variables defined on a common probability space such that $P(B_1 = 1) = P(B_1 = -1) = 1/2$. For each positive integer n, let $X_n = B_n/n$. Then note that since

$$\sum_{n=1}^{\infty} E[(X_n)^2] < \infty,$$

it follows that

$$\sum_{k=1}^{n} X_k$$

converges almost surely unconditionally to a real-valued random variable. However, note that

$$\sum_{k=1}^{n} |X_k| = n,$$

which converges to ∞. □

Example 9.11. *There exists a sequence $\{X_n\}_{n\in\mathbb{N}}$ of mutually independent, zero mean random variables defined on a common probability space such that*

$$\frac{1}{n}\sum_{i=1}^{n} X_i$$

converges to $-\infty$ with probability one.

Proof: Let the mutually independent random variables be defined on a common probability space such that $P(X_n = -n) = 1-(1/n^2)$ and $P(X_n = n^3 - n) = 1/n^2$. Then it follows that $\mathrm{E}[X_n] = 0$ for each positive integer n. Further, it follows that X_n/n converges to -1 almost surely. From this it follows that $(1/n)\sum_{i=1}^{n} X_i$ converges to $-\infty$ with probability one. This example might make one wonder what is fair about a fair game. □

Example 9.12. *The necessary conditions of a weak law of large numbers due to Kolmogorov are not necessary.*

Proof: This example is from Breiman [1957]. In 1928, Kolmogorov [cf. Breiman, 1957] presented the following now widely propagated version of the weak law of large numbers. Let $X_1, X_2, \ldots$ be mutually independent zero mean random variables defined on the same probability space, and let $X_{n,k} = X_k$ if $|X_k| < n$ and zero otherwise. Then $(1/n)\sum_{k=1}^{n} X_k$ converges in probability to zero if and only if

$$\text{(1)}\quad \sum_{k=1}^{n} P(|X_k| \geq n) \to 0, \text{ as } n \to \infty$$

$$\text{(2)}\quad \frac{1}{n}\sum_{k=1}^{n} \mathrm{E}[X_{n,k}] \to 0, \text{ as } n \to \infty, \text{ and}$$

$$\text{(3)}\quad \frac{1}{n^2}\sum_{k=1}^{n} \mathrm{Var}\,[X_{n,k}^2] \to 0, \text{ as } n \to \infty.$$

Presented without proof in the same paper was a sharpened version of the above theorem with condition (3) replaced by

$$\text{(3}')\quad \frac{1}{n^2}\sum_{k=1}^{n} \mathrm{E}\,[X_{n,k}^2] \to 0, \text{ as } n \to \infty.$$

A proof of this last theorem was given in Gnedenko and Kolmogorov [1954] and appears elsewhere in the literature as well.

On a suitable probability space, define mutually independent random variables $X_1, X_2, \ldots$ such that $P(X_1 = 0) = 1$, $P(X_k = (-1)^k k^{5/2}) = 1/k^2$ for $k \geq 2$, and $P(X_k = (-1)^{k+1}\sqrt{k}[1 - (1/k^2)]^{-1}) = 1 - (1/k^2)$ for $k \geq 2$. Note that $\mathrm{E}[X_k] = 0$ for each positive integer k. We will now show that conditions (1), (2), and (3) are satisfied and yet (3′) is not satisfied. In the following, n is taken to be at least 4.

Proof of (1): If $k^{5/2} < n$, then $X_{n,k} = X_k$, and if $k \leq n$, then $\sqrt{k}[1 - (1/k^2)]^{-1} < n$. Thus,

$$P(|X_k| \geq n) = \begin{cases} 0, & \text{if } 1 \leq k^{5/2} < n \\ 1/k^2, & \text{if } n \leq k^{5/2} \text{ and } k \leq n, \end{cases}$$

and consequently,

$$\sum_{k=1}^{n} P(|X_k| \geq n) = \sum_{k=G(n^{2/5})}^{n} \frac{1}{k^2} \to 0$$

as $n \to \infty$, where $G(\cdot)$ is the greatest integer function.

Proof of (2): Note that

$$\mathrm{E}[X_{n,k}] = \begin{cases} 0, & \text{if } 1 \leq k^{5/2} < n \\ (-1)^{k+1}\sqrt{k}, & \text{if } n \leq k^{5/2} \text{ and } k \leq n. \end{cases}$$

Hence,

$$\frac{1}{n}\sum_{k=1}^{n} \mathrm{E}[X_{n,k}] = \frac{1}{n} \sum_{k=G(n^{2/5})}^{n} (-1)^{k+1}\sqrt{k},$$

where, as above, $G(\cdot)$ is the greatest integer function. For a real number $s > 1$, note that $\sqrt{s} - \sqrt{s-1} < 1/(2\sqrt{s-1})$. Using this inequality, we see that

$$\left|\frac{1}{n}\sum_{k=1}^{n} \mathrm{E}[X_{n,k}]\right| \leq \frac{1}{2^n}\sum_{k=1}^{n-1} \frac{1}{\sqrt{k}} + \frac{1}{n}\sqrt{G(n^{2/5})} \to 0$$

as $n \to \infty$.

Proof of (3): Note that

$$\mathrm{Var}[X_{n,k}] = \begin{cases} k^3 + k(1-k^{-2})^{-1}, & \text{if } 2 \leq k^{5/2} < n \\ k(1-k^{-2})^{-1} - k, & \text{if } n \leq k^{5/2} \text{ and } k \leq n. \end{cases}$$

For $k \geq 2$, $k^3 + k(1-k^{-2})^{-1} \leq k^3 + (4/3)k \leq 2k^3$, and $k(1-k^{-2})^{-1} - k = k^{-1}(1-k^{-2})^{-1} \leq (4/3)k^{-1}$. Hence,

$$\begin{aligned} \frac{1}{n^2}\sum_{k=1}^{n} \mathrm{Var}[X_{n,k}^2] &\leq \frac{1}{n^2}\sum_{k=2}^{G(n^{2/5})} 2k^3 + \frac{1}{n^2}\sum_{k=G(n^{2/5})}^{n} \frac{4}{3k} \\ &\leq \frac{(G(n^{2/5}))^4}{2n^2} + \frac{4}{3n^2}\ln(n) \to 0 \end{aligned}$$

as $n \to \infty$.

Now, for the final step, we show that (3′) is not satisfied. Note that

$$\frac{1}{n^2}\sum_{k=1}^{n} \mathrm{E}[X_{n,k}^2] \geq \frac{1}{n^2} \sum_{k=G(n^{2/5})}^{n} \mathrm{E}[X_{n,k}^2]$$

$$= \frac{1}{n^2} \sum_{k=G(n^{2/5})}^{n} k(1-k^{-2})^{-1} \geq \frac{1}{n^2} \sum_{k=G(n^{2/5})}^{n} k \to \frac{1}{2}$$

as $n \to \infty$, where, as above, $G(\cdot)$ is the greatest integer function. □

Example 9.13. *There exists a martingale with a constant positive mean that converges almost surely to zero in finite time and yet with positive probability exceeds any real number.*

Proof: The following example shows that a martingale may have a constant positive mean, converge almost surely to zero in finite time, and yet with positive probability exceed any real number.

Let $\{X_n\colon n \in \mathbb{N}\}$ be a sequence of mutually independent, identically distributed random variables such that $P(X_1 = 0) = P(X_1 = 2) = 1/2$. For each positive integer n, define $Y_n = X_1X_2 \cdots X_n$ and note that $\{Y_n\colon n \in \mathbb{N}\}$ is a martingale and that $\mathrm{E}[Y_n] = 1$ for all $n \in \mathbb{N}$. Further, notice that not only does the sequence $\{Y_n\colon n \in \mathbb{N}\}$ converge almost surely to zero, but, with probability one, only a finite number of terms of the sequence $\{Y_n\colon n \in \mathbb{N}\}$ are nonzero. Even so, it follows easily that Y_n exceeds any real number with positive probability since $P(Y_n = 2^n) > 0$ for all $n \in \mathbb{N}$. □

Example 9.14. *A last element of a martingale need not exist.*

Proof: The following example from Ash [1972] shows that a martingale need not possess a last element. Let $\{Y_i\colon i \in \mathbb{N}\}$ be a family of mutually independent random variables defined on a probability space $(\Omega, \mathcal{F}, P)$ such that, for $0 < p < 1$, $P(Y_i = 1) = p$ and $P(Y_i = 0) = 1 - p$. Define $X_n = (1/p^n)\prod_{i=1}^{n} Y_i$ and $\mathcal{G}_n = \sigma(Y_1, \ldots, Y_n)$. Then $\mathrm{E}[X_{n+1} \mid \mathcal{G}_n] = (1/p^{n+1})Y_1 \cdots Y_n \mathrm{E}[Y_{n+1}] = (p/p^{n+1})Y_1 \cdots Y_n = (1/p^n)Y_1 \cdots Y_n = X_n$ a.s. Hence, it is clear that $\{X_n, \mathcal{G}_n\}$ is a martingale. Note that, with probability one, X_n is nonzero for only finitely many integers n since $P(Y_j = 1$ for every $j) = \lim_{k\to\infty} p^k = 0$. Thus, X_n converges to 0 a.s.

Assume that the X_n's are uniformly integrable. Then $X_n \to 0$ a.s. implies that $X_n \to 0$ in $\mathbf{L}_1$, or that $\mathrm{E}[X_n] \to 0$. But, $\mathrm{E}[X_n] = 1$ for each n, which thus implies that the X_n's cannot be uniformly integrable.

Assume now that this martingale possesses a last element X_∞. Then it is required that $X_i = \mathrm{E}[X_\infty \mid \mathcal{G}_i]$ a.s. for every positive integer i. But then

$$\int_{\{|X_i| \geq c\}} |X_i|\, dP \leq \int_{\{|X_i| \geq c\}} \mathrm{E}[|X|_\infty \mid \mathcal{G}_i]\, dP = \int_{\{|X_i| \geq c\}} |X_\infty|\, dP$$

since $\{|X_i| \geq c\}$ is an element of $\mathcal{G}_i$. Further, Chebyshev's inequality implies that $P(\{|X_i| \geq c\}) \leq \mathrm{E}[|X_i|]/c = 1/c \to 0$ as $c \to \infty$ uniformly in i. Hence, the X_i's are uniformly integrable. This contradiction implies that no last element X_∞ exists for this martingale. □

Example 9.15. *A martingale may converge without being* $\mathbf{L}_1$ *bounded.*

Proof: Many martingale convergence theorems require that the martingale be $\mathbf{L}_1$ bounded. The following example from Ash [1972] shows that a martingale may converge without being $\mathbf{L}_1$ bounded.

Consider a Markov chain defined as follows. Let $X_1 = 0$ and, if $X_n = 0$, let

$$X_{n+1} = \begin{cases} a_{n+1}, & \text{with probability } p_{n+1} \\ -a_{n+1}, & \text{with probability } p_{n+1} \\ 0, & \text{with probability } 1 - 2p_{n+1}, \end{cases}$$

and, if $X_n \neq 0$, let $X_{n+1} = X_n$, where, for each $n \in \mathbb{N}$, $0 < p_{n+1} < 1/2$, $a_n > 0$, $\sum_{k=1}^{\infty} p_k < \infty$, and $\sum_{k=1}^{\infty} a_k p_k = \infty$. Note that if $X_n \neq 0$, then $X_j = X_n$ for every $j \geq n$. Further, notice that for each positive integer n, $\mathrm{E}[X_{n+1} \,|\, X_n = 0] = a_{n+1}p_{n+1} - a_{n+1}p_{n+1} + 0(1 - 2p_{n+1}) = 0$, $\mathrm{E}[X_{n+1} \,|\, X_n = a_n] = a_n$, and $\mathrm{E}[X_{n+1} \,|\, X_n = -a_n] = -a_n$. Hence, the Markov chain $\{X_n\colon n \in \mathbb{N}\}$ defined above is a martingale. Further, assuming that all of the above random variables are defined on a probability space $(\Omega, \mathcal{F}, P)$, for each $\omega \in \Omega$, $X_n(\omega) = 0$ for every n or there exists some m such that for $n \geq m$, $X_n(\omega) = X_m(\omega)$. Hence, X_n converges for every $\omega \in \Omega$. Notice, however, that $\mathrm{E}[|X_k|] = 2p_2a_2 + (1 - 2p_2)2p_3a_3 + (1 - 2p_2)(1 - 2p_3)2p_4a_4 + \cdots + (1 - 2p_2)\cdots(1 - 2p_{k-1})2p_ka_k$. Further, since $a_n > 0$ and $0 < p_{n+1} < 1/2$ for each positive integer n, it follows that $\mathrm{E}[|X_k|] \geq 2[(1 - 2p_1)(1 - 2p_2)\cdots(1 - 2p_{k-1})](p_2a_2 + p_3a_3 + \cdots + p_ka_k)$. Hence, it follows that

$$\lim_{k\to\infty} \mathrm{E}[|X_k|] \geq 2\left[\prod_{j=1}^{\infty}(1 - 2p_j)\right]\left[\sum_{j=2}^{\infty} p_j a_j\right].$$

Note that the infinite product in this expression is not equal to zero since $\sum_{k=1}^{\infty} p_k < \infty$ [Gradshteyn and Ryzhik, 1980, p. 11, 0.252]. Thus, it is clear that $\lim_{k\to\infty} \mathrm{E}[|X_k|] = \infty$, or in other words, it is clear that this martingale, which was previously shown to converge everywhere, is not $\mathbf{L}_1$ bounded. □

Example 9.16. *There exists a nonnegative martingale* $\{(X_n, \mathcal{F}_n)\colon n \in \mathbb{N}\}$ *such that* $\mathrm{E}[X_n] = 1$ *for all* $n \in \mathbb{N}$ *and such that* X_n *converges to zero pointwise as* $n \to \infty$ *yet such that* $\sup_{n\in\mathbb{N}} X_n$ *is not integrable.*

Proof: In this example from Neveu [1975], let $\Omega = \mathbb{N}$ and $\mathcal{F} = \mathbb{P}(\mathbb{N})$. Further, define a probability measure P on $(\Omega, \mathcal{F})$ via $P(\{\omega\}) = 1/[\omega(\omega + 1)]$. For $n \in \mathbb{N}$, let $\mathcal{F}_n = \sigma(\{1\}, \{2\}, \ldots, [n + 1, \infty) \cap \mathbb{N})$. Finally, for $n \in \mathbb{N}$, let $X_n(\omega) = (n + 1)\mathrm{I}_{S_n}(\omega)$, where $S_n = [n + 1, \infty) \cap \mathbb{N}$. Now, note that

$$\mathrm{E}[X_n] = (n + 1) \sum_{k=n+1}^{\infty} \frac{1}{k(k + 1)} = 1$$

for all $n \in \mathbb{N}$. Further, for positive integers k and n with $k > n$, note that $\mathrm{E}[X_k \mid \mathcal{F}_n] = X_n$ pointwise on Ω, since X_n is $\mathcal{F}_n$-measurable and since

$$\int_A X_n \, dP = \int_A X_k \, dP$$

for all $A \in \mathcal{F}_n$. That is, this equality reduces to

$$\int_{A \cap [n+1, \infty)} X_n \, dP = \int_{A \cap [n+1, \infty)} X_k \, dP$$

for all $A \in \mathcal{F}_n$. Thus, if $A \cap [n+1, \infty) = \emptyset$, then each side of the equality is zero. Otherwise, we have that

$$\begin{aligned}\int_{[n+1, \infty)} X_n \, dP &= \mathrm{E}[X_n] = 1 = \mathrm{E}[X_k] \\ &= \int_{[k+1, \infty)} X_k \, dP = \int_{[n+1, \infty)} X_k \, dP.\end{aligned}$$

Thus, $\{(X_n, \mathcal{F}_n): n \in \mathbb{N}\}$ is a nonnegative martingale with a constant mean of one. Further, note that as $n \to \infty$, $X_n \to 0$ pointwise. However, we see that $\sup_{n \in \mathbb{N}} X_n = \omega$, and this random variable is not integrable. □

Example 9.17. *A martingale that is indexed by a directed set can be* $\mathbf{L}_2$ *bounded and converge in* $\mathbf{L}_2$ *and yet not converge for any point off a null set.*

Proof: This example is from Walsh [1986]. Recall that a directed set is a nonempty set S with a relation R on S such that $(a, a) \in R$ for all $a \in S$, such that if $(a, b) \in R$ and $(b, c) \in R$ then $(a, c) \in R$, and such that if a and b are two elements of S, there exists an element c of S such that $(a, c) \in R$ and $(b, c) \in R$. Further, we denote $(a, b) \in R$ by the notation $a \leq b$. We will next briefly consider an aspect of a martingale indexed by a directed set. Let I be a directed set and let $\{X_t: t \in I\}$ be an indexed family of random variables. Further, let $\{\mathcal{F}_t: t \in I\}$ be a family of σ-algebras such that for $s \leq t$, $\mathcal{F}_s \subset \mathcal{F}_t$. We say that $\{X_t, \mathcal{F}_t: t \in I\}$ is a martingale if $\mathrm{E}[|X_t|] < \infty$ for each $t \in I$, if X_t is $\mathcal{F}_t$-measurable for each $t \in I$, and if for $s \leq t$, $\mathrm{E}[X_t \mid \mathcal{F}_s] = X_s$ almost surely.

Let $\{X_n\}_{n \in \mathbb{N}}$ be a sequence of mutually independent random variables such that for each $n \in \mathbb{N}$, $P(X_n = -1) = P(X_n = 1) = 1/2$. First note that $\sum_{n=1}^{\infty} X_n/n$ converges almost surely. Let I be the family of all finite subsets of $\mathbb{N}$, partially ordered by set inclusion. Thus I is a directed set. If $t \in I$, let $M_t = \sum_{n \in t} X_n/n$. Then, with respect to its own filtration, $\{M_t: t \in I\}$ is a martingale. Also, this martingale converges in probability. Further, $\mathrm{E}[M_t^2] = \sum_{n=1}^{\infty} 1/n^2 < 2$. Thus, this martingale is $\mathbf{L}_2$-bounded. However, note that $M_t(\omega)$ converges only if it converges regardless of the order of summation, that is, only if it converges absolutely. However, $\sum_{n=1}^{\infty} |X_n/n| = \sum_{n=1}^{\infty} 1/n = \infty$. Thus, here we have an example of an $\mathbf{L}_2$-bounded martingale indexed by a directed set that converges in probability but does not converge pointwise for any point on the underlying probability space off a null set. □

Example 9.18. *There exists a martingale $\{X(t), \mathcal{F}_t\colon t \in \mathbb{R}\}$ such that for any positive real number p, $\mathrm{E}[|X(t)|^p] < \infty$ for all real numbers t and yet for almost no point on the underlying probability space does $X(t)$ converge as $t \to \infty$.*

Proof: Consider the underlying probability space and the random process $\{X(t)\colon t \in \mathbb{R}\}$ from Example 7.31. For real numbers t, let $\mathcal{F}_t = \sigma(X(s)\colon s \leq t)$. Then $\{X(t), \mathcal{F}_t\colon t \in \mathbb{R}\}$ is a martingale such that for any positive real number p, $\mathrm{E}[|X(t)|^p] < \infty$ for all nonnegative real numbers t. However, since for any point on the underlying probability space, $X(\cdot)$ is the indicator function of a saturated non-Lebesgue measurable set, it follows that $X(t)$ does not converge as $t \to \infty$. □

Example 9.19. *If $\{X(t)\colon t \in \mathbb{R}\}$ is a random process defined on a probability space such that for each $t \in \mathbb{R}$, $X(t) = 0$ a.s., the set on which $\lim_{t\to\infty} X(t) = 0$ need not be an event.*

Proof: Let the probability space be given by $(0, 1]$, the Borel subsets, and Lebesgue measure, and let the random process $\{X(t)\colon t \in \mathbb{R}\}$ be defined via $X(t, \omega) = f_t(\omega)$, where f_t is as defined in Example 2.27. □

Example 9.20. *If $\{Y_n\}_{n\in\mathbb{N}}$ is a sequence of random variables defined on a common probability space such that $Y_n \to 0$ in probability as $n \to \infty$, and if $\{X(t) : t \in [0, \infty)\}$ is an $\mathbb{N}$-valued random process defined on the same probability space such that $X(t) \to \infty$ a.s. as $t \to \infty$, then $Y_{X(t)}$ need not converge to zero in probability as $t \to \infty$.*

Proof: This example is from Gut [1988]. Consider the probability space given by $[0, 1]$, $\mathcal{B}([0, 1])$, and $\lambda|_{\mathcal{B}([0,1])}$. For each positive integer n, let $Y_n(\omega) = I_{B_j}(\omega)$, where B_j is the interval $[j2^{-m}, (j+1)2^{-m}]$ with $n = 2^m + j$ for a nonnegative integer j such that $0 \leq j \leq 2^m - 1$. Further, let $X(t) = \min\{k \in \mathbb{N} : k \geq 2^t \text{ and } Y_k > 0\}$. It straightforwardly follows that $Y_n \to 0$ in probability (but not a.s.) as $n \to \infty$. Note, however, that $Y_{X(t)} = 1$ a.s. for all nonnegative t. □

References

Ash, R. B., *Real Analysis and Probability,* Academic Press, New York, 1972.

Billingsley, P., *Probability and Measure,* Second edition, John Wiley, New York, 1986.

Breiman, L., "A counterexample to a theorem of Kolmogorov," *The Annals of Mathematical Statistics,* Vol. 28, pp. 811–814, September, 1957.

Feller, W., *An Introduction to Probability Theory and its Applications,* Vol. 2, Second edition, John Wiley, New York, 1971.

Gnedenko, B., and A. Kolmogorov, *Limit Distributions for Sums of Independent Random Variables,* Addison–Wesley, Cambridge, 1954.

Gradshteyn, I. S., and I. M. Ryzhik, *Table of Integrals, Series, and Products,* corrected and enlarged edition, Academic Press, New York, 1980.

Gut, A., *Stopped Random Walks: Limit Theorems and Applications,* Springer–Verlag, New York, 1988.

Neveu, J., *Discrete–Parameter Martingales,* North Holland, Amsterdam, 1975.

Walsh, J. B., "Martingales with a multidimensional parameter and stochastic integrals in the plane," *Lecture Notes in Mathematics,* No. 1215, Springer–Verlag, New York, 1986, pp. 329–491.

10. Applications of Probability

In this chapter we will consider counterexamples concerning applications from engineering and statistics in which probability theory is used.

Example 10.1. *There exist two complex-valued, uncorrelated, mutually Gaussian random variables that are not independent.*

Proof: This example is from Ash and Gardner [1975]. Consider four mutually Gaussian, zero mean random variables X_1, X_2, X_3, and X_4 with covariance matrix

$$\begin{bmatrix} 1 & 0 & 0 & -1 \\ 0 & 1 & -1 & 0 \\ 0 & -1 & 1 & 0 \\ -1 & 0 & 0 & 1 \end{bmatrix}.$$

Let $X = X_1 + \imath X_2$ and $Y = X_3 + \imath X_4$. Note that $\mathrm{E}[XY^*] = \mathrm{E}[X_1X_3 + X_2X_4] + \imath\mathrm{E}[X_2X_3 - X_1X_4] = 0$ and hence X and Y are uncorrelated, mutually Gaussian, complex-valued random variables. However, X and Y are not independent, since, if they were, then X_1 and X_4 would be independent. This, however, is not the case, since $\mathrm{E}[X_1X_4] = -1$. □

Example 10.2. *There exists a linear filter described via convolution with a nowhere zero function that is a no-pass filter for a certain family of input random processes.*

Proof: This example describes a situation in which a linear filter described via convolution with a nowhere zero function is a no-pass filter for a large family of input random processes. In particular, recall that Example 5.43 presented two bounded linear operators on $\mathbf{L}_1(\mathbb{R})$ to $\mathbf{L}_1(\mathbb{R})$ denoted by $T_\alpha(\cdot)$ and $T_\beta(\cdot)$ for which $T_\alpha(f) * T_\beta(g)$ is zero for any elements f and g in $\mathbf{L}_1(\mathbb{R})$. Recall also that $h(t) = T_\beta(\mathrm{I}_{(-1,\,1]}(t))$, for $t \in \mathbb{R}$, is bounded and nowhere zero.

Consider a linear filter that is described via convolution with the fixed bounded nowhere zero integrable function $h(t)$. For a probability space $(\Omega, \mathcal{F}, P)$, let $\mathcal{A}$ denote the family of all random processes $\{X(t)\colon t \in \mathbb{R}\}$ defined on $(\Omega, \mathcal{F}, P)$ for which every sample path of $\{X(t)\colon t \in \mathbb{R}\}$ is in $\mathbf{L}_1(\mathbb{R})$. [For an interesting element of $\mathcal{A}$ when $(\Omega, \mathcal{F}, P)$ is given by $\mathbb{R}$, $\mathcal{B}(\mathbb{R})$, and standard Gaussian measure on $(\mathbb{R}, \mathcal{B}(\mathbb{R}))$, see Example 7.26.] Let $\mathcal{G}$ denote the family of random processes $\{Y(t)\colon t \in \mathbb{R}\}$ defined on $(\Omega, \mathcal{F}, P)$ that are of the form $Y(t) = T_\alpha(X(t))$, where $X(t)$ is an element of $\mathcal{A}$. The result in Example 5.43 implies that $Y(t) * h(t) = 0$ pointwise on the underlying probability space. That is, the linear filter under consideration is a no-pass filter for the family of random processes given by $\mathcal{G}$ even though the linear filter is described via convolution with a bounded integrable function that is nowhere zero. □

Example 10.3. *There exists a bounded random variable X and a symmetric, bounded function C mapping $\mathbb{R}$ into $\mathbb{R}$ that is nondecreasing on $[0, \infty)$ such that there exists no one-level quantizer Q with the property that $\mathrm{E}[C(X - Q(X))]$ is minimized over all one-level quantizers.*

Proof: This example is from Abaya and Wise [1984]. Let

$$C(t) = \begin{cases} 0, & \text{if } 0 \le |t| < 2 \\ 1, & \text{if } 2 \le |t| \end{cases}$$

and let X be a random variable having a discrete distribution such that $P(X = 2) = P(X = -2) = 1/4$, and $P(X = 2 - (1/m)) = P(X = -2 + (1/m)) = 2^{-m-2}$ for positive integers m. The claimed result then follows straightforwardly. □

Example 10.4. *A random variable X having a symmetric probability density function may be such that no two-level quantizer $\tilde{Q}$ such that $\tilde{Q}(x) = -\tilde{Q}(-x)$ for all real x minimizes the quantity* $\mathrm{E}[[X - Q(X)]^2]$ *over all two-level quantizers Q.*

Proof: This example is from Abaya and Wise [1981]. Let X be a random variable having a probability density function given by

$$f(x) = \begin{cases} (-|x|/3) + (5/12), & \text{if } |x| < 1 \\ (7 - |x|)/72, & \text{if } 1 \le |x| < 7 \\ 0, & \text{if } 7 \le |x|. \end{cases}$$

It is easily verified that, up to almost everywhere equivalence with respect to Lebesgue measure, there are only two optimum two-level quantizers Q_1 and Q_2, where $Q_1(x) = -1$ if $x < 1$ and $Q_1(x) = 3$ if $x \ge 1$ and where $Q_2(x) = Q_1(-x)$. Further, the minimum mean-square error is given by 2.61. By comparison, the best odd symmetric quantizer Q has output levels at $\pm 61/36$ and has a mean-square error of 2.74. □

Example 10.5. *For any positive integer k and any integer $N > 1$, a probability space, a Gaussian random vector X defined on the space taking values in $\mathbb{R}^k$ and having a positive definite covariance matrix, and an N-level quantizer Q exist such that the random vector $Q(X)$ takes on each of the N values in its range with equal probability and such that X and $Q(X)$ are independent.*

Proof: This example is from Hall and Wise [1993]. For a positive integer k, a k-dimensional quantizer of a random variable X defined on a probability space $(\Omega, \mathcal{F}, P)$ is any function Q: $\mathbb{R}^k \to F$ such that F is a finite subset of $\mathbb{R}^k$, such that $Q(x) = x$ for all x in F (i.e., such that Q restricted to F is the identity map on F), and such that $Q(X)$ is itself a random variable defined on $(\Omega, \mathcal{F}, P)$. If F is a finite subset of $\mathbb{R}^k$ with cardinality N, then a quantizer Q: $\mathbb{R}^k \to F$ of a random variable X is said to be an N-level quantizer.

Let $S_1, \ldots, S_N$ be a partition of $\mathbb{R}^k$ as provided by Example 6.9, and for these N subsets of $\mathbb{R}^k$ let $(\Omega, \mathcal{F}, \mu)$ be a probability space as provided by Example 7.7, where P is chosen to be the product measure induced by placing standard Gaussian measure on each factor of $(\mathbb{R}^k, \mathcal{B}(\mathbb{R}^k))$. For each positive integer $i \le N$, let α_i be an element from S_i. Let F denote the set $\{\alpha_1, \ldots, \alpha_N\}$. Define an N-level k-dimensional quantizer Q: $\mathbb{R}^k \to F$ via

$Q(x) = \sum_{i=1}^{N} \alpha_i I_{S_i}(x)$. Further, notice that the random vector $X(\omega) = \omega$, $\omega \in \Omega$, is a zero mean Gaussian random vector defined on $(\Omega, \mathcal{F}, \mu)$ whose covariance matrix is the $k \times k$ identity matrix. Also, notice that for $1 \leq i \leq N$, $\mu(Q(X(\omega)) = \alpha_i) = \mu(\omega \in S_i) = 1/N$. Finally, notice that X and $Q(X)$ are independent via the result of Example 7.7.

The result of this example may be used to answer many questions that frequently arise when quantization is employed. For example, if X is a random variable and Q is an N-level quantizer with $N > 1$, then must X and $Q(X)$ be dependent random variables? (No.) If X is a random variable and Q is an N-level quantizer with N large, then must the mean-square error between X and $Q(X)$ be small? (No. In fact, the mean-square error can be arbitrarily large for any value of $N > 1$.) If X is a random variable and Q is an N-level quantizer with $N > 1$, then must $E[Q(X) \mid X] = Q(X)$ a.s.? (No.) If X is a second-order random variable and Q_n is an n-level quantizer for each positive integer n, then must $Q_n(X)$ converge in mean-square to X as $n \to \infty$? (No.) If X and Y are independent nondegenerate random variables, then is it possible that X is equal to a function of Y pointwise on the underlying probability space? (Yes.) □

Example 10.6. *The martingale convergence theorem need not be a useful estimation technique even when every random variable of concern is Gaussian.*

Proof: This example is from Wessel and Wise [1988]. Consider the probability space $(\mathbb{R}, \mathcal{B}(\mathbb{R}), P)$, where P denotes standard Gaussian measure on $(\mathbb{R}, \mathcal{B}(\mathbb{R}))$. Let P_* denote the inner P measure on $(\mathbb{R}, \mathcal{B}(\mathbb{R}))$ and let S be a subset of $\mathbb{R}$ such that $P_*(S) = P_*(S^c) = 0$. [To obtain such a set, see Example 1.46 and recall that P is equivalent to Lebesgue measure on $\mathcal{B}(\mathbb{R})$.] Further, let $\mathcal{W} = \{(S \cap B_1) \cup (S^c \cap B_2): B_1, B_2 \in \mathcal{B}(\mathbb{R})\}$ and note that $\mathcal{W}$ is a σ-algebra on $\mathbb{R}$ that includes $\mathcal{B}(\mathbb{R})$. Define a probability measure μ on $(\mathbb{R}, \mathcal{W})$ via $\mu((S \cap B_1) \cup (S^c \cap B_2)) = [P(B_1) + P(B_2)]/2$. (As in Example 7.7, that μ is well defined follows from the properties of S.) Note that the restriction of μ to $\mathcal{B}(\mathbb{R})$ is P.

Consider now the probability space $(\mathbb{R}, \mathcal{W}, \mu)$. Note first that S and S^c are each independent of $\mathcal{B}(\mathbb{R})$, since, for any Borel set B, $\mu(S \cap B) = \mu(B)/2 = \mu(S)\mu(B)$ and $\mu(S^c \cap B) = \mu(B)/2 = \mu(S^c)\mu(B)$. Next, define a random variable X on $(\mathbb{R}, \mathcal{W}, \mu)$ via $X(x) = x I_S(x) - x I_{S^c}(x)$ and notice that, for any Borel set B, $\mu(X \in B) = \mu((S \cap B) \cup (S^c \cap -B)) = P(B)$ since P is symmetric. Hence, X is a Gaussian random variable with zero mean and unit variance. Further, note that $E[X \mid \mathcal{B}(\mathbb{R})] = x E[I_S - I_{S^c} \mid \mathcal{B}(\mathbb{R})] = x E[I_S - I_{S^c}] = 0$ a.s. since the identity map is Borel measurable, S and S^c are independent of $\mathcal{B}(\mathbb{R})$, and $P(S) = P(S^c) = 1/2$.

Now, let $\{Y_k: k \in \mathbb{N}\}$ be any sequence of Borel measurable functions mapping $\mathbb{R}$ into $\mathbb{R}$ and note that $\{Y_k: k \in \mathbb{N}\}$ is a sequence of random variables on $(\mathbb{R}, \mathcal{W}, \mu)$. Consider the martingale $\{X_k = E[X \mid Y_1, \ldots, Y_k]: k \in \mathbb{N}\}$. Since, given any $k \in \mathbb{N}$, $\mathcal{B}(\mathbb{R})$ includes $\sigma(Y_1, \ldots, Y_k)$, it follows that $X_k = E[X \mid Y_1, \ldots, Y_k] = E[E[X \mid \mathcal{B}(\mathbb{R})] \mid Y_1, \ldots, Y_k] = 0$ a.s. using the previous result. Hence, the martingale convergence theorem is completely useless in estimating the random variable X in terms of the random variables $\{Y_k\colon k \in \mathbb{N}\}$. Furthermore, note that for any sequence $\{s_k\colon k \in \mathbb{N}\}$ of positive real numbers, one could set $Y_k(x) = s_k x$. In this case, the above phenomenon is exhibited when all of the random variables of concern are Gaussian, and even though, for any positive integer k, there exists a function $f\colon \mathbb{R} \to \mathbb{R}$ such that $X = f(Y_k)$ pointwise on the underlying probability space. □

Example 10.7. *For any positive integer N, a perfectly precise Kalman filtering scheme may fail completely when based upon observations subjected to an N-level maximum entropy quantizer.*

Proof: This example is from Hall and Wise [1990]. Let $\mathbb{Z}^+$ denote the nonnegative integers and consider a setting for the Kalman filter in which for each $k \in \mathbb{Z}^+$, $X_{k+1} = a_k X_k + b_k \eta_k$ and $V_k = c_k X_k + d_k \xi_k$, where X_0 is a Gaussian random variable and the sequences $\{\xi_k\colon k \in \mathbb{Z}^+\}$ and $\{\eta_k\colon k \in \mathbb{Z}^+\}$ are uncorrelated zero mean Gaussian random processes such that $\mathrm{E}[X_0\xi_k] = \mathrm{E}[X_0\eta_k] = 0$ for all $k \in \mathbb{Z}^+$. Recall that the Kalman filtering problem seeks a minimum mean-square, affine, recursive estimate of X_k based upon the random variables $\{V_0, \ldots, V_k\}$. In this example, we will consider the problem of minimum mean-square, affine, recursive estimation of X_k based upon the random variables $\{Q(V_0), \ldots, Q(V_k)\}$, where Q is a quantizer.

Let $(\Omega, \mathcal{F}, \mu)$ be a probability space as described by Example 7.7 with $k = 1$, $N > 1$, and standard Gaussian measure P on $(\mathbb{R}, \mathcal{B}(\mathbb{R}))$. In the setting for the Kalman filter detailed above, let $a_k = b_k = d_k = 1$ for all $k \in \mathbb{Z}^+$, let $X_0 = 0$, let $\xi_k = 0$ for all k, and let $\eta_k(\omega) = s_k\omega$, where $\omega \in \Omega$ and where $\{s_k\colon k \in \mathbb{Z}^+\}$ is a sequence of positive real numbers. Define a sequence $\{m_k\colon k \in \mathbb{Z}^+\}$ via $m_0 = 0$ and $m_k = s_0 + \cdots + s_{k-1}$ for $k \in \mathbb{N}$. For $A > 0$, define a sequence $\{c_k\colon k \in \mathbb{Z}^+\}$ via $c_0 = 0$ and $c_k = \sqrt{A}/m_k$ for $k \in \mathbb{N}$. Note that $\mathrm{E}[X_0\xi_k] = \mathrm{E}[X_0\eta_k] = 0$ for all $k \in \mathbb{Z}^+$ and, also, that $X_k(\omega) = m_k\omega$ for $k \in \mathbb{Z}^+$ and $\omega \in \Omega$, $V_0 = 0$, and $V_k(\omega) = \sqrt{A}\,\omega$ for $k \in \mathbb{N}$ and $\omega \in \Omega$. Hence, $\mathrm{E}[X_k \mid V_0, \ldots, V_k] = X_k$ a.s. for $k \in \mathbb{Z}^+$. However, for the partition $S_1, \ldots, S_N$ of $\mathbb{R}$ associated with $(\Omega, \mathcal{F}, \mu)$, consider a quantizer Q given by $Q(x) = \sum_{i=1}^{N} \alpha_i \mathrm{I}_{S_i}\left(x/\sqrt{A}\right)$, where α_i is some fixed element of S_i for each i. Note that $\mathrm{E}[X_k \mid Q(V_0), \ldots, Q(V_k)] = 0$ for each integer $k \geq 0$ via the results in Example 7.7, even though the probability of the set $\{\omega\colon Q(V_k(\omega)) = x\}$ is the same positive constant for each point x in the range of the N-level quantizer Q and for each integer $k > 0$. Note that for this quantization of the data any affine, recursive estimation scheme is doomed to failure. Indeed, any Borel measurable, recursive estimation scheme is doomed to failure. Also, note that the number of levels of the quantizer Q could have been any preselected integer greater than 1. □

Example 10.8. *A Kalman filter may fail to provide a useful estimate based upon observations subjected to an arbitrarily high-level quantizer even when the quantity being estimated may be reconstructed precisely from a single observation.*

Proof: This example is from Hall and Wise [1990]. Consider a probability space $(\Omega, \mathcal{F}, \mu)$ as provided by Example 7.7 for $k = 1$, $N = 2$, and standard Gaussian measure P on $(\mathbb{R}, \mathcal{B}(\mathbb{R}))$. For the partition S_1 and S_2 provided by that example, let $S = S_1$ and note that $S^c = S_2$. Recall from Example 7.7 that the set S is independent of $\mathcal{B}(\mathbb{R})$. Further, notice that the identity map on Ω is a Gaussian random variable with zero mean and unit variance. Since standard Gaussian measure P is symmetric [that is, $P(A) = P(-A)$ for each $A \in \mathcal{B}(\mathbb{R})$], it follows that $X(\omega) = \omega[\mathrm{I}_S(\omega) - \mathrm{I}_{S^c}(\omega)]$ is also a Gaussian random variable with zero mean and unit variance. That is, for any $B \in \mathcal{B}(\mathbb{R})$, $\mu(X \in B) = \mu((B \cap S) \cup ((-B) \cap S^c)) = [P(B) + P(-B)]/2 = P(B)$.

In the setting for the Kalman filter detailed in Example 10.7, let $b_k = c_k = 0$ for all $k \in \mathbb{Z}^+$ and let $a_k = d_k = 1$ for all $k \in \mathbb{Z}^+$. Further, let $X_0(\omega) = \omega[\mathrm{I}_S(\omega) - \mathrm{I}_{S^c}(\omega)]$. For a sequence $\{s_k\colon k \in \mathbb{Z}^+\}$ of nonzero real numbers, let $\xi_k(\omega) = s_k\omega$ for all $k \in \mathbb{Z}^+$. Also, for all $k \in \mathbb{Z}^+$, let $\eta_k(\omega) = 0$. Note that the sequences $\{\xi_k\colon k \in \mathbb{Z}^+\}$ and $\{\eta_k\colon k \in \mathbb{Z}^+\}$ are uncorrelated zero mean Gaussian random processes. Also, note that $\mathrm{E}[X_0\eta_k] = 0$ for all $k \in \mathbb{Z}^+$ and that $\mathrm{E}[X_0\xi_k] = \mathrm{E}[s_k\omega^2[\mathrm{I}_S(\omega) - \mathrm{I}_{S^c}(\omega)]] = s_k\mathrm{E}[\omega^2\mathrm{E}[\mathrm{I}_S - \mathrm{I}_{S^c} \,|\, \mathcal{B}(\mathbb{R})]] = s_k\mathrm{E}[\omega^2\mathrm{E}[\mathrm{I}_S - \mathrm{I}_{S^c}]] = 0$ for all $k \in \mathbb{Z}^+$ since ω^2 is Borel measurable since S is independent of $\mathcal{B}(\mathbb{R})$ and since $\mu(S) = 1/2$. For all $k \in \mathbb{Z}^+$, note that $X_k(\omega) = \omega[\mathrm{I}_S(\omega) - \mathrm{I}_{S^c}(\omega)]$ and $V_k = \xi_k$. Also, note that $\mathrm{E}[X_k \,|\, V_0, \ldots, V_k] = 0$ a.s. That is, $\sigma(V_0, \ldots, V_k) = \mathcal{B}(\mathbb{R})$ and $\mathrm{E}[X_k \,|\, V_0, \ldots, V_k] = \mathrm{E}[\omega(\mathrm{I}_S - \mathrm{I}_{S^c}) \,|\, \mathcal{B}(\mathbb{R})] = \omega\mathrm{E}[\mathrm{I}_S - \mathrm{I}_{S^c} \,|\, \mathcal{B}(\mathbb{R})] = \omega\mathrm{E}[\mathrm{I}_S - \mathrm{I}_{S^c}] = 0$ a.s. since the identity map is Borel measurable, since S is independent of $\mathcal{B}(\mathbb{R})$, and since $\mu(S) = 1/2$. Now, consider any Borel measurable quantizer Q, let $\mathcal{G} = \sigma(Q(V_0), \ldots, Q(V_k))$, and note that $\mathrm{E}[X_k \,|\, \mathcal{G}] = 0$ a.s. since $\mathcal{G}$ is a σ-subalgebra of $\mathcal{B}(\mathbb{R})$ and $\mathrm{E}[X_k \,|\, \mathcal{G}] = \mathrm{E}[\mathrm{E}[X_k \,|\, \mathcal{B}(\mathbb{R})] \,|\, \mathcal{G}] = 0$ a.s. Since the best mean-square estimate of X_k based upon a Borel measurable function of $Q(V_0), \ldots, Q(V_k)$ is a.s. zero, the best mean-square estimate of X_k based upon an affine function of $Q(V_0), \ldots, Q(V_k)$ is also a.s. zero. Thus we note in this case the failure of the Kalman filter, using quantized data, to provide any useful estimate of X_k. In fact, we have in this case the failure of the Kalman filter, using precise data, to provide any useful estimate of X_k. However, consider what can be done with judiciously chosen recursive estimates. Letting $f_k\colon \mathbb{R} \to \mathbb{R}$ via $f_k(x) = (x/s_k)[\mathrm{I}_S(x/s_k) - \mathrm{I}_{S^c}(x/s_k)]$ for $k \in \mathbb{Z}^+$, we easily produce a "recursive" estimate $\hat{X}_k$ of X_k as a function of V_k and $\hat{X}_{k-1}$ given by $\hat{X}_k = f_k(V_k) = X_k$ pointwise on the underlying probability space. Thus, the nonlinear recursive filter given above works precisely, whereas conditional expectation and the Kalman filter in this case are useless. Further, the nonlinear recursive filter in this situation is able to extract the signal from the noise precisely. □

Example 10.9. *For an integer $n > 2$, there exists a set $\{Y_1, \ldots, Y_n\}$ of random variables such that Y_n is a strict sense random variable that is equal to a polynomial function of Y_i for some positive integer $i < n$, yet the best $\mathbf{L}_2$ linear unbiased estimator of Y_n based on $\{Y_1, \ldots, Y_{n-1}\}$ is zero.*

Proof: Consider a probability space $(\Omega, \mathcal{F}, P)$ and a random variable X defined on this space such that X has a zero mean, unit variance Gaussian distribution. Let n be an integer greater than one. Apply the Gram–Schmidt orthogonalization procedure to the family $\{1, X, X^2, \ldots, X^n\}$ to obtain an orthogonal family $\{Y_1, \ldots, Y_{n+1}\}$, where it follows that $Y_1 = 1$, $Y_2 = X$, and that $\mathrm{E}[Y_{n+1}] = 0$. Let $\hat{\mathrm{E}}[Y_{n+1} \mid Y_1, \ldots, Y_n]$ denote the best $\mathbf{L}_2$ linear estimator of Y_{n+1} based on elements from the set $\{Y_1, \ldots, Y_n\}$. Notice that $\hat{\mathrm{E}}[Y_{n+1} \mid Y_1, \ldots, Y_n] = a_1 + a_2 Y_2 + \cdots + a_n Y_n$ a.s. for some choice of the a_i's. Since, for each positive integer $i \le n$, $(Y_{n+1} - \hat{\mathrm{E}}[Y_{n+1} \mid Y_1, \ldots, Y_n])$ is orthogonal to Y_i and since the elements from $\{Y_1, \ldots, Y_{n+1}\}$ are orthogonal, it follows that $a_i = 0$ for each positive integer $i \le n$. Hence, the best $\mathbf{L}_2$ linear unbiased estimator of Y_{n+1} based on elements from the set $\{Y_1, \ldots, Y_n\}$ is zero a.s. However, Y_{n+1} is equal pointwise to a polynomial function of Y_2. □

Example 10.10. *There exists a probability density function for which conventional importance sampling is a useless variance reduction technique.*

Proof: This example is from Wessel, Hall, and Wise [1990]. Let $P_0 = \int_T^\infty f(x)\,dx$, where f is a probability density function and T is a positive real number, and let f^* be a probability density function that is positive on $[T, \infty)$. In an effort to evaluate P_0 based on Monte Carlo techniques, P_0 is often estimated by the sample mean

$$\overline{P_0} = \frac{1}{N} \sum_{i=1}^{N} h(X_i),$$

where h denotes the indicator function of $[T, \infty)$ and $X_1, \ldots, X_N$ are mutually independent random variables, each with the probability density function f. The standard approach to importance sampling begins by noticing that

$$P_0 = \int_T^\infty \frac{f(x)\, f^*(x)}{f^*(x)}\, dx.$$

Interpreting this integral as an expectation with respect to f^*, a new estimate of P_0, denoted by $\overline{P_0^*}$ and given by

$$\overline{P_0^*} = \frac{1}{N} \sum_{i=1}^{N} \frac{h(X_i^*)\, f(X_i^*)}{f^*(X_i^*)},$$

is obtained where the X_i^*'s are mutually independent random variables, each having the probability density function f^*. The objective of importance sampling is to choose f^* in such a manner as to reduce the variance of $h(X_1^*)/f(X_1^*)\, f^*(X_1^*)$ compared with the variance of $h(X_1)$.

Consider a sequence of real numbers $\{a_n \colon n \in \mathbb{N}\}$ defined via

$$a_n = \begin{cases} 1, & \text{if } n = 1 \\ n(1 + a_{n-1}), & \text{if } n \in \mathbb{N} \text{ and } n > 1, \end{cases}$$

and define a probability density function via

$$f(x) = \begin{cases} f(-x), & \text{if } x < 0 \\ 2^{-(n+1)}, & \text{if } x \in [a_n, 1 + a_n) \text{ for } n \in \mathbb{N} \\ 0, & \text{otherwise.} \end{cases}$$

Let T be a fixed positive real number. Conventional importance sampling dictates that the choice for f^* should be $\alpha f(\alpha x)$, where α is some fixed element of (0, 1). Ideally, α should be chosen so as to minimize the variance of $h(X_1^*)/f(X_1^*)\, f^*(X_1^*)$, which is given by

$$\int_T^\infty \frac{f^2(x)}{\alpha f(\alpha x)}\, dx - P_0^2.$$

This variance will now be shown to be infinite for any value of α in (0, 1).

Fix α in (0, 1) and choose $m \in \mathbb{N}$ so that $m > 1/\alpha$ and $a_m > T$. Recall that $f(x) = 2^{-(m+1)}$ for $x \in [a_m, 1 + a_m)$ and note that $f(\alpha x) = 0$ for $x \in [(1+a_{m-1})/\alpha, a_m/\alpha)$. Also, $(1+a_{m-1})/\alpha < a_m$ from the definition of a_m and since $m > 1/\alpha$. Further, there exists a real number β such that $a_m < \beta < a_m/\alpha$ and $\beta \leq 1 + a_m$. Thus, it follows that $f(x)$ is nonzero on $[a_m, \beta)$ and that $f(\alpha x)$ is zero on $[(1 + a_{m-1})/\alpha, a_m/\alpha)$, a proper superset of $[a_m, \beta)$, which implies that

$$\int_T^\infty \frac{f^2(x)}{\alpha f(\alpha x)}\, dx = \infty.$$

Therefore, conventional importance sampling techniques are useless when applied to the problem of calculating tail probabilities for the density f given above. Notice that any density that is nonzero on the support of the above density f and zero off the support of f will exhibit the same phenomenon. Thus, for instance, such a probability density function could be chosen to be infinitely differentiable. □

Example 10.11. *There exists a probability density function for which improved importance sampling is a useless variance reduction technique.*

Proof: This example is from Wessel, Hall, and Wise [1990]. In the setting of Example 10.10, improved importance sampling is performed using a shifted version of the density $f(x)$, say $f(x + \alpha)$, where $\alpha \in \mathbb{R}$. Let $\zeta_n = \sum_{j=1}^n j$ and $\xi_n = \sum_{j=1}^n j^{-2}$, and define two real-valued sequences as follows:

$$a_n = \begin{cases} 0, & \text{if } n = 1 \\ \zeta_{n-1} + \xi_{n-1}, & \text{if } n \in \mathbb{N} \text{ and } n > 1, \end{cases}$$

and

$$b_n = \begin{cases} 1, & \text{if } n = 1 \\ \zeta_n + \xi_{n-1}, & \text{if } n \in \mathbb{N} \text{ and } n > 1. \end{cases}$$

Further, define a probability density function $f(x)$ via:

$$f(x) = \begin{cases} f(-x), & \text{if } x < 0 \\ 45/2\pi^4 n^5, & \text{if } x \in [a_n, b_n) \text{ for } n \in \mathbb{N} \\ 45/2\pi^4 n^2, & \text{if } x \in [b_n, a_{n+1}) \text{ for } n \in \mathbb{N}. \end{cases}$$

To see that $f(x)$ is indeed a probability density function, simply note that $\sum_{n=1}^{\infty} n^{-4} = \pi^4/90$.

Now let α be any fixed nonzero real number and let T be any fixed positive real number. Define $M_\alpha = \{n \in \mathbb{N}: n^{-2} < |\alpha| < n \text{ and } b_n > T\}$. Note that M_α is nonempty and, in fact, is an infinite subset of N since $j \in M_\alpha$ implies that $j+1 \in M_\alpha$. Let $m_\alpha = \min\{n \in M_\alpha\}$.

Note that $b_n - a_n = n$ and $a_{n+1} - b_n = n^{-2}$ for all $n \in \mathbb{N}$. Consider the case when $\alpha < 0$. Then, $b_n - a_n = n > |\alpha|$ if $n \geq m_\alpha$. Thus, $a_n - \alpha < b_n$ if $n \geq m_\alpha$. Further, $a_{n+1} - b_n = n^{-2} < |\alpha|$ if $n \geq m_\alpha$. Thus, $a_{n+1} < b_n - \alpha$ if $n \geq m_\alpha$. Therefore, $a_n - \alpha < b_n < a_{n+1} < b_n - \alpha$ if $\alpha < 0$ and $n \geq m_\alpha$. Consider the case when $\alpha > 0$. Then, $a_{n+1} - b_n = n^{-2} < \alpha$ if $n \geq m_\alpha$. Thus, $a_{n+1} - \alpha < b_n$ if $n \geq m_\alpha$. Also, $b_{n+1} - a_{n+1} = n+1 > n > \alpha$ if $n \geq m_\alpha$. Thus, $a_{n+1} < b_{n+1} - \alpha$ if $n \geq m_\alpha$. Therefore, $a_{n+1} - \alpha < b_n < a_{n+1} < b_{n+1} - \alpha$ if $\alpha > 0$ and $n \geq m_\alpha$.

In summary, it has been shown that the interval $[b_n, a_{n+1})$ is a proper subset of $[a_n - \alpha, b_n - \alpha)$ if $n \geq m_\alpha$ and $\alpha < 0$ and is a proper subset of $[a_{n+1} - \alpha, b_{n+1} - \alpha)$ if $n \geq m_\alpha$ and $\alpha > 0$.

Recall that $f(x) = 45/2\pi^4 n^2$ if $x \in [b_n, a_{n+1})$. Further, note that $f(x+\alpha) = 45/2\pi^4 n^5$ if $x \in [a_n - \alpha, b_n - \alpha)$, and that $f(x+\alpha) = 45/2\pi^4(n+1)^5$ if $x \in [a_{n+1} - \alpha, b_{n+1} - \alpha)$. Thus, using the previous result, it follows that if $n \geq m_\alpha$, then $f(x+\alpha) = 45/2\pi^4 n^5$ for $x \in [b_n, a_{n+1})$ and $\alpha < 0$, and $f(x+\alpha) = 45/2\pi^4(n+1)^5$ for $x \in [b_n, a_{n+1})$ and $\alpha > 0$.

Therefore, if $\alpha < 0$, then

$$\int_T^\infty \frac{f^2(x)}{f(x+\alpha)}\,dx \geq \sum_{n=m_\alpha}^{\infty} \int_{b_n}^{a_{n+1}} \frac{f^2(x)}{f(x+\alpha)}\,dx = \sum_{n=m_\alpha}^{\infty} \frac{45}{2n\pi^4} = \infty,$$

and, if $\alpha > 0$, then

$$\int_T^\infty \frac{f^2(x)}{f(x+\alpha)}\,dx \geq \sum_{n=m_\alpha}^{\infty} \int_{b_n}^{a_{n+1}} \frac{f^2(x)}{f(x+\alpha)}\,dx$$

$$= \sum_{n=m_\alpha}^{\infty} \frac{45(n+1)^5}{2\pi^4 n^6} \geq \sum_{n=m_\alpha}^{\infty} \frac{45}{2n\pi^4} = \infty.$$

Thus, since this density causes the variance of $h(X_1^*)/f(X_1^*)\,f^*(X_1^*)$ to be infinite for any nonzero choice of α, improved importance sampling techniques are useless in this situation. Again, notice that using standard techniques, the above phenomenon could be exhibited by using an infinitely differentiable density. □

Example 10.12. *There exists a situation involving a Gaussian signal of interest and mutually Gaussian observations in which methods of fusion are useless.*

Proof: This example is from Hall, Wessel, and Wise [1990]. Consider a random variable X and a set of random variables $\{Y_1, \ldots, Y_n\}$ all defined on the same probability space. A commonly considered problem in the area of distributed estimation is that of how best to fuse or combine estimators of the form $E[X \mid \mathcal{D}]$, where $\mathcal{D}$ is a nonempty proper subset of $\{Y_1, \ldots, Y_n\}$, in order to obtain a single good estimator of $E[X \mid Y_1, \ldots, Y_n]$. In the following example a situation is described, using common distributions, in which any such method of fusion is useless.

For an integer $n > 1$, let $\{X, Y_1, \ldots, Y_n\}$ be a set of random variables with a joint probability density function given by

$$f(x, y_1, \ldots, y_n) = \left(\frac{1}{\sqrt{2\pi}}\right)^{n+1} \exp\left[-\frac{1}{2}\left(x^2 + \sum_{i=1}^{n} y_i^2\right)\right] \times \left[1 + x \exp\left(-\frac{x^2}{2}\right) \prod_{i=1}^{n}\left[y_i \exp\left(-\frac{y_i^2}{2}\right)\right]\right].$$

As shown in Example 7.15, it follows that the set $\{X, Y_1, \ldots, Y_n\}$ is not mutually Gaussian and not mutually independent, yet any proper subset of $\{X, Y_1, \ldots, Y_n\}$ containing at least two random variables is mutually independent, mutually Gaussian, and identically distributed with each random variable having zero mean and unit variance. For any nonempty proper subset $\mathcal{D}$ of $\{Y_1, \ldots, Y_n\}$, notice that $E[X \mid \mathcal{D}] = 0$ a.s., since X is independent of $\mathcal{D}$. However, it follows easily that

$$E[X \mid Y_1, \ldots, Y_n] = \frac{1}{2\sqrt{2}} Y_1 \cdots Y_n \exp\left[-\frac{1}{2}\left(Y_1^2 + \cdots + Y_n^2\right)\right] \text{ a.s.}$$

Thus, since any Borel measurable function of the estimators $E[X \mid \mathcal{D}]$, where $\mathcal{D}$ ranges over all nonempty proper subsets of $\{Y_1, \ldots, Y_n\}$, would be constant a.s., it would be futile to attempt to estimate $E[X \mid Y_1, \ldots, Y_n]$ based on a combination of these estimators. □

Example 10.13. *A popular Markovian model for bit error rate estimation may lead to misleading estimates in practical applications.*

Proof: The bit error rate is an important parameter associated with the analysis of digital channels. Attempts to estimate the bit error rate of a digital channel are often hypothesized upon the assumption of mutually independent errors. Such an assumption, however, is not compatible with many digital systems. Consequently much work in this area has recently centered around a model of the channel given by a finite-state Markov chain where one of the states symbolizes an error. The performance of a relative frequency approach to bit error rate estimation based on such a model will now be briefly considered.

Consider the following situation. Assume that a digital channel is modeled by a three-state Markov chain $\{Z_i\colon i \in \mathbb{N}\}$ with a transition matrix Q given by

$$Q = \begin{bmatrix} \varepsilon & 0 & 1-\varepsilon \\ 0 & 1-\varepsilon & \varepsilon \\ 1-\varepsilon & \varepsilon/2 & \varepsilon/2 \end{bmatrix},$$

where $0 < \varepsilon < 1$. Further, define the initial probability vector $v = [p, 1 - p, 0]^T$, where $0 < p < 1$. State three will be associated with an error, and the remaining two states will be associated with error-free operation. Also, for each positive integer i, define $E_i = 1$ if $Z_i = 3$ and $E_i = 0$ otherwise. Note then that $R_n = E_1 + \cdots + E_n$ denotes the number of errors that occur during the first n transmissions. The relative frequency estimator of the bit error rate is thus given by $P_n = R_n/n$. Now let n be an even positive integer and note that $P(Z_2 = 2, \dots, Z_n = 2 \mid Z_1 = 2) = (1 - \varepsilon)^{n-1}$, and that $P(Z_2 = 3, Z_3 = 1, Z_4 = 3, Z_5 = 1, \dots, Z_n = 3 \mid Z_1 = 1) = (1-\varepsilon)^{n-1}$. Thus, given any fixed even positive integer n, ε can be chosen so that $P(P_n = 1/2 \mid Z_1 = 2)$ is almost equal to 1 and $P(P_n = 0 \mid Z_1 = 1)$ is almost equal to 1. Hence, depending upon the initial state, the relative frequency estimate of the bit error rate may vary greatly. That is, if the initial state is 2, then with high probability the Markov chain will remain in state 2 up to step n and hence P_n will with high probability be equal to zero. On the other hand, if the initial state is 1, then with high probability the Markov chain will alternate between states 1 and 3 up to step n, and hence P_n will with high probability be equal to 1/2.

This example exposes a basic problem with much of the research in this area. In particular, some concept of ergodicity is often called upon to justify neglecting the initial probabilities in an asymptotic analysis of a relative frequency estimator for a Markovian model. Unfortunately, one is always limited to taking a finite number of samples in such an estimation scheme, and therefore the initial probabilities of the Markov chain cannot be neglected. In fact, in the previous example, one could choose n to be beyond the realm of computation and still have a situation where the relative frequency bit error rate estimate strongly depends upon the initial state. □

Example 10.14. *Sampling a strictly stationary random process at a random time need not preserve measurability.*

Proof: Consider the probability space $(\mathbb{R}, \mathcal{B}(\mathbb{R}), P)$, where P is standard Gaussian measure on $(\mathbb{R}, \mathcal{B}(\mathbb{R}))$. Recalling that this measure space is not complete, it follows that there exists some Borel set B with P measure zero and a subset A of B such that A is not a Borel set. For $t \in \mathbb{R}$ and $\omega \in \mathbb{R}$, let $X(t, \omega) = \mathrm{I}_A(t)\mathrm{I}_B(\omega)$. Further, let $U(\omega) = \omega$ and note that U is a Gaussian random variable defined on $(\mathbb{R}, \mathcal{B}(\mathbb{R}), P)$. Since $P(X(t) = 0) = 1$ for each $t \in \mathbb{R}$, it is clear that $X(t)$ is a strictly stationary random process. However, $X(U(\omega), \omega) = \mathrm{I}_A(\omega)$, which is not a random variable, since the inverse image of the Borel set $\{1\}$ is given by A, which is not a Borel set. Thus, sampling a strictly stationary random process at random times does not imply that measurability is preserved. □

Example 10.15. *There exists a random process $\{X(t)\colon t \in \mathbb{R}\}$ that is not stationary yet is such that for any positive real number Δ the random process $\{X(n\Delta)\colon n \in \mathbb{Z}\}$ is strictly stationary.*

Proof: This example is from Wise [1991]. In many practical problems, one is interested in knowing whether or not a random process $\{X(t)\colon t \in \mathbb{R}\}$ is stationary. Due to the current digital trend in signal processing, one might instead attempt to determine the stationarity of $\{X(n\Delta)\colon n \in \mathbb{Z}\}$. If $\{X(n\Delta)\colon n \in \mathbb{Z}\}$ is stationary for every positive real number Δ, then must $\{X(t)\colon t \in \mathbb{R}\}$ be stationary?

Let $\{Y(t)\colon t \in \mathbb{R}\}$ be a stationary zero mean Gaussian random process defined on some underlying probability space such that $\mathrm{E}[Y(t)Y(t+\tau)] = \exp(-|\tau|)$. Define a stationary zero mean Gaussian random process $\{Z(t)\colon t \in \mathbb{R}\}$ via $Z(t) = Y(2t)$. Also, define a zero mean Gaussian random process $\{X(t)\colon t \in \mathbb{R}\}$ via $X(t) = Y(t)$ if t is rational and $X(t) = Z(t)$ if t is irrational. Note that $\{X(t)\colon t \in \mathbb{R}\}$ is not stationary, since, if t and τ are rational, then $\mathrm{E}[X(t)X(t+\tau)] = \exp(-\tau)$, yet if t is irrational and τ is rational, then $\mathrm{E}[X(t)X(t+\tau)] = \exp(-2\tau)$. Next, let Δ be any positive real number. Note that if Δ is rational, then $n\Delta$ is rational for all integers n, and if Δ is irrational, then $n\Delta$ is irrational for all nonzero integers n. Further, note that $Y(0) = Z(0) = X(0)$. Thus, for all integers n, $X(n\Delta) = Y(n\Delta)$ if Δ is rational and $X(n\Delta) = Z(n\Delta)$ if Δ is irrational. Thus, for any positive real number Δ, $\{X(n\Delta)\colon n \in \mathbb{Z}\}$ is a stationary Gaussian random process even though $\{X(t)\colon t \in \mathbb{R}\}$ is a Gaussian random process that is not stationary. □

Example 10.16. *Estimating an autocorrelation function via a single sample path can be futile.*

Proof: Consider a wide sense stationary random process $\{X(t)\colon t \in \mathbb{R}\}$ having locally integrable sample paths and its autocorrelation function $R(\tau) = \mathrm{E}[X(t)X(t+\tau)]$. Let $x(t)$ denote a sample path of the random process $X(t)$. One often proposed estimator for $R(\tau)$ is based on a single sample path $x(t)$ restricted to some interval $[-T, T]$ and is given by

$$R_T(\tau) = \frac{1}{2T}\int_{-T+(|\tau|/2)}^{T-(|\tau|/2)} x\left(t+\frac{\tau}{2}\right) x\left(t-\frac{\tau}{2}\right) dt.$$

The following example shows how easily such a scheme can fail to provide a good estimate for $R(\tau)$.

Let $\{X_1(t)\colon t \in \mathbb{R}\}$ and $\{X_2(t)\colon t \in \mathbb{R}\}$ be wide sense stationary random processes with autocorrelation functions $R_1(\tau)$ and $R_2(\tau)$, respectively. Further, define a random process $X(t)$ via

$$X(t) = \begin{cases} X_1(t), & \text{with probability } 1/2 \\ X_2(t), & \text{with probability } 1/2. \end{cases}$$

Then, the autocorrelation function R of $X(t)$ is given by $R(\tau) = [R_1(\tau) + R_2(\tau)]/2$. Notice that a sample path of $X(t)$ is either a sample path of $X_1(t)$ or $X_2(t)$, which could be greatly different random processes. Thus, choices for $R_1(\tau)$ and $R_2(\tau)$ abound such that $R(\tau)$ could in no way be estimated well by a single sample path. □

Example 10.17. *A method of moments estimator need not be unbiased.*

Proof: The following example points out that a method of moments estimator is not in general an unbiased estimator. Let $X_1, X_2, \dots, X_n$ be a sequence of mutually independent, identically distributed second-order random variables. Let $S_n = (X_1 + \cdots + X_n)$ and $V_n = (X_1^2 + \cdots + X_n^2)/n$. Recalling that the variance σ^2 of X_1 is equal to $\mathrm{E}[X_1^2] - (\mathrm{E}[X_1])^2$, it follows that a method of moments estimator for σ^2 is given by

$$M = \frac{1}{n}\sum_{k=1}^{n} X_k^2 - \frac{1}{n^2}\left[\sum_{k=1}^{n} X_k\right]^2.$$

Let $\mu = \mathrm{E}[X_1]$ and notice that

$$\begin{aligned}
\mathrm{E}[M] &= \frac{1}{n}\sum_{k=1}^{n}\mathrm{E}[X_k^2] - \frac{1}{n^2}\mathrm{E}\left[\left(\sum_{k=1}^{n} X_k\right)^2\right] \\
&= \frac{1}{n}\left[\sum_{k=1}^{n}(\sigma^2+\mu^2)\right] - \frac{1}{n^2}\left[n\mathrm{E}[X_1^2] + (n^2-n)\mathrm{E}[X_1]^2\right] \\
&= \sigma^2 + \mu^2 - \frac{1}{n^2}\left[n(\sigma^2+\mu^2) + (n^2-n)\mu^2\right] = \left(1 - \frac{1}{n}\right)\sigma^2.
\end{aligned}$$

Thus, since $\mathrm{E}[M] \neq \sigma^2$, it is clear that the method of moments estimator given by M is not unbiased. □

Example 10.18. *A method of moments estimator need not be a reasonable estimator.*

Proof: The following example presents a situation in which a method of moments estimator is a ridiculous estimator. Let N be a positive integer and consider a collection of N items with each item uniquely identified by a distinct positive integer less than or equal to N. Sample this family of items n times with replacement and let X_i denote the number associated with the ith item selected. Note that $\mathrm{E}[X_i] = (N+1)/2$. Hence, if $\overline{X} = (X_1 + \cdots + X_n)/n$, then $2\overline{X} - 1$ is a method of moments estimator for the parameter N. This estimator of N, however, is clearly unreasonable if $\max_{1\le i\le n} X_i$ is larger than $2\overline{X} - 1$, since $\max_{1\le i\le n} X_i$ is never greater than N. □

Example 10.19. *An unbiased estimator need not exist.*

Proof: The following example shows that an unbiased estimator need not exist. Consider a random variable X with a Poisson distribution given by $P(X = k) = e^{-\lambda}\lambda^k/k!$ for a nonnegative integer k and $\lambda > 0$. Assume that a statistic $T(X)$ is an unbiased estimator for the value of $1/\lambda$. In this case it follows that $\mathrm{E}[T(X)] = \sum_{k=0}^{\infty} T(k)(e^{-\lambda}\lambda^k/k!) = 1/\lambda$, or that $\sum_{k=0}^{\infty} T(k)(\lambda^k/k!) = e^{\lambda}/\lambda$. Note, however, that the function e^{λ}/λ does not admit a Maclaurin expansion. This contradiction implies that no such statistic T exists. In other words, there exists no unbiased estimator of $1/\lambda$ for the above situation. □

Example 10.20. *A situation exists in which a biased constant estimator of the success probability in a binomial distribution has a smaller mean-square error than an unbiased relative frequency estimator.*

Proof: The following example, inspired by Rao [1962], describes a situation in which a biased constant estimator has a smaller mean-square error than an unbiased relative frequency estimator. Consider a random variable X with a binomial distribution given by $P(X = r) = \binom{n}{r} p^r(1-p)^{n-r}$ for $r = 0, 1, \ldots, n$. Assume that $p \in (1/4, 3/4)$ and that $n = 3$. Further, consider a relative frequency estimator $\hat{p}$ of p given by $\hat{p} = X/n$. Notice immediately that $\hat{p}$ is an unbiased estimator of p and that $\mathrm{E}[(\hat{p}-p)^2] = (p/3) - (p^2/3)$.

Consider now an estimate of p given by the constant $1/2$. Note that this estimate is biased if $p \neq 1/2$. Further, notice that $\mathrm{E}[[(1/2) - p]^2] = p^2 - p + (1/4)$. Hence it follows that $\mathrm{E}[(\hat{p}-p)^2] - \mathrm{E}[[(1/2)-p]^2] = (-4p^2/3) + (4p/3) - (1/4) > 0$ for $p \in (1/4, 3/4)$. Thus, the constant estimator $1/2$ may be biased yet still have a smaller mean-square error than the unbiased relative frequency estimator $\hat{p}$. □

Example 10.21. *A minimum variance unbiased estimator need not exist.*

Proof: Consider a random variable X possessing a probability density function denoted by $f_\theta(x)$, where θ is some element from a nonempty subset Θ of $\mathbb{R}$. Assume that one wishes to find an unbiased estimator $T(X)$ of θ that minimizes $\mathrm{E}_{\theta_0}[[T(X)]^2]$, where $\theta_0 \in \Theta$, $T\colon \mathbb{R} \to \mathbb{R}$ is Borel measurable, and $\mathrm{E}_\theta[g(X)] = \int_{\mathbb{R}} g(x) f_\theta(x)\, dx$, subject to the constraint that $\mathrm{E}_\theta[T(X)] = \theta$ for every in $\theta \in \Theta$. The following example, inspired by Stein [1950] and Lehmann [1983], shows that even when unbiased estimators exist, it might so happen that no unbiased estimator exists that is a minimum variance unbiased estimator.

Consider the case in which $\Theta = \{0, 1\}$ and $\theta_0 = 0$. Further, assume that $f_0(x) = \mathrm{I}_{(0,1)}(x)$ and $f_1(x) = [1/(2\sqrt{x})]\mathrm{I}_{(0,1)}(x)$. For this situation, given any $\varepsilon > 0$, there exists an unbiased estimator $d_\varepsilon(X)$ of θ for which $0 < \mathrm{E}_0[(d_\varepsilon(X))^2] < \varepsilon$, yet there exists no unbiased estimator $g(X)$ for which $\mathrm{E}_0[[g(X)]^2] = 0$. In particular, let $\varepsilon > 0$ be fixed and consider the estimator given by $d_\varepsilon(x) = a\,[b + (1/\sqrt{x})]\,\mathrm{I}_{(c,1)}(x)$, where $0 < c < 1$. Note that there always exists a positive number $c < 1$ such that $(1-\sqrt{c})/(1+\sqrt{c}) + [\ln(c)/4] < -1/\varepsilon$. Choose such a value for c and let $a = -\left([\ln(c)/2] + [2(1-\sqrt{c})/(1+\sqrt{c})]\right)^{-1}$ and $b = -2/(1+\sqrt{c})$. After some algebraic manipulations, it follows that for these choices of a, b, and c, the estimator $d_\varepsilon(x)$ is such that $\mathrm{E}_0[d_\varepsilon(X)] = 0$, $\mathrm{E}_1[d_\varepsilon(X)] = 1$, and $\mathrm{E}_0[[d_\varepsilon(X)]^2] < \varepsilon$. Nevertheless, it is clear that there exists no unbiased estimator $g(X)$ for which $\mathrm{E}_0[[g(X)]^2] = 0$. That is, there exists no minimum variance unbiased estimator for this situation, even though, given any positive number ε, there exists an unbiased estimator of θ that has a variance less than ε. □

Example 10.22. *A unique minimum variance unbiased estimator need not be a reasonable estimator.*

Proof: For a nonempty subset Θ of $\mathbb{R}$, consider a family of distributions given by $\mathcal{P} = \{P_\theta : \theta \in \Theta\}$. Further, consider a random vector X composed of mutually independent, identically distributed random variables each with the same, but unknown, distribution from the family $\mathcal{P}$ and a statistic T; that is, T is a real-valued Borel measurable function of X. Recall that the statistic T is said to be a complete statistic if $\mathrm{E}_\theta[f(T(X))] = 0$ for a Borel measurable function $f : \mathbb{R} \to \mathbb{R}$ and for all $\theta \in \Theta$ implies that $f(t) = 0$ a.e. with respect to $\mathcal{P}$ where $\mathrm{E}_\theta[\cdot]$ denotes expectation with respect to the distribution P_θ. Let $g : \Theta \to \mathbb{R}$ be such that a nonconstant function of T exists that is an unbiased estimator of $g(\theta)$. Assume that $S_1(T)$ and $S_2(T)$ are two such unbiased estimators of $g(\theta)$. Then, since $S_1(T)$ and $S_2(T)$ are unbiased, $\mathrm{E}_\theta[S_1(T) - S_2(T)] = 0$ for all $\theta \in \Theta$. If T is a complete statistic, then it follows immediately that $S_1(T) = S_2(T)$ a.e. with respect to $\mathcal{P}$. Thus, an unbiased estimator of $g(\theta)$ that is a nonconstant function of the complete statistic T is the a.s. unique such unbiased estimator.

The following example demonstrates how a uniformly minimum variance unbiased estimator may completely fail to provide a reasonable estimate. Let X be a random variable possessing a Poisson density function with a fixed but unknown parameter $\theta > 0$. Choose $\lambda > 0$ and define $g(\theta) = \exp(-\lambda\theta)$. Assume that $T(X)$ is an unbiased estimator of $g(\theta)$. Then it follows that

$$\sum_{j=0}^{\infty} \frac{T(j)\theta^j}{j!} = \exp(\theta - \lambda\theta) = \sum_{j=0}^{\infty} \frac{(1-\lambda)^j\theta^j}{j!},$$

and hence it follows that $T(X) = (1-\lambda)^X$. Next, assume that $\mathrm{E}[f(X)] = 0$ for some Borel measurable function $f : \mathbb{R} \to \mathbb{R}$. Then it follows that

$$\sum_{k=0}^{\infty} f(k) \frac{e^{-\theta}\theta^k}{k!} = 0,$$

which in turn implies that $f(k) = 0$ for each nonnegative integer k, or in other words, that $f(k) = 0$ a.e. with respect to the measure induced by X. Hence, the statistic X is complete, and $T(X)$, a measurable function of the complete statistic X, is the a.s. unique unbiased estimator of $g(\theta)$ (that is a nonconstant function of T).

Finally, notice that if, for example, $\lambda = 3$, then the a.s. unique unbiased estimator of $g(\theta)$ as a nonconstant function of X is given by $(-2)^X$. This estimator oscillates wildly between positive and negative values and is virtually worthless in any attempt to estimate $\exp(-\lambda\theta)$. □

Example 10.23. *There exists a situation in which a single observation from a distribution with a finite variance is a better estimator of the mean than is a sample mean of two such observations.*

Proof: Consider a probability space $(\Omega, \mathcal{F}, P)$ and let $\{X_1, \ldots, X_n\}$ be a collection of mutually independent, identically distributed random variables defined on $(\Omega, \mathcal{F}, P)$. Assume that X_1 has a probability density function given by $f(x-\theta)$, where θ is a real constant and f is continuous and even. Note that the sample mean $\overline{X}_n = (X_1 + \cdots + X_n)/n$ provides an unbiased estimator of the parameter θ. A widely held belief is that $\overline{X}_n$ always provides at least as good of an estimator as $\overline{X}_m$ if $n > m$. The following example, suggested by Stigler [1980] and based on a result by Edgeworth, shows that this claim is false in general. In particular, an example is given in which X_1 is a better estimator of θ than is $\overline{X}_2 = (X_1 + X_2)/2$.

The following result will be used in the example that follows. Let X and Y be random variables with probability density functions $f_1(x-\theta)$ and $f_2(x-\theta)$, respectively, where $f_i(x)$ is continuous and even for $i \in \{1, 2\}$ and θ is a fixed real number. Assume that $f_1(0) > f_2(0)$. Then, since $f_1(x) - f_2(x)$ is continuous and positive at the origin, there exists some positive number ε such that $f_1(x) - f_2(x)$ is positive on the interval $[0, \varepsilon]$. Since $f_i(x)$ is even for $i \in \{1, 2\}$, the previous observation implies that

$$\int_{\theta-\varepsilon}^{\theta+\varepsilon} f_1(x-\theta)\,dx > \int_{\theta-\varepsilon}^{\theta+\varepsilon} f_2(x-\theta)\,dx.$$

Hence, for this choice of ε, $P(|X-\theta| \le \varepsilon) > P(|Y-\theta| \le \varepsilon)$.

Consider now the probability density function given by

$$f(x) = \frac{k-1}{2(1+|x|)^k},$$

where k is an integer greater than 3. Notice that the first through $k-2$ moments of this density are each finite, and, in particular, the variance is finite. Further, consider a cost function that assigns unit cost to errors that are larger than some fixed positive constant ε and assigns zero cost to smaller errors. In particular, a good estimator $\hat{\theta}$ is one for which $P(|\hat{\theta} - \theta| \le \varepsilon)$ is large. Note that the density of $\overline{X}_2$ evaluated at θ is given by $2\int_{\mathbb{R}} f^2(x)\,dx$. Hence, the ratio of the density of $\overline{X}_2$ at θ to that of X_1 at θ is given by

$$D = \frac{2}{f(0)} \int_{\mathbb{R}} f^2(x)\,dx = 1 - \frac{1}{2k-1}.$$

Notice that $D < 1$ for k as above. Hence, the result of the previous paragraph implies that there exists some ε for which $P(|X_1 - \theta| \le \varepsilon) > P(|\overline{X}_2 - \theta| \le \varepsilon)$. Therefore, for this situation and for the cost function as above associated with this value of ε, a single observation will yield a superior estimate of the mean than will the sample mean of two observations. □

Example 10.24. *Ancillary statistics need not be location invariant.*

Proof: Consider an arbitrary fixed probability density function $f(\cdot)$ and a family of probability density functions given by $\{g(x, \theta)\colon \theta \in \Theta\}$, where $g(x, \theta) = f(x+\theta)$ and Θ is some nonempty subset of $\mathbb{R}$. For a positive integer n, let $\{X_1, X_2, \ldots, X_n\}$ be a set of mutually independent, identically distributed random variables, each with a probability density function

$g(x, \theta)$, where θ is assumed to be some fixed but unknown element from Θ. Recall that a statistic S for the family $\{X_1, X_2, \ldots, X_n\}$ is said to be location invariant if, for any real constant α, $S(X_1 + \alpha, \ldots, X_n + \alpha) = S(X_1, \ldots, X_n)$. Also, recall that a statistic S for the family $\{X_1, X_2, \ldots, X_n\}$ is said to be ancillary if the distribution of S does not depend on θ. It is clear that any location invariant statistic is ancillary. The following example from Padmanabhan [1977] demonstrates that the converse statement is false. That is, an ancillary statistic is presented that is not location invariant.

Let X_1 and X_2 be independent, identically distributed random variables each defined on a probability space $(\Omega, \mathcal{F}, P)$ and each with a Gaussian distribution with mean θ and unit variance where θ is some fixed but unknown real number. Fix a real number K and define a statistic T as follows:

$$T(X_1, X_2) = \begin{cases} X_1 - X_2, & \text{if } X_1 + X_2 \geq K \\ X_2 - X_1, & \text{if } X_1 + X_2 < K. \end{cases}$$

It follows easily that $X_1 - X_2$ and $X_2 - X_1$ each has a Gaussian distribution with mean zero and variance equal to 2. Further, it follows quickly that $X_1 - X_2$ and $X_2 - X_1$ are each independent of $X_1 + X_2$. Hence, it follows that $T(X_1, X_2)$ has a Gaussian distribution with mean zero and variance equal to 2 for any value of θ, and thus it follows that the statistic T is ancillary.

Consider, however, the set $A = \{\omega \in \Omega\colon -\alpha < X_1 < 0,\ K - \alpha < X_2 < K\}$, where α is some fixed positive number. It follows easily that on the set A, $T(X_1, X_2) = X_2 - X_1$ yet $T(X_1 + \alpha, X_2 + \alpha) = X_1 - X_2$ since $X_1 + X_2 < K$, whereas $X_1 + X_2 + 2\alpha > K$. Hence, $T(X_1, X_2) \neq T(X_1 + \alpha, X_2 + \alpha)$ on the set A, which implies that the ancillary statistic T is not location invariant. □

Example 10.25. *The Fisher information may stress small values of a parameter yet stress large values when estimating a strictly increasing function of the parameter.*

Proof: Consider a random variable X that possesses a probability density function $f_\theta(x)$, where θ is an unknown parameter from a nonempty open subset Θ of $\mathbb{R}$. A standard problem in estimation involves determining the parameter based on X. In this regard the quantity

$$\mathrm{E}\left[\left(\frac{\partial}{\partial\theta}\ln[f_\theta(X)]\right)^2\right],$$

denoted by $I(\theta)$ and called the Fisher information, is often used, when it exists, to indicate the "information" that X contains about the parameter θ. In particular, the Fisher information is frequently employed as an indicator of how easily a specific element $\theta_0 \in \Theta$ may be distinguished from neighboring values; that is, if $I(\theta_0) > I(\theta_1)$, then it is often claimed that θ can be estimated more accurately at $\theta = \theta_0$ than at $\theta = \theta_1$. The following example from Lehmann [1983] demonstrates some curious behavior that the Fisher information may exhibit when a strictly increasing function is composed with the parameter of interest.

Consider a Poisson random variable with parameter $\lambda > 0$. For this distribution it follows that $I(\lambda) = \mathrm{E}[[(X/\lambda) - 1]^2] = 1/\lambda$ since $\mathrm{E}[X] = \lambda$ and $\mathrm{E}[X^2] = \lambda^2 + \lambda$, so that the Fisher information might suggest that the parameter λ can be estimated most accurately when λ is small. Assume now that one wishes to estimate the parameter $\theta = \ln(\lambda)$. It is important to note that the Fisher information depends on the particular parameterization under consideration. In particular, if $I(\alpha)$ is the Fisher information that X contains about the parameter α and if $\alpha = g(\xi)$, where g is differentiable, then the Fisher information that X contains about ξ is given by $I^*(\xi) = I(g(\xi))[g'(\xi)]^2$ [Lehmann, 1983, p. 118]. Hence, in this case where $\lambda = \exp(\theta)$, it follows that $I^*(\theta) = I(\exp(\theta))[\exp(\theta)]^2 = \exp(\theta)$. Hence, the Fisher information that X contains about $\ln(\lambda)$ is given by λ, and thus, the Fisher information might suggest that the parameter $\ln(\lambda)$ can be estimated most accurately when λ is large. Thus, although the Fisher information stresses small values of the parameter λ when estimating λ, it stresses large values of λ when estimating $\ln(\lambda)$, a strictly increasing function of the parameter λ. □

Example 10.26. *There exists a sequence of estimators with an asymptotic variance that is never greater than the Cramer–Rao lower bound and that is less than that bound for certain values of the parameter of interest.*

Proof: Consider a sequence $\{X_1, X_2, \ldots\}$ of mutually independent, identically distributed random variables, each possessing a Gaussian distribution with unit variance and mean $\theta \in \mathbb{R}$. Let S_n denote the sum of the first n of these random variables and note that S_n/n is a maximum likelihood estimator of θ. Further, notice that S_n/n is Gaussian with mean θ and variance $1/n$. Now, for a real number α, the Hodges–Le Cam estimator is given by:

$$T_n\left(\frac{S_n}{n}\right) = \begin{cases} S_n/n, & \text{if } |S_n|/n \geq \sqrt[4]{\dfrac{1}{n}} \\ \alpha S_n/n, & \text{if } |S_n|/n < \sqrt[4]{\dfrac{1}{n}}. \end{cases}$$

Notice first that $\lim_{n\to\infty} P\left(|S_n|/n < \sqrt[4]{1/n}\right)$ is equal to 0 if θ is nonzero via the strong law of large numbers and the dominated convergence theorem and is equal to 1 if $\theta = 0$ via Chebyshev's inequality. Hence, if θ is nonzero, then it follows that T_n and S_n/n each converge in distribution to the constant θ. Now notice if $\theta = 0$, then

$$\mathrm{Var}(T_n) \leq \frac{\alpha^2}{n} + \frac{1}{n} P\left(\frac{|S_n|}{n} \geq \sqrt[4]{\frac{1}{n}}\right) \leq \frac{\alpha^2}{n} + \sqrt{\frac{1}{n^3}},$$

and the variance of S_n/n is equal to $1/n$. Thus, for $|\alpha| < 1$ and for n sufficiently large, the variance of T_n is strictly smaller than the variance of

S_n/n. The point $\theta = 0$ is called a point of superefficiency. Recalling that the maximum likelihood estimator S_n/n achieves the Cramer–Rao lower bound, it follows that, at the point $\theta = 0$, and for n sufficiently large, the variance of $T_n(S_n/n)$ is in fact smaller than the Cramer–Rao lower bound when $|\alpha| < 1$. □

Example 10.27. *A maximum likelihood estimator need not exist.*

Proof: Consider a collection $\{X_1, \ldots, X_n\}$ of mutually independent, identically distributed random variables, each possessing a probability density function given by

$$f_\theta(x) = \frac{1-\varepsilon}{\sigma} h\left(\frac{x-\mu}{\sigma}\right) + \varepsilon h(x-\mu),$$

where $\theta = (\mu, \sigma^2) \in \Theta = \mathbb{R} \times (0, \infty)$, where $h(\cdot)$ is the continuous zero mean, unit variance Gaussian probability density function, and where ε is a fixed number in $(0, 1)$. The likelihood function for this situation is given by

$$L(\theta, x_1, \ldots, x_n) = \prod_{i=1}^{n} \left[\frac{1-\varepsilon}{\sigma} h\left(\frac{x_i-\mu}{\sigma}\right) + \varepsilon h(x_i-\mu)\right].$$

Notice, in particular, that

$$L(\theta, x_1, \ldots, x_n) > \frac{1-\varepsilon}{\sigma} h\left(\frac{x_1-\mu}{\sigma}\right) \prod_{j=2}^{n} \varepsilon h(x_j-\mu).$$

Further, notice that this lower bound is unbounded for θ in Θ, since it approaches infinity as $\sigma \to 0$ when $\mu = x_1$. Thus a maximum likelihood estimate for θ does not exist since the supremum of the likelihood function over all θ in Θ is not finite. Further, notice that when ε is sufficiently small, f_θ looks very much like a Gaussian density function. □

Example 10.28. *A maximum likelihood estimator need not be unique.*

Proof: Consider a family of probability density functions $\{f_\theta(\cdot)\colon \theta \in \mathbb{R}\}$, where $f_\theta(x) = I_{[\theta-(1/2), \theta+(1/2)]}(x)$. Further, for a fixed positive integer n, consider a collection $X_1, X_2, \ldots, X_n$ of mutually independent, identically distributed random variables, each with a probability density function given by $f_\theta(x)$ for some fixed, yet unknown, value of θ. Further, for $1 \leq i \leq n$, let Y_i denote the ith-order statistic of the X_i's. Notice that the likelihood function $L(\theta, x_1, x_2, \ldots, x_n) = 1$ if $x_i \in [\theta - (1/2), \theta + (1/2)]$ for each positive integer $i \leq n$ and is equal to zero otherwise. That is, $L(\theta, x_1, x_2, \ldots, x_n) = 1$ if $\max_{i\leq n} x_i - (1/2) \leq \theta \leq \min_{i\leq n} x_i + (1/2)$ and is equal to zero otherwise. Therefore, any statistic $T(X_1, X_2, \ldots, X_n)$ for which $Y_n - (1/2) \leq T(X_1, X_2, \ldots, X_n) \leq Y_1 + (1/2)$ holds is a maximum likelihood estimator of the parameter θ. It follows easily that for any $\lambda \in [0, 1]$, $(1-\lambda)Y_n + \lambda Y_1 + [\lambda - (1/2)]$ is a maximum likelihood estimator for θ. Hence, not only are maximum likelihood estimators not in general unique, but, as in this case, there may exist uncountably many distinct maximum likelihood estimators for a single parameter. □

Example 10.29. *A maximum likelihood estimator need not be unbiased or admissible.*

Proof: This example is from Kiefer [1987]. Consider a random variable X that has a Bernoulli distribution such that $P(X = 1) = \theta$ and $P(X = 0) = 1 - \theta$, where $\theta \in \Theta = [1/3, 2/3]$. The likelihood function resulting from a single observation is then given by $L(\theta, x) = \theta^x(1-\theta)^{1-x}$, which implies that a maximum likelihood estimator of θ is given by $\hat{\theta}(x) = (x+1)/3$ for $x = 0, 1$. Note that $\mathrm{E}[\hat{\theta}(X)] = (\theta+1)/3$, which implies that this maximum likelihood estimator is not unbiased. Consider for a moment a family of estimators each of the form

$$\theta_\alpha(x) = \begin{cases} \alpha, & \text{if } x = 0 \\ 1-\alpha, & \text{if } x = 1, \end{cases}$$

where $1/3 \le \alpha \le 1/2$ and note that $\hat{\theta}(x) = \theta_{1/3}(x)$. Further, note that $R(\alpha, \theta) = \mathrm{E}[[\theta_\alpha(X) - \theta]^2] = (\alpha-\theta)^2(1-\theta) + [(1-\alpha)-\theta]^2\theta$. For the maximum likelihood estimator $\hat{\theta}$ given above, we see that $R(1/3, \theta) = (\theta^2/3) - (\theta/3) + (1/9)$. Further,

$$R(\alpha, \theta) - R\left(\frac{1}{3}, \theta\right) = \left(\alpha - \frac{1}{3}\right)\left[(2\theta-1)^2 + \left(\alpha - \frac{2}{3}\right)\right] < 0$$

for all $\alpha \in (1/3, 1/2]$ and all $\theta \in \Theta$, since θ in Θ implies that $(2\theta-1)^2 \le 1/9$ and $\alpha \in (1/3, 1/2]$ implies that $[\alpha - (2/3)] \le -1/6$ and $[\alpha - (1/3)] \ge 0$. Thus, not only is this maximum likelihood estimator not admissible, but it has the uniformly largest mean-square error over all of the uncountably many estimators in the family $\{\theta_\alpha\colon \alpha \in [1/3, 1/2]\}$. □

Example 10.30. *A maximum likelihood estimator need not be consistent.*

Proof: This example is suggested in Bickel and Doksum [1977]. Fix an integer $k > 1$ and consider a collection $\{X_{ij}\colon i \in \mathbb{N}, 1 \le j \le k\}$ of mutually independent random variables with X_{ij} having a Gaussian distribution with mean μ_i and positive variance σ^2 for each $i \in \mathbb{N}$ and $1 \le j \le k$. Let p be an integer greater than 1 and consider a subset of $\{X_{ij}\colon i \in \mathbb{N}, 1 \le j \le k\}$ given by $\{X_{ij}\colon 1 \le i \le p, 1 \le j \le k\}$. Likelihood functions for $\mu_1, \dots, \mu_p$ and σ^2 based upon elements from this subset are given by

$$L_1(\mu_i, x_{i1}, \dots, x_{ik}) = \sqrt{\frac{1}{(2\pi\sigma^2)^k}} \exp\left(\frac{-1}{2\sigma^2}\left[(x_{i1}-\mu_i)^2 + \cdots + (x_{ik}-\mu_i)^2\right]\right)$$

and

$$L_2(\sigma^2, x_{11}, \dots, x_{pk}) = \sqrt{\frac{1}{(2\pi\sigma^2)^{kp}}} \exp\left(\frac{-1}{2\sigma^2}\left[(x_{11}-\mu_1)^2 + \cdots + (x_{pk}-\mu_p)^2\right]\right),$$

respectively. It follows quickly that a maximum likelihood estimator for μ_i, where $1 \le i \le p$, is given by $T(X_{i1}, \ldots, X_{ik}) = (1/k)(X_{i1} + \cdots + X_{ik})$. Notice that

$$\begin{aligned}&\ln[L_2(\sigma^2, x_{11}, \ldots, x_{pk})]\\ &\quad = \ln\left(\sqrt{\frac{1}{(2\pi\sigma^2)^{kp}}}\right) + \left(\frac{-1}{2\sigma^2}\left[(x_{11} - \mu_1)^2 + \cdots + (x_{pk} - \mu_p)^2\right]\right)\end{aligned}$$

and that

$$\begin{aligned}&\frac{\partial}{\partial\sigma^2}\ln[L_2(\sigma^2, x_{11}, \ldots, x_{pk})]\\ &\quad = \frac{-kp}{2\sigma^2} + \left(\frac{1}{2\sigma^4}\left[(x_{11} - \mu_1)^2 + \cdots + (x_{pk} - \mu_p)^2\right]\right)\end{aligned}$$

from which it follows that a maximum likelihood estimator for σ^2 is given by

$$S(X_{11}, \ldots, X_{pk}) = \frac{1}{kp}\sum_{i=1}^{p}\sum_{j=1}^{k}[X_{ij} - T(X_{i1}, \ldots, X_{ik})]^2.$$

For a moment, consider a collection $\{Z_1, \ldots, Z_n\}$ of mutually independent Gaussian random variables each with mean μ and unit variance. Let $\overline{Z} = (Z_1 + \cdots + Z_n)/n$. In addition, let A be an $n \times n$ matrix (a_{ij}) such that the columns of A form an orthonormal basis for $\mathbb{R}^n$ and such that $a_{i1} = 1/\sqrt{n}$ for $1 \le i \le n$. Define Y_i for $1 \le i \le n$ via $(Y_1 \cdots Y_n) = [(Z_1 \cdots Z_n)\,A]$ and note that, since the columns of A are orthonormal, the Y_i's are mutually independent Gaussian random variables with variances equal to one and means given by $\mathrm{E}[Y_1] = \mu\sqrt{n}$ and $\mathrm{E}[Y_i] = 0$ for $1 < i \le n$. Finally notice that

$$\begin{aligned}\sum_{k=1}^{n}(Z_k - \overline{Z})^2 &= \left(\sum_{k=1}^{n} Z_k^2\right) - n\overline{Z}^2 = \left(\sum_{k=1}^{n} Z_k^2\right) - Y_1^2\\ &= \left(\sum_{k=1}^{n} Y_k^2\right) - Y_1^2 = \sum_{k=2}^{n} Y_k^2,\end{aligned}$$

which has a χ^2 density with $n-1$ degrees of freedom. Hence, we see that

$$\frac{1}{\sigma^2}\sum_{j=1}^{k}(X_{ij} - T(X_{i1}, \ldots, X_{ik}))^2$$

has a χ^2 distribution with $k-1$ degrees of freedom for each positive integer $i \le p$. Thus, $\mathrm{E}[S(X_{11}, \ldots, X_{pk})] = [(k-1)/k]\,\sigma^2$. Letting $p \to \infty$ and applying the weak law of large numbers to the sequence

$$\left\{\frac{1}{k}\sum_{j=1}^{k}[X_{ij} - T(X_{i1}, \ldots, X_{ik})]^2 : i \in \mathbb{N}\right\}$$

implies that $S(X_{11}, \ldots, X_{pk})$ converges in probability to $[(k-1)/k]\,\sigma^2$. Thus, the maximum likelihood estimator $S(X_{11}, \ldots, X_{pk})$ is not consistent. □

Example 10.31. *A maximum likelihood estimator of a real parameter may be strictly less than the parameter with probability one.*

Proof: Consider a collection $\{X_1, X_2, \ldots, X_n\}$ of mutually independent random variables, each with a probability density function given by $f_\theta(x) = (1/\theta)\mathrm{I}_{(0,\theta]}(x)$, where $\theta \in \Theta = (0, \infty)$. Notice that the likelihood function is given by $L(\theta, x_1, x_2, \ldots, x_n) = (1/\theta^n)\mathrm{I}_{(0,\theta]}(\max_{i\leq n} x_i)$. This function possesses a unique maximum at $\theta = \max_{i\leq n} x_i$. Hence, it follows that the maximum likelihood estimator of θ is given by $\hat{\theta}(x_1, x_2, \ldots, x_n) = \max_{i\leq n} x_i$. Notice that this estimator of θ is less than θ with probability one. □

Example 10.32. *The computational demands of a maximum likelihood estimator may increase dramatically with the number of observations.*

Proof: Consider a collection $\{X_1, X_2, \ldots, X_n\}$ of mutually independent random variables, each with a Cauchy density function given by $f_\theta(x) = \left[\pi[1 + (x - \theta)^2]\right]^{-1}$, where $\theta \in \Theta = \mathbb{R}$. For this case it follows that the likelihood ratio is given by

$$\ln[L(\theta, x_1, x_2, \ldots, x_n)] = -n\ln(\pi) - \sum_{i=1}^{n} \ln[1 + (x_i - \theta)^2],$$

and, hence that

$$\frac{\partial}{\partial\theta}\ln[L(\theta, x_1, x_2, \ldots, x_n)] = \sum_{i=1}^{n} \frac{2(x_i - \theta)}{1 + (x_i - \theta)^2}.$$

Notice that this expression may be written as a fraction whose numerator is a polynomial in θ of degree $2n - 1$. As the number of observations n increases, the computational demands of finding the roots of this polynomial increase dramatically. Not only must $2n - 1$ roots in general be considered, but a large number of roots must be checked to determine whether they correspond to local or global extrema. □

Example 10.33. *A maximum likelihood estimator may be unique for an odd number of observations yet fail to be unique for an even number of observations.*

Proof: Let $\Theta = \mathbb{R}$ and let $f_\theta(x) = \exp(-|x - \theta|)/2$ for each θ in Θ and x in $\mathbb{R}$. Note that in this case the log of the likelihood function is given by

$$\ln[L(\theta, x_1, \ldots, x_n)] = -n\ln(2) - \sum_{i=1}^{n} |x_i - \theta|,$$

which thus implies that a maximum likelihood estimator for θ is given by a median of $\{X_1, \ldots, X_n\}$. Also, notice that the observations $X_1, \ldots, X_n$ are almost surely distinct. Thus, we see that for an even number of observations, there exist almost surely uncountably many distinct maximum likelihood estimators for θ, yet for an odd number of observations, there exists almost surely a unique maximum likelihood estimator. □

Example 10.34. *A maximum likelihood estimator of a real parameter may increase without bound as the number of observations upon which the estimate is based approaches infinity.*

Proof: This example is from Le Cam [1986]. Let $\alpha_0 = 1$ and define a decreasing sequence of positive numbers via

$$\int_{\alpha_{k+1}}^{\alpha_k} \left[\exp\left(\frac{1}{x^2}\right) - \frac{1}{2}\right] dx = \frac{1}{2}$$

for each nonnegative integer k. Further, define a family of probability density functions $\{f_\theta(x)\colon \theta \in \Theta\}$ on $(0, 1]$, where $\Theta = \mathbb{N}$ and

$$f_\theta(x) = \frac{1}{2} + \left[\exp\left(\frac{1}{x^2}\right) - \frac{1}{2}\right] \mathrm{I}_{(\alpha_\theta, \alpha_{\theta-1}]}(x),$$

and let $\{X_n\colon n \in \mathbb{N}\}$ be a sequence of mutually independent random variables, each with the same probability density function $f_\theta(x)$ for some fixed yet unknown value of θ from Θ. Note that $\ln[L(\theta, x_1, \ldots, x_n)] = \sum_{i=1}^{n} \Lambda(\theta, x_i)$, where

$$\Lambda(\theta, x_i) = \begin{cases} (x_i)^{-2}, & \text{if } x_i \in (\alpha_\theta, \alpha_{\theta-1}] \\ -\ln(2), & \text{otherwise.} \end{cases}$$

Hence, a maximum likelihood estimator $\hat{\theta}_n$ exists and is equal to an integer j for which the following term is maximized:

$$\sum_{\{m\colon \alpha_j < x_m \le \alpha_{j-1}\}} \frac{1}{x_m^2}.$$

Note that for any $\beta > 0$, it follows from the Borel–Cantelli lemma that infinitely many terms from the sequence $\{X_n\colon n \in \mathbb{N}\}$ will fall in the interval $(0, \beta]$ almost surely. Thus it follows immediately that $\hat{\theta}_n$ tends almost surely to infinity even though the actual value of θ is a fixed positive integer. □

Example 10.35. *There exists a probability space $(\Omega, \mathcal{F}, P)$, a stationary Gaussian random process $\{X(t)\colon t \in \mathbb{R}\}$ defined thereon and composed of mutually independent standard Gaussian random variables, and a function $f\colon \mathbb{R}^2 \to \mathbb{R}$ such that for any fixed but unknown real number θ, any real number t, and any point ω from Ω, $f(t, X(t-\theta, \omega)) = \theta$.*

Proof: Consider the partition of the real line $\{S_t\colon t \in \mathbb{R}\}$, the probability space $(\Omega, \mathcal{F}, P)$, and the stochastic process $\{X(t)\colon t \in \mathbb{R}\}$ defined thereon that were given in Example 7.30. As before, let $\omega = \{\omega_t\}_{t \in \mathbb{R}}$ denote a generic element of Ω. Since $X(t-\theta, \omega) = \omega_{t-\theta} \in S_{t-\theta}$, the result follows immediately. □

Example 10.36. *There exists a probability space $(\Omega, \mathcal{F}, P)$, and a stationary Gaussian random process $\{X(t): t \in \mathbb{R}\}$ defined thereon and composed of mutually independent standard Gaussian random variables, and a function $f: \mathbb{R} \to \mathbb{R}$ such that for any $t \in \mathbb{R}$ and any $\omega \in \Omega$, $f(X(t, \omega)) = t$.*

Proof: Consider the partition of the real line $\{S_t: t \in \mathbb{R}\}$, the probability space $(\Omega, \mathcal{F}, P)$, and the stochastic process $\{X(t): t \in \mathbb{R}\}$ defined thereon that were given in Example 7.30. As before, let $\omega = \{\omega_t\}_{t\in\mathbb{R}}$ denote a generic element of Ω. Let the function f be given by $f(x) = \sum_{t\in\mathbb{R}} t\mathrm{I}_{S_t}(x)$ and note that for any $t \in \mathbb{R}$ and any $\omega \in \Omega$, $f(X(t, \omega)) = t$. □

Example 10.37. *A most powerful statistical hypothesis test need not correspond to an optimal detection procedure.*

Proof: As in Example 1.46, let S_1 and S_2 be a partition of $\mathbb{R}$ into saturated non-Lebesgue measurable subsets. Corresponding to this partition, consider a probability space $(\Omega, \mathcal{F}, P)$ as provided by Example 7.7, where P agrees with standard Gaussian measure on $\mathcal{B}(\mathbb{R})$. Let $S = S_1$ and note that $S^c = S_2$. Let X, Y_0, and Y_1 be random variables defined on this space via $X(\omega) = \omega$, $Y_0(\omega) = \mathrm{I}_S(\omega)$, and $Y_1(\omega) = 2\mathrm{I}_{S^c}(\omega)$.

Consider a detection problem in which we wish to decide between a null detection hypothesis given by the ordered pair of random variables (X, Y_0) and an alternative detection hypothesis given by (X, Y_1). Notice that the statistical hypothesis problem that corresponds to this detection problem is given as follows: For a positive real number h define a probability measure $Q_h: \mathcal{B}(\mathbb{R}) \to [0, 1]$ via $Q_h(B) = [\mathrm{I}_B(0) + \mathrm{I}_B(h)]/2$, and consider two simple statistical hypotheses $H_0 = \{P_0\}$ and $H_1 = \{P_1\}$, where P_0 and P_1 are probability measures defined on $\mathcal{B}(\mathbb{R}^2)$ given by $P_0(A) = (\mu \times Q_1)(A)$ and $P_1(A) = (\mu \times Q_2)(A)$. Now, let $\alpha_0 \in \left(0, \frac{1}{2}\right)$ be a given upper bound on the false alarm probability. It is a simple exercise to verify that a Neyman–Pearson test for this statistical hypothesis testing situation is given by the following: If the second member of the ordered pair of random variables is equal to 2, then announce H_1; if the second member of the ordered pair of random variables is equal to 0, then announce H_1 with probability $2\alpha_0$; otherwise announce H_0. The detection probability for this statistical hypothesis test is $\frac{1}{2} + \alpha_0$. However, this statistical test does not correspond to the best detection scheme.

In particular, consider the detection scheme given by the following procedure: Announce the null detection hypothesis if and only if either the first member of the ordered pair of random variables is in S and the second member of the ordered pair of random variables is positive, or the first member of the ordered pair of random variables is not in S and the second member of the ordered pair of random variables is zero. Notice that the false alarm probability associated with this detection scheme is zero and the detection probability is one. Further, recall that the scheme derived from the Neyman–Pearson lemma did not exhibit such performance. Indeed, the detection scheme given here is a perfect detector, whereas the one associated with a most powerful statistical hypothesis test is not. To see further that the detection and statistical hypothesis problems are in fact distinct, simply note that if we let $Y_1(\omega) = 2I_S(\omega)$, then the statistical hypothesis testing problem is unchanged, whereas the detection hypothesis problem changes dramatically. Also, note that the random variables under consideration possess commonplace distributions. Nevertheless, there exists a detection scheme that greatly out performs the detector corresponding to a Neyman–Pearson test. □

Example 10.38. *A Neyman–Pearson detector for a known positive signal corrupted by additive noise possessing a unimodal and symmetric probability density function may announce the absence of the signal for large values of the observation.*

Proof: This example was inspired by Ferguson [1967]. Consider a one sample detection problem modeled by a hypothesis testing problem in which H_0 states that X is equal in law to N and in which H_1 states that X is equal in law to $s+N$, where the signal s is a known constant and the noise N is a random variable whose distribution is known. A Neyman–Pearson detector is a detector whose false alarm probability does not exceed a set value and whose detection probability is maximized subject to the constraint on the false alarm probability. If the known signal s is positive and if the noise N has a probability density function that is unimodal and symmetric, then will a Neyman–Pearson detector announce that the signal is present when the observation is large? If the distribution of the noise were zero mean Gaussian, the answer to this question would be yes. However, in general, there exist situations consistent with the above question, such that the Neyman–Pearson detector announces that the signal is absent for large values of the observation. For example, assume that $s = 2$ and that the probability density function of the noise is given by $f(x) = [\pi(1+x^2)]^{-1}$, which is unimodal and symmetric. Then $f(x-2)/f(x)$ approaches one as $x \to \infty$, equals one at $x = 1$, is strictly greater than one for $x \in (1, \infty)$, and is less than one for $x \in (-\infty, 1)$. Further, note that $\int_{-\infty}^{1} f(x)\,dx = 3/4$. Thus, for a Neyman–Pearson detector such that the false alarm probability is fixed to be not greater than some number $\alpha < 1/4$, the threshold for the likelihood ratio will be some number greater than one, and consequently, the detector will announce the signal as absent when the observation is sufficiently large. □

Example 10.39. *A uniformly most powerful test of a simple hypothesis against a composite alternative may exist for one size yet not exist for another size.*

Proof: Consider three probability distributions P_0, P_1, and P_2 on $\{0, 1, 2\}$ defined as follows:

$$P_0(\{0\}) = P_0(\{1\}) = P_0(\{2\}) = \frac{1}{3},$$

$$P_1(\{1\}) = 1 - P_1(\{2\}) = \frac{1}{3},$$

$$P_2(\{0\}) = 1 - P_2(\{2\}) = \frac{1}{3}.$$

Further, consider the simple hypothesis H given by $\{P_0\}$ and the composite alternative K given by $\{P_1, P_2\}$. Although the test given by

$$\phi(x) = \begin{cases} 1, & \text{if } x > 3/2 \\ 0, & \text{if } x < 3/2 \end{cases}$$

is uniformly most powerful for testing H against K at a size given by $1/3$, it follows quickly that no such test exists for any size greater than $1/3$. □

References

Abaya, E. F., and G. L. Wise, "Some notes on optimal quantization," *Proceedings of the 1981 International Conference on Communications,* Denver, Colorado, June 14–18, 1981, pp. 30.7.1–30.7.5.

Abaya, E. F., and G. L. Wise, "Some remarks on the existence of optimal quantizers," *Statistics and Probability Letters,* Vol. 2, December 1984, pp. 349–351.

Ash, R. B., and M. F. Gardner, *Topics in Stochastic Processes,* Academic Press, New York, 1975.

Bickel, P. J., and K. A. Doksum, *Mathematical Statistics: Basic Ideas and Selected Topics,* Holden–Day, Oakland, California, 1977.

Ferguson, T. S., *Mathematical Statistics: A Decision Theoretic Approach,* Academic Press, New York, 1967.

Hall, E. B., A. E. Wessel, and G. L. Wise, "Some aspects of fusion in estimation theory," *IEEE Transactions on Information Theory,* Vol. 37, No. 2, pp. 420–422, March 1991.

Hall, E. B., and G. L. Wise, "An analysis of convergence problems in Kalman filtering which arise from numerical effects," *Proceedings of the 33rd Midwest Symposium on Circuits and Systems,* Calgary, Alberta, Canada, pp. 315–318, August 12–14, 1990.

Hall, E. B., and G. L. Wise, "A result on multidimensional quantization," *Proceedings of the American Mathematical Society,* Vol. 118, June 1993, pp. 609–613.

Kiefer, J. C., *Introduction to Statistical Inference,* Springer-Verlag, New York, 1987.

Le Cam, L., *Asymptotic Methods in Statistical Decision Theory,* Springer-Verlag, New York, 1986.

Lehmann, E. L., *Theory of Point Estimation,* John Wiley, New York, 1983.

Padmanabhan, A. R., "Ancillary statistics which are not invariant," *The American Statistician,* Vol. 31, August 1977, p. 124.

Rao, C. R., "Apparent anomalies and irregularities in maximum likelihood estimation," *Sankhya,* Series A, Vol. 24, 1962, pp. 73–86.

Stein, C., "Unbiased estimates with minimum variance," *The Annals of Mathematical Statistics,* Vol. 21, September 1950, pp. 406–415.

Stigler, S. M., "An Edgeworth curiosum," *The Annals of Statistics,* Vol. 8, 1980, pp. 931–934.

Wessel, A. E., E. B. Hall, and G. L. Wise, "Importance sampling via a simulacrum," *The Journal of the Franklin Institute,* Vol. 327, No. 5, 1990, pp. 771–783.

Wessel, A. E., and G. L. Wise, "On estimation of random variables via the martingale convergence theorem," *Systems and Control Letters,* Vol. 11, 1988, pp. 61–64.

Wise, G. L., "A cautionary aspect of stationary random processes," *IEEE Transactions on Circuits and Systems,* Vol. 38, November 1991, pp. 1409–1410.

Index

absolute continuity, xiv, 9, 10, 12, 13, 18, 20, 26, 53, 63, 72, 81, 89, 90, 115, 138, 159, 175
admissible estimator, 29, 201
analytic set, 7, 33, 47, 102
ancillary statistic, 29, 197, 198
arc-length measure, 100, 142
atomic measure, 13, 26, 81, 90, 94, 95, 175
autocorrelation function, 23, 28, 151, 152, 154, 193
axiom of choice, vii, 33

Bernstein set, 6, 14, 33, 41–43, 62, 101
binomial distribution, 28, 195
Borel–Cantelli lemma, 204
Brownian motion, 24, 156–158

Cantor ternary set, 5, 8, 21, 33, 36–38, 41, 54, 57, 60, 61, 73, 137, 138, 148
Cantor–Lebesgue function, 38, 54, 55, 62, 76, 106
Cantor-like set, 5, 33, 35, 36, 39, 42, 62, 72
cardinal, vii, 7, 33, 50, 51
Cartesian product, 100
Cauchy distribution, 163, 203
central limit theorem, 171, 173
Chapman–Kolmogorov equation, 22, 147, 148
characteristic function, 21, 25, 141, 142, 152, 171–173
Chebyshev's inequality, 179, 199
co-countable set, 57, 82, 83, 89, 90, 114, 130, 134, 150, 160, 162
completely normal space, 13, 14, 17, 19, 82, 92, 95–97, 113, 131
complex-valued random variable, 27, 183
composite hypothesis, 30, 206, 207
condensation point, 40
conditional expectation, 25, 103, 159, 162–164, 166, 169, 187
conditional independence, 24, 25, 161, 162
conditional probability, 24, 103, 159–161, 163
connected set, 10, 65
continuous function, xiii, 5, 7–11, 14, 15, 17, 18, 20, 33, 38, 41, 47, 53–57, 59, 61, 62, 66, 68, 73–77, 99, 104, 106–108, 111, 113, 114, 116, 127, 132–134, 142, 146, 162, 173, 174, 197
continuum hypothesis, vii, xiii, 6, 7, 10, 11, 19, 44, 45, 49, 68, 84, 129, 130
convergence, 10, 25, 26, 58, 61, 66, 67, 169, 171, 175, 179
convex function, 21, 53, 59, 60, 127, 142
convex set, 18, 21, 126, 127, 142
convolution, 18, 27, 116, 118, 132, 173, 183
correlation, 21, 23, 27, 140–143, 183, 186, 187
cost function, 197
counting measure, 57, 87, 88, 90, 95, 112, 115, 117, 134
covariance, 28, 159, 183–185
Cramer–Rao lower bound, 29, 199, 200
critical point, 73
critical value, 10, 73

Darboux function, 9–11, 53, 64, 65, 74, 75
derivative, 10, 11, 23, 71–75, 106, 107, 152
detection theory, 29, 205, 206
diffuse measure, 11, 13, 22, 81, 83, 84, 94, 146, 147
Dini derivative, 11, 71, 75
Dirac measure, 98, 99, 114
directed set, 27, 109, 180
dominated convergence theorem, 104, 151, 199
Dynkin system theorem, 161

entropy, 28, 138, 186
equipotence, vii, 130

equivalence relation, 34, 39, 67, 76, 133
ergodicity, 192
extrema, 203

Fatou's lemma, 25, 169
filter, 27, 164, 183, 187
filtration, 22, 23, 25, 149, 150, 169, 180
first category, 5, 7, 8, 33, 37, 42, 43, 49, 50, 56, 78, 79
Fisher information, 29, 198, 199
Fourier series, 117
Fourier transform, 18, 116
Fubini's theorem, 55, 112, 117, 118, 129–132, 135, 149
fusion, 28, 170, 190, 191, 207

Gaussian distribution, 21–24, 26–29, 139–146, 149, 150, 152, 154–157, 163, 164, 172–174, 183–188, 190–193, 198–202, 204–206
Gram–Schmidt procedure, 188
graph, 58
group, 67, 121

Hahn–Jordan decomposition theorem, 88
Hamel basis, 6, 7, 33, 43–49, 51, 57, 59, 63, 65, 67
Hausdorff space, 12–14, 17, 19, 20, 82, 85, 86, 92, 95–97, 99, 101, 113, 114, 131, 132, 134
Hodges–Le Cam estimator, 199

importance sampling, 28, 188–190, 208
increments, 24, 151, 156
independence, 24, 25, 158, 161, 162
indiscrete topology, 82
inner measure, xiii, 6, 33, 41, 43–45, 96, 160, 164, 185
inner product, 117, 157
inner regular, 13, 14, 82, 95–97, 101
isolated point, 93

Jensen's inequality, 142

Kalman filter, 28, 186, 187, 207

Lebesgue integral, 103, 110, 137, 138
Lebesgue–Stieltjes measure, 12, 89
limit point, 33, 40, 86, 92, 98
linear function, 17, 113, 114
Lipschitz property, 9, 53, 62
locally compact space, 12–14, 19, 20, 85, 92, 96–98, 131–133
lower semicontinuous, 8, 53, 55

Maclaurin series, 194
Markov property, 22, 28, 147, 148, 161, 179, 191, 192
martingale, 25, 27, 169, 171, 178–181, 185
martingale convergence theorem, 28, 179, 185, 208
maximum likelihood estimation, 29, 199–204, 207
mean-square continuity, 23, 154
mean-square derivative, 23, 152
mean-square error, 25, 28, 164, 184, 185, 195, 201
median, 25, 172, 203
method of moments, 28, 194
metric space, 5, 8, 12, 14, 20, 21, 33, 34, 58, 59, 82, 86, 97–99, 104, 132, 133, 139, 171
midpoint convexity, 8, 53, 59, 60
minimization, 164
modification, 22, 149
Monte Carlo estimation, 188

nesting property, 163
Neyman–Pearson test, 29, 205, 206
noise, 29, 187, 206
nowhere dense, 6, 16, 20, 33, 36, 40, 42, 49, 50, 57, 68, 73, 78, 79, 98, 107, 108, 138

ordinal, vii, xiii, 51, 92, 93, 96, 99, 109, 110, 113, 130, 132, 134
orthogonal, 121, 151, 188
orthonormal, 202
oscillation, 56
outer measure, xiii, 13, 14, 48, 83, 90, 96, 101, 144, 160, 164

outer regular, 13, 14, 82, 95–98

pairwise independence, 147, 148
partial derivative, 18, 127
partition, 6, 7, 18, 19, 21, 23, 33, 34, 45, 46, 48, 50, 51, 65, 67, 77, 85, 87, 121–126, 129, 139, 142, 145, 150, 155, 156, 184, 186, 187, 204, 205
perfect set, 6, 9, 24, 33, 36, 40, 46, 47, 51, 62, 64, 98, 155
periodic function, 8, 23, 53, 57–59, 152
Poisson distribution, 194, 196, 199
Polish space, 33, 47
product space, 19, 20, 131–134, 145, 150, 155
property of Baire, 6, 33, 42, 43
pseudometric, 23, 153, 154

quantization, 27, 28, 158, 183–187, 207

Radon–Nikodym, 17, 18, 103, 114, 115, 137, 159
regression function, 25, 165, 170
regular measure, 13, 14, 82, 94, 96, 97
Riemann integral, 10, 14, 15, 20, 72, 73, 103–105, 137, 138
Riemann–Stieltjes integral, 15, 106
rigid motion, 121, 125, 126
rotation, 121, 123–125

sample path, 22–24, 148–156, 183, 193
saturated measure, 13, 83, 90, 91
saturated nonmeasurable set, 7, 19, 21, 22, 24, 33, 50, 51, 61, 90, 129, 139, 144, 150, 151, 155, 181, 205
second category, 6, 7, 33, 43, 49
separability, 21, 23, 33, 139, 153, 154
signed measure, 12, 17, 81, 87, 88, 114, 159
Skorohod imbedding theorem, 24, 156
Sorgenfrey plane, 86
spectral measure, 23, 154
spectral representation, 23, 151, 152
stationarity, 23, 24, 28, 29, 150–152, 154–156, 192, 193, 204, 205
Stieltjes integral, 15, 105
stochastic integral, 151, 182
stopping time, 25, 156, 157, 169, 170
strong law of large numbers, 172, 199
sufficiency, 25, 159, 165–168
superefficiency, 200

Taylor's series, 10, 71
threshold, 206
tightness, 21, 139, 172
Tonelli's theorem, 104
topological space, xiii, 33, 81, 82, 85, 86, 96, 97, 121
transfinite induction, 43, 44, 47, 49–51, 64
translate, 48, 98, 99, 121, 126, 174

unbiased estimator, 28, 29, 187, 188, 194–197, 201
uniform distribution, 138, 144, 152, 165, 175
upper semicontinuous, 53

Vitale set, 39, 66

weak law of large numbers, 27, 176, 202
weak* topology, 26, 171, 174, 175

Zermelo–Fraenkel axioms, vii